MICROSTRUCTURES AND STRUCTURAL DEFECTS IN HIGH-TEMPERATURE SUPERCONDUCTORS

MICROSTRUCTURES AND STRUCTURAL DEFECTS IN HIGH-TEMPERATURE SUPERCONDUCTORS

Zhi-Xiong Cai & Yimei Zhu

Brookhaven National Laboratory
Upton, New York, USA

World Scientific
Singapore • New Jersey • London • Hong Kong

Published by

World Scientific Publishing Co. Pte. Ltd.
P O Box 128, Farrer Road, Singapore 912805
USA office: Suite 1B, 1060 Main Street, River Edge, NJ 07661
UK office: 57 Shelton Street, Covent Garden, London WC2H 9HE

QC
611.98
.H54
C35
1998

British Library Cataloguing-in-Publication Data
A catalogue record for this book is available from the British Library.

MICROSTRUCTURES AND STRUCTURAL DEFECTS IN HIGH-TEMPERATURE SUPERCONDUCTORS

Copyright © 1998 by World Scientific Publishing Co. Pte. Ltd.

All rights reserved. This book, or parts thereof, may not be reproduced in any form or by any means, electronic or mechanical, including photocopying, recording or any information storage and retrieval system now known or to be invented, without written permission from the Publisher.

For photocopying of material in this volume, please pay a copying fee through the Copyright Clearance Center, Inc., 222 Rosewood Drive, Danvers, MA 01923, USA. In this case permission to photocopy is not required from the publisher.

ISBN 981-02-3285-3

Printed in Singapore by Uto-Print

For

Danning and Eriko

Preface

This book is geared toward upper-level undergraduate students and entry-level graduate students majoring in physics or materials science who are interested in crystal structure and structural defects, particularly those related to high-temperature superconductors. Researchers involved in studying high-T_c materials should also find the book useful as it consistently summarizes the recent progress in this field.

The book has an extensive introduction to microstructures and structural defects in high-temperature superconductors. It emphasizes the application of experimental as well as theoretical modeling techniques to the study of these complex materials, rather than simply presenting the results of current research. The reader is given an overview of the complexities in understanding structure-sensitive properties of bulk high-T_c materials, such as the transport properties, and in developing large-scale (high current, high field) applications. Whenever feasible, the effects of defects on the superconducting properties of these materials are described to put the study of the microstructure in a proper perspective.

We do not attempt to cover all aspects of high-temperature superconductivity research; instead, only the bare-bone knowledge of superconductivity theory is introduced. To link the structure to the properties, we focus on the interpretation of the microstructure and structural defects based on observations with advanced transmission electron microscopy coupled with theoretical modeling techniques. We illustrate experimental and theoretical techniques to study microstructures and structural defects with examples drew from various high-T_c superconductors. The usefulness of most of the techniques is not unique to high-temperature superconductors; they can be easily applied to metals, ceramics, and other complex materials.

This book is mainly derived from our own researches, and is our joint effort. In Chapters 1 (Cai) we briefly review the history of research in high-temperature superconductivity. In Chapter 2 (Cai) we introduced the basic notion of superconductivity and its relationship to microstructure and defects, especially in high-T_c superconductors. Chapter 3 (Cai and Zhu) is an overview of experimental techniques and theoretical modeling, mainly those used in the researches described in this book to study defects in materials. Chapter 4 (Zhu and Cai) summarizes the crystal structure of three major families of high-temperature superconductors, the La-, Y-, and Bi-based cuprates. In Chapter 5 (Zhu), we introduce the basic notions of various types of defects, both intra-granular and inter-granular that appear in superconducting oxides, and the geometrical approach (interfacial dislocations and Coincidence-Site-Lattice model) to studying grain boundaries. In the following chapters, we discuss experimental and theoretical studies of various types of microstructural features and defects in high-temperature superconductors. Chapters 6 and 7 are devoted to oxygen disorder/order related defects in $YBa_2Cu_3O_{7-\delta}$. Chapter 6 (Cai and Zhu) describes in detail the "tweed" structure, a displacive modulation due to oxygen ordering, and its structural origin. Chapter 7 (Zhu) offers observations and interpretations of the atomic structures of twinning variants and varied twin boundaries, another type of oxygen ordering associated with high-symmetrical lattice planes in $YBa_2Cu_3O_{7-\delta}$. Chapter 8 (Cai and Zhu) discusses structural modulations and the related phase transformations in $La_{2-x}Ba_xO_4$. Chapter 9 (Cai and Suenaga) covers the kinetics of grain alignment and formation of Bi-2223 in Bi-2223/Ag tapes, which is very important to large-scale applications of the high-T_c superconductors. Chapter 10 (Zhu) presented a novel diffraction technique to study charge distribution in $YBa_2Cu_3O_{7-\delta}$ and Bi-2212 superconductors. Chapter 11 (Zhu) and Chapter 12 (Zhu) extensively discuss the structure of grain boundaries. The former describes various experimental methods based on nano-probe electron microscopy that is required to study grain boundaries, such as determining boundary crystallography and measuring local hole concentration and lattice parameters. The latter provides a detailed account of grain boundary structure including misorientation distribution in practical Y- and Bi-based superconducting materials. We emphasize the correlation between the boundary hole concentration and the boundary lattice constraint for YBCO, and address the structure-property relation by examining the structure of the electromagnetically characterized Bi-2212 bicrystals. In the last section (12.4) of the chapter (Cai), we propose a theoretical model for understanding superconductivity in polycrystalline materials. In Chapter 13 (Zhu, Sec.13.4: Cai) we present our study of columnar defects produced by heavy-ion irradiation, the effects of crystal orientation, stochiometry, and imperfection on these columnar defects, and their structural nature and formation mechanism. Brief conclusions are given in the final chapter (Cai), and suggestions made on the future directions of research.

We have tried to treat the subject in an elementary way. While the inclusion of some mathematical formulae is unavoidable, we have emphasized the physics, instead of fancy mathematical treatment. Thus, the reader needs only basic knowledge of statistical mechanics, solid-state physics and calculus to follow the discussions in this book.

It is our hope that this book will draw attention to the area of materials science in which fruitful researches are being conducted and in which many challenging problems remain.

This book has developed over a number of years of research in the Materials Science Division, Department of Applied Science at Brookhaven National Laboratory. We have benefited greatly from the input of our colleagues, visiting scientists, collaborators, students, and friends, who have directly and indirectly contributed to this book. In particular we wish to thank M. Suenaga and D.O. Welch, BNL, who encouraged us to take up this project, for their knowledge, and stimulating discussions. We thank J. Tafto, Univ. of Oslo, for his creative and fruitful contribution during his short but productive summer visits in the recent years. We gratefully acknowledge the generous assistance and cooperation of Q. Li, A.R. Moodenbaugh, R.L. Sabatini and L. Wu. We are also indebted to W. Bian, R.C. Budhani, Y. Corcoran, J. Cowley, Y. Fukumoto, A.G. Khachaturyan, A. King, T.R. Thurston, Y.N. Tsay, J.-Y. Wang, Y.-L. Wang, Z.L. Wang, H.J. Wiesmann, Y. Xu, H. Zhang and J.M. Zuo (in alphabetical order), who are the co-authors of our published researches. We thank F. Ling, Y. Koh, and S. Patt from World Scientific Publishing Inc. for their help in preparing the manuscript. Last, but not least, we thank our families, Danning and Yida, and Eriko and Anna. This book would never have been written without their understanding and support.

Zhi-Xiong Cai
Yimei Zhu

Long Island, New York
February 1998

Contents

Preface	vi
1 Introduction	**1**
2 Defects and Superconductivity	**7**
2.1 What is Superconductivity?	7
2.2 The Electromagnetic Properties of Superconductor	9
2.3 The London Model	10
2.4 Ginzburg-Landau Theory	14
2.5 Two Types of Superconductors	18
2.6 Critical Current in Type-II Superconductors	23
2.7 Flux-line Pinning and Microstructure	24
2.8 Good and Bad Defects	25
3 Tools Used to Study the Defects	**27**
3.1 Introduction	27
3.2 Experimental Tools	27
3.2.1 Transmission Electron Microscopy	27
3.2.2 X-ray and Neutron Diffraction	30
3.3 Theoretical Modeling and Computer Simulation	31
3.3.1 Interaction Between Atoms	31
3.3.2 Computer Simulation	34
4 The Structure Characteristics	**41**
4.1 Introduction	41
4.2 Crystal Structure	41
4.2.1 La214 and Related Structures	42
4.2.2 $YBa_2Cu_3O_{7-\delta}$ (Y123) and Related Structures	44

	4.2.3	$Bi_2Sr_2Ca_{n-1}Cu_nO_{2n+6}$ and Related Structures	48
4.3	Bonding and Structural Features		52

5 Types of Defects — 55

5.1 Intragranular Defects ... 55
    5.1.1 Point Defects ... 55
    5.1.2 Line Defects-Dislocations ... 58
    5.1.3 Planar Defects ... 61
    5.1.4 Volume Defects - Inclusions ... 62
    5.1.5 Structural Modulations ... 64
    5.1.6 Artificially Created Defects ... 65
5.2 Intergranular Defects – Grain Boundaries ... 69
    5.2.1 The Concept of Coincidence Site Lattice ... 70
    5.2.2 The Concept of O-lattice ... 72
    5.2.3 Frank Formula ... 73

6 Oxygen Ordering Related Defects in $YBa_2Cu_3O_7$ — 75

6.1 Introduction ... 75
6.2 Models for Oxygen Ordering in $YBa_2Cu_3O_{7-\delta}$ ... 77
6.3 Effect of Thermal Quenching on Oxygen Ordering ... 89
6.4 [110] Tweedy Modulation ... 96
6.5 Theoretical Model for Fe Doped $YBa_2Cu_3O_{7-\delta}$ and Tweed ... 107
    6.5.1 Model for Oxygen Ordering in Fe Doped $YBa_2Cu_3O_{7-\delta}$ ... 107
    6.5.2 Modification of the Lattice Gas Model to Include Elastic Strain ... 113
6.6 Reduction and Reoxidation in $YBa_2(Cu_{1-x}Fe_x)_3O_{7-\delta}$ ... 120
6.7 Three Dimensional Structural Modulation ... 131
    6.7.1 $0.02 < x < 0.10$ ... 131
    6.7.2 $0.10 < x < 0.33$ ... 134
    6.7.3 Simulation of Tweed Image Contrast ... 142
6.8 Summary ... 145

7 The Twin Boundary Structure in $YBa_2Cu_3O_{7-\delta}$ — 147

7.1 Introduction ... 147
7.2 Gross Features of the Twin Boundary ... 149
    7.2.1 Conventional TEM Observations ... 149

	7.2.2	HREM Observations 153
	7.2.3	The Effects of Cation Substitution 156
	7.2.4	The Effects on Superconducting Properties 159
7.3	Studies of the Displacement at the Twin Boundary . . . 160
	7.3.1	Evolution of the Twin-Boundary Structure During Electron-Beam Irradiation 160
	7.3.2	The Variability of the Twin Boundaries 162
	7.3.3	Analyses of Fringe Contrast of the Twin Boundary 166
	7.3.4	Analyses of the Streaks of Diffuse Scattering . . 170
	7.3.5	Interfacial Energy of the Twin Boundaries 174
7.4	Crystallographic Analysis of the Twin Boundary 177
	7.4.1	A $\Sigma 64$ Coincidence Boundary 177
	7.4.2	Twin-Boundary Steps and Twinning Dislocations 179
	7.4.3	A Displacement-Shift-Complete-Lattice (DSCL) Treatment 183
	7.4.4	Structure of Mixed Twin Boundaries 185
7.5	Twin Tip 188
	7.5.1	Four Twinning Variants 190
	7.5.2	The Shape of a Tapered Twin 195
	7.5.3	Interfaces of the Orthogonally Oriented Twins . . 199
7.6	Summary 201

8 Structural Modulation in $La_{2-x}Ba_xCuO_4$ 203

8.1	Introduction 203
8.2	Consecutive Structural Transformation and Lattice Dynamical Model 205
8.3	Low-Temperature Microstructure 215
8.4	Theoretical Model of Twin Boundary Structure 224
	8.4.1	Landau Model 225
	8.4.2	The Structure of Twin Boundary 227
	8.4.3	The Size of the Twin Domains 233
8.5	Summary 236

9 Kinetics of the Alignment and the Formation of Bi-2223   237

9.1 Introduction 237

	9.2	The Fabrication Procedure 241
	9.3	Evolution at Early Stages of Heat Treatment 244
		9.3.1 *In situ* X-ray Diffraction Measurements 245
		9.3.2 Electron Microscopy 248
		9.3.3 The Formation of Bi-2223 262
	9.4	Dislocation and Bismuth 2212-to-2223 Transformation . 271
		9.4.1 Introduction 271
		9.4.2 Edge-dislocations as Channels for Fast Ion Diffusion 274
		9.4.3 The Layer Rigidity Model 275
		9.4.4 Kinetics of Bi-2212 Conversion to Bi-2223 282
	9.5	Summary 287

10 Charge Distribution   289

    10.1 Introduction 289
    10.2 $Bi_2Sr_2CaCu_2O_8$ 292
         10.2.1 A General Nano-scale Description 292
         10.2.2 Direct Imaging of Charge Modulation 295
    10.3 $YBa_2Cu_3O_7$ 301
         10.3.1 A Novel Diffraction Method 301
         10.3.2 Distribution of Electron Holes 308
    10.4 Conclusion 312

11 Experimental Techniques for Grain Boundary Studies   313

    11.1 Analyses of Grain Boundary Crystallography 314
         11.1.1 Determination of Grain Boundary Geometry ... 314
         11.1.2 Case Study of $YBa_2Cu_3O_{7-\delta}$ 320
    11.2 Electron Energy-Loss Spectroscopy of Oxygen K-edge .. 328
         11.2.1 Experimental Considerations of TEM-EELS ... 329
         11.2.2 Orientational Dependence of the Oxygen Pre-peak 334
         11.2.3 Quantitative Analysis of Oxygen/hole Concentration 337
    11.3 Lattice Parameter from CBED Measurement of HOLZ Pattern 340

Contents   xv

12 The Structure of Grain Boundaries   345
12.1 Introduction 345
12.2 Grain Boundaries in $YBa_2Cu_3O_{7-\delta}$ 346
    12.2.1 Misorientation Distribution in Textured Bulks . . 347
    12.2.2 Grain Boundary Dislocations 352
    12.2.3 Oxygen Content and CCSL Boundaries 355
    12.2.4 Variation of Oxygen/hole at the Boundaries . . . 358
    12.2.5 Cation Segregation at the Boundary 368
    12.2.6 Strain Energy of the Grain Boundaries 371
12.3 Grain Boundaries in $Bi_2Sr_2CaCu_2O_{8+\delta}$ 376
    12.3.1 Misorientation Distribution in Bi-2212 and Bi-2223 Composite Tapes 378
    12.3.2 Superconducting Properties of Twist Boundaries of Bi-2212 383
    12.3.3 Structural Features of the Twist Boundaries . . . 387
12.4 RSJ Model 403
12.5 Summary 417

13 Artificially Created Defects   419
13.1 Introduction 419
13.2 Experimental Methods 420
    13.2.1 Sample Preparation 420
    13.2.2 Heavy-ion Irradiation 421
13.3 Structure of the Defects 421
    13.3.1 General Morphology 421
    13.3.2 Radiation Damage Induced by Au, Ag, Cu, and Si Ions 424
    13.3.3 The Effect of Crystallographic Orientation 430
    13.3.4 The Effects of Oxygen Concentration in $YBa_2Cu_3O_{7-\delta}$ 431
    13.3.5 The Effect of the Pre-existing Crystal Imperfections 435
    13.3.6 Creation of Stacking Faults 437
    13.3.7 EELS Measurements 439
13.4 Thermal Spike Model 441
    13.4.1 Stopping Power of Heavy Ions 442
    13.4.2 The Size of the Defect Area 443
    13.4.3 Anisotropy of the Size and Shape of the Defects . 448

13.5 Strain Field and Strain Contrast of Columnar Defects . . 450
    13.5.1 The Displacement Fields of Columnar Defects . . 451
    13.5.2 Intensity of the Diffraction Contrast of the Columnar Defects 455
    13.5.3 Simulation of the Strain Contrast of the Columnar Defects 457
    13.5.4 The Radial Displacement and Strain of the Columnar Defects 462
13.6 Planar Defects Induced by Heavy-ion Irradiation 465
    13.6.1 Morphology of the Planar Defects 465
    13.6.2 Microchemical Analysis of the Planar Defects . . 467
    13.6.3 Structure of the Planar Defects 469
13.7 Summary 474

14 Conclusions   477

Bibliography   479

Index   505

Chapter 1

Introduction

The discovery of superconductivity in La-Ba-Cu-O compounds with critical temperature (T_c) >30K by Bednorz and Müller immediately touched off one of the most intense research efforts in the history of materials research [Bednorz and Müller 1986, Bednorz et al. 1987]. Several groups quickly reproduced their results and identified the superconducting phase as $La_{2-x}Ba_xCuO_4$. Many researchers then began to determine the effects of elemental substitutions and different processing conditions on the structure and superconducting properties of this oxide. Following this approach, in 1987, several groups discovered superconductivity above 90K in the Y-Ba-Cu-O system [Wu et al. 1987, Zhao et al. 1987, Cava et al. 1987a]. As before, the discovery was quickly reproduced around the world. The 90K phase was identified as $YBa_2Cu_3O_7$, and studies of the effects of substitution and processing conditions were initiated. This cycle has been repeated twice in 1988 with the independent discoveries of superconductivity at $110K$ in the $Bi_2Sr_2Ca_2Cu_3O_{10}$ systems by Maeda et al. [Maeda et al. 1988] and $120K$ in the Tl-Ba-Ca-Cu-O systems by Sheng and Hermann [Sheng and Hermann 1988a, Sheng and Hermann 1988b]. In 1993, the world record of T_c was again raised to 134K in $HgBa_2Ca_2Cu_3O_8$ systems [Putilin et al. 1993, Schilling et al. 1993] under ambient pressure and to 164K under the high pressure of 30GPa [Chu et al. 1993, Gao et al. 1994]. There are now hundreds of compounds discovered that become superconducting at above liquid nitrogen temperature (77K).

The rapid pace of discovery in higher temperature superconductors produced wide spread euphoria among the physics and materials science communities. High-temperature superconductors was expected to have applications in many areas of our life, be it for energy transport and high-power generators, transformers, current limiters, or for ultra-fast computers and communication technology, or in medicine for NMR (Nuclear Mangetic Resonance) tomography and for SQUID (Superconducting QUantum Interference Device) in magnetic encephalography. Large-scale uses were also foreseen, such as levitation for trains, magnetohydrodynamic propulsion in ships, and applications for fusion reactors. There is also great excitement about the fundamental physics of superconductivity. The figures showing the rapid rise of T_c such as Fig.1.1 were shown in almost every meeting related to superconductivity, implying that the days of "room-temperature superconductor" will not be far off.

Soon after, however, the euphoria was dampened by the discovery of the "weak-link" behavior of the high-T_c superconductors. It has become clear that the major barrier to the practical applications of these wonderful materials is not the critical temperature but their low critical current density, especially at high magnetic field. Improving these properties depends heavily on our detailed knowledge of the microstructures of these compounds and whether we can apply this knowledge to improve the properties of the materials.

Superconducting properties depend sensitively on the structures of the material in the scales larger than its coherence length. The conventional superconductors (such as Nb_3Sn) have coherent length in the order of several thousand anstroms. The high-T_c materials, on the other hand, have coherence length in the order of 10 to 20 anstroms. Therefore for high-T_c materials, a knowledge of the average, or ideal structure is not sufficient; often, it is the structural defects that determine their superconducting properties.

We have learned that some defects destroy superconductivity, while others promote it [Jorgensen 1991, Raveau 1992]. In high-temperature superconductors, defects play an important role concerning two aspects. On the one hand, high critical current densities in these materials can only be achieved by the presence of a high density of defects providing pinning centers for the magnetic flux lines. It is well known that ideal defects for flux line pinning should have a size equal to about

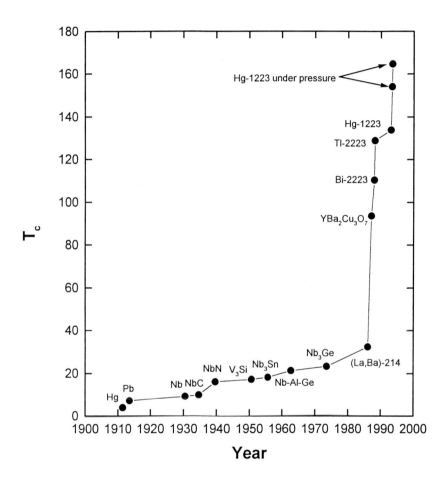

Figure 1.1 The highest superconducting transition temperature T_c over time.

the coherence length (10–20 Å). Furthermore, their density should be high enough to provide sufficient flux pinning at high magnetic fields. Due to the short coherence length of these superconductors, the superconductivity is depressed considerably by local defects. In this way the point defects can be effective pinning sites for magnetic flux lines, thus increasing the critical current density in single-crystalline materials. The strongly layered structure of some high-T_c materials (such as Bi-Sr-Ca-Cu-O) make the flux lines disassociate itself into individual vortices known as "pancakes", which makes line defects more appropriate pinning sites. On the other hand, extended defects in the superconductors such as large-angle grain boundaries form weak-links in high-T_c materials, which are key elements limiting their uses as bulk superconductors. However, properly engineered, their weak-link nature enables the fabrication of high-T_c Josephson junctions, which are useful as fast electronic switches and sensors. Therefore an understanding of the structures of such defects is critical to the application of high-T_c superconductors.

The intensive research effort in the past decade has not been in vain. In recent years, better understanding of the microstructure has helped us to make great improvement on the properties of high-T_c materials. The pioneering work by Heine, Tenbrink and Thoener in 1989 paved the way to make $Bi_2Sr_2CaCu_2O_x$ wires with reasonably high critical current densities over long lengths [Heine et al. 1989]. The processing technique they discovered has been effectively exploited by the large industrial labs, particularly at Sumitomo Electric Industries in Japan and at American Superconductor Corporation in USA, both of whom are producing multifilamentary Bi-2223 wires and tapes over 1 km in length with reasonable consistency. The critical current density of the bulk materials have been rising steadily, from several hundred A/cm^2 in 1980's to 70,500 A/cm^2 in 1997 (at 77K with no magnetic field). As Fig.1.2 shows, the critical current density of the best samples of the high-T_c materials at the liquid nitrogen temperature increases on average at the rate about 11 kA/cm^2 per year. Moreover, there seems to be no decrease in the recent rate of improvement, which is very encouraging. Various textured growth techniques have been developed that produces multifilament wires as long as 10 km that is superconducting at liquid nitrogen temperature. New techniques such

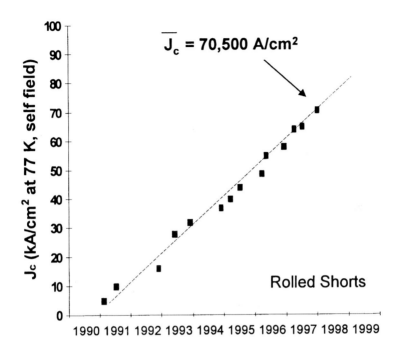

Figure 1.2 Plot of critical current density J_c performance vs. time for rolled multifilamentary $Bi_2Sr_2Ca_2Cu_3O_{10+\delta}$ conductors at American Superconductor Corp. [Qi Li, American Superconductor Corp., private communication]. See also Fig.1 in [Li et al. 1997a].

as heavy ion irradiation also improve the superconducting properties of these materials at high magnetic field by orders of magnitudes. On the other hand, high-quality $YBa_2Cu_3O_7$ thick films were manufactured using biaxially textured growth techiques such as Ion-Beam-Assited-Deposition (IBAD) [Iijima et al. 1992] and Rolling-Assisted-Biaxial Texturing (RABiTS) techinques [Wu et al. 1995] that largely eliminated the weak-links (large-angle grain boundaries) in these materials.

In the area of superconducting thin film applications the situations are even brighter, with the first commercial high-T_c sensors appearing in 1994.

This book is devoted to the applications of modern experimental as well as theoretical techniques in elucidating various structural defects in superconducting oxides. These techniques are complementary to each other, and very often several techniques have to be used to solve one structural problem. Thanks to the world-wide interest in their potential applications, rich experimental data has been obtained in recent years and sample quality has improved continuously. Furthermore, the richness of the microstructural variations in these materials and the close relationship between superconductivity and microstructure gives us insights of the microstructure of these compounds that are unavailable for other materials. This has set the stage for the physical realization and experimental accessibility of the entire field of solid state physics as well as materials science. Given the rapid progress and divserse research effort involved in this field, it goes without saying that this book cannot cover every aspect of these materials, nor does it represent the final knowledge in this field of research. Rather, the aim of this book is to describe (what we, the authors, believe to be) the major aspect of microstructures that are most important for improving the superconducting properties of high-T_c materials.

Although we have come a long way since the discovery of the first high-T_c superconductor in 1986, there has been a lack of consistent treatment of the materials science's perspective of this material. The information on the microstructures of high-T_c superconductors is scattered among review articles, conference reports, and papers in scientific journals. We hope this book will give the readers a consistent picture of our present knowledge in this field of research.

Chapter 2

Defects and Superconductivity

2.1 What is Superconductivity?

In 1911 H. Kamerlingh Onnes [Onnes 1911] discovered that when cooled to below 4.2 K, the resistance of mercury suddenly disappeared. He called the temperature at which this transition occurs the *critical temperature* T_c. The extremely small resistance (very high conductivity) below T_c suggested the name of the phenomena "superconductivity". When the temperature is below T_c the mercury is in *superconducting state*. As the temperature rise above T_c the superconductivity of the mercury is destroyed and it re-entered *normal state* that has electrical resistance.

Further experiments by Onnes established two other thresholds besides T_c that can destroy superconductivity. One, as Onnes indicated in 1913 [Onnes 1913], is large current density, which has a "threshold value" J_c above which the resistance increases very rapidly. This "threshold value" was later called *critical current density* which became one of the most important parameters for designing superconducting magnets.

Below the critical temperature T_c, the superconductivity can also be destroyed by a high external magnetic field. The minimum magnetic field H_c needed to destroy superconductivity is called the *critical field*, which is a function of temperature. For the majority of elemental metal superconductors the relationship between the critical field and temperature can be approximated by [Onnes 1913]

$$H_c(T) = H_{c0}[1 - (\frac{T}{T_c})^2], \qquad (2.1)$$

where H_{c0} is the critical field at zero temperature.

In the years followed the many other materials were found to be superconducting at low temperatures. These include most metallic elements, intermetallic compounds such as Ni_3Ge, even some organic compounds, and of course, the superconducting oxides which are the subject of this book. Of particular importance to the application of superconductivity is a group of intermetallic compounds with the so-called A-15 structure. Among these intermetalic compounds, Nb_3Ge, with $T_c \approx 23$ K, held the record of the highest T_c before the discovery of high temperature superconducting oxides; while Nb_3Sn and Nb_3Ti, with their high critical current density ($J_c \sim 10^6$ A/cm^2 at 4.2 K), was widely used to make superconducting wires and magnet. These so-called "low-T_c" superconductors have been used in a variety of applications such as high field magnets and radio-frequency cavities in particle accelerators, as well as nuclear magnetic resonance and magnetic resonance imaging in medical facilities.

The research into the properties and mechanisms of the superconductivity soon followed its discovery. In 1933 Meissner and Ochsenfeld discovered the perfect diamagnetism of the materials in superconducting state (see the next section for detailed discussions). Based on this findings, F. and H. London proposed a two-fluid model in 1934. The London model (discussed later in this chapter) explained the Meissner effect and predicted the penetration length λ, which is a characteristic length of penetration of magnetic flux into a superconductor.

In 1950 V. Ginzburg and Lev Landau proposed a phenomenological theory of superconductivity based on the thermodynamics of phase transitions. In 1957 A. Abrikosov studied the behavior of superconductor in an external magnetic field and classified superconductors into two types. Type-I superconductor has only one critical magnetic field, below which the magnetic flux is completely expelled from its interior. Type-II superconductor, however, has two critical magnetic fields. It expels magnetic flux from its interior only when the external field is below its lower critical field. Above this lower critical field the magnetic

2.2. The Electromagnetic Properties of Superconductor

flux can partially penetrate into the interior of a type-II superconductor and forms so-called "mixed states". The superconductivity in type-II superconductor can be sustained at very high external field, making it ideal for large-scale (high-field, high current) applications.

The microscopic theory of superconductivity was not established until nearly 50 years since the discovery of superconductivity. In 1957 John Bardeen, Leon Cooper and Robert Schrieffer proposed a quantum mechanical theory of superconductivity, which is usually referred to as BCS theory.

2.2 The Electromagnetic Properties of Superconductor

As we discussed in the previous section, the most striking property of a superconductor is its zero electrical resistance. It is natural to consider superconductor as a perfect conductor with electrical conductivity σ to be infinite. The current density J relates to the electric field ϵ by Ohm's law

$$J = \sigma \epsilon \tag{2.2}$$

so that the current density in a superconductor can remain constant in absence of the electric field ϵ. Then according to Maxwell equation

$$\nabla \times \epsilon = -\frac{\partial \mathbf{B}}{\partial t}, \tag{2.3}$$

The magnetic induction $\mathbf{B}$ will be independent of time since ϵ is zero everywhere. In other words, The magnetic induction $\mathbf{B}$ in a perfect conductor should depend on its history of magnetization.

When a perfect conductor is cooled below T_c and is then applied a weak magnetic field, the magnetic induction inside the material $\mathbf{B} = 0$, i.e. there is no magnetic flux line. If, however an external magnetic field is applied when $T > T_c$, and then its temperature is lowered to below T_c, one would expect the magnetic induction $B \neq 0$ at the interior of the perfect conductor.

In 1933 Meissner and Ochsenfeld discovered that this is not the case for superconductors. They found that the magnetic induction becomes zero inside the superconductor when it is cooled below T_c in a weak magnetic field. In other words, the magnetic flux is expelled from the interior of the superconductor. Therefore regardless of its initial conditions before it became superconducting, the magnetic induction insider a superconductor is always zero (as long as the external field is less than H_c) when $T < T_c$. i.e. The superconductor is a perfect diamagnet. This effect was later called "Meissner Effect".

The Meissner effect shows that superconductor cannot be simply regarded as a perfect conductor. It also shows that superconductivity is a thermodynamically equilibrium state, independent of the path it takes to arrive at the superconducting state.

Thus the two properties, zero resistance and Meissner effect, become the signature of the superconductivity. Any true superconductors have to have these two properties below the critical temperature.

2.3 The London Model

It is obvious from the discussion in the previous section that the conventional electrodynamic equations [Eqs.(2.2) and (2.3)] cannot explain the Meissner effect. Since Eqn.(2.3) (the Maxwell equation) is always valid, we need to modify Eqn.(2.2) to take into account the special properties of the superconductors.

The simplest model of superconductivity is the two-fluid model, which is established by London brothers in 1935 [London and London 1935]. They propose that there are two types of charge-carriers in the superconducting state, the normal electrons and superconducting electrons which have zero resistivity. While the normal electrons obey Ohm's law [Eqn.(2.2)] with finite σ, the superconducting electrons are not scattered by the defects thus they do not contribute to resistivity. The whole system shows no resistivity because the current is carried by the superconducting electrons. The equation of motion of the superconducting electrons can then be deduced as follows.

Let us assume the superconducting electrons in the system have charge of e^*, effective mass of m^* and density of n_s, the superconducting current density is then

$$\mathbf{j_s} = n_s e^* \mathbf{v} \tag{2.4}$$

where $\mathbf{v}$ is the velocity of the superconducting electrons. In a electrical

2.3. The London Model

field ϵ the equation of motion of the superconducting electron is

$$m^* \frac{d\mathbf{v}}{dt} = e^*\epsilon. \tag{2.5}$$

From the above two equations we can establish the relationship

$$\frac{d\mathbf{j_s}}{dt} = \frac{n_s e^{*2}\epsilon}{m*} = \frac{\epsilon}{\Lambda^2}, \tag{2.6}$$

where

$$\Lambda^2 = \frac{m^{*2}}{n_s e^{*2}}. \tag{2.7}$$

Putting Eqn.(2.6) into the Maxwell equation [Eqn.(2.3)] and expressing the magnetic induction $\mathbf{B}$ in terms of the vector potential $\mathbf{A}$, $\mathbf{B} = \nabla \times \mathbf{A}$, we have

$$\frac{\partial}{\partial t}\left[\nabla \times \left(\mathbf{j_s} + \frac{1}{\Lambda^2}\mathbf{A}\right)\right] = 0. \tag{2.8}$$

In order to explain the Meissner effect, F. and H. London took the following solution of Eqn.(2.8)

$$\mathbf{j_s} + \frac{1}{\Lambda^2}\mathbf{A} = 0,$$

i.e.

$$\mathbf{j_s} = -\frac{1}{\Lambda^2}\mathbf{A}. \tag{2.9}$$

Eqn.(2.9) is called London equation which describes the electrodynamics of a superconductor, From which the magnetic induction $\mathbf{B}$ is given as

$$\mathbf{B} = \nabla \times \mathbf{A} = -\nabla \times \Lambda^2 \mathbf{j_s}. \tag{2.10}$$

Applying the Maxwell equation

$$\nabla \times \mathbf{B} = \mu_0 \mathbf{j_s}, \tag{2.11}$$

to Eqn.(2.10) yields

$$\begin{aligned}\mathbf{B} &= -\frac{\Lambda^2}{\mu_0}\nabla \times \nabla \times \mathbf{B} \\ &= -\lambda_L^2[\nabla(\nabla \cdot \mathbf{B}) - \nabla^2 \mathbf{B}].\end{aligned}$$

Again from Maxwell equation $\nabla \cdot \mathbf{B} = 0$, we obtain

$$\nabla^2 \mathbf{B} = \frac{1}{\lambda_L^2} \mathbf{B}, \tag{2.12}$$

where

$$\lambda_L = \frac{\Lambda}{\mu_0^{1/2}} = \left(\frac{m^*}{\mu_0 n_s e^{*2}}\right)^{1/2} \tag{2.13}$$

is called the *London penetration depth*.

The physical meaning of the London penetration depth can be illustrated using a one-dimensional system. If we consider a uniform superconductor in the region $x > 0$ and apply the magnetic field H_0 parallel to the superconductor-vacuum interface, from Eqn.(2.12) we get

$$\frac{d^2 B_y(x)}{dx^2} = \frac{B_y(x)}{\lambda_L^2} \tag{2.14}$$

which has the solution

$$B_y(x) = \mu_0 H_0 \exp(-x/\lambda_L). \tag{2.15}$$

Fig.2.1 shows the space profile of the magnetic induction near the surface of the superconductor. We can see clearly that the magnetic induction approaches zero in the interior of the superconductor (when $x \gg \lambda_L$). However unlike the simple picture we have in the previous section, the magnetic flux can penetrate into the superconductor in the region $0 < x < \lambda_L$. From Maxwell equation we find

$$j_{sz} = \frac{1}{\mu_0}(\nabla \times \mathbf{B})_z = -\frac{H_0}{\lambda_L} \exp(-x/\lambda_L). \tag{2.16}$$

In other words, there is a superconducting current along z-axis near the surface of the superconductor which decays exponentially along x axis, with characteristic decay length λ_L. The superconducting current density j_{sz} creates a magnetic field along negative y axis at $x > \lambda_L$ which cancels the the external magnetic field. Therefore the magnetic field inside the superconductor is zero, which is exactly what Meissner and Ochsenfeld observed. Because of this, the superconducting current along the surface of a superconductor is called *shielding current*.

2.3. The London Model

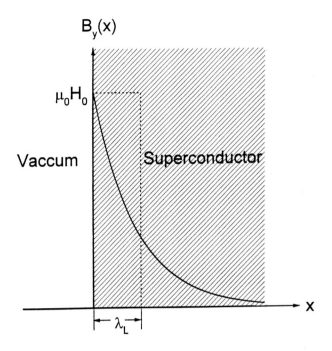

Figure 2.1 The distribution of magnetic induction near the superconductor-vacuum interface according to London equation.

We can see from the examples above that the London theory describes the Meissner effect very well. We can also see that the London penetration depth describes the characteristic length over which magnetic induction and shielding current decays.

Although London theory has great success in describing many phenomena in superconductivity, it has one serious flaw. As we can see from the discussions in this section, the London theory assumes that the density of the superconducting electrons n_s is uniform. It also does not give the temperature-dependence of n_s. Surely n_s will decreases as the temperature increases, and eventually goes to zero at T_c. Near the superconductor-normal interface, n_s should also varies depending on the distance from the interface. For such information we have to turn to Ginzburg-Landau theory, which is the subject of our next section.

2.4 Ginzburg-Landau Theory

While the mechanism of superconductivity was not discovered until the invention of BCS theory in 1960s, it was well-established that the superconducting current (without energy dissipation) is a quantum phenomena in macroscopic scale, just as the eternal current created by electrons in atoms (which can be regarded as quantum phenomena in a microscopic scale). Therefore without discussion of the detailed mechanism of superconductivity, we can nevertheless write down the wave functions of superconducting electron as

$$\psi(\mathbf{r}) = \sqrt{n_s(\mathbf{r})} \exp[i\phi(\mathbf{r})], \tag{2.17}$$

and the current density can be expressed as

$$\mathbf{j_s}(\mathbf{r}) = \frac{ie^*\hbar}{2m^*}(\psi^*\nabla\psi - \psi\nabla\psi^*) - \frac{e^{*2}}{m^*}\mathbf{A}\psi\psi^*, \tag{2.18}$$

where $\mathbf{A}$ is the magnetic vector potential, and $\mathbf{B} = \nabla \times \mathbf{A}$. Combining the above equations we get

$$\mathbf{j_s}(\mathbf{r}) = -n_s(\mathbf{r})\frac{e^{*2}}{m^*}\left[\frac{\hbar}{e^*}\nabla\phi - \mathbf{A}(\mathbf{r})\right], \tag{2.19}$$

taking curl on both side of the equation and noticing $\nabla \times \nabla\phi = 0$ we have

$$\nabla \times \left[\frac{m^*}{n_s(\mathbf{r})e^{*2}}\mathbf{j_s}(\mathbf{r})\right] = -\nabla \times \mathbf{A}. \tag{2.20}$$

We can see that only when $n_s(\mathbf{r})$ is constant can we get London equation [Eqn.(2.10)] which is

$$\frac{m^*}{n_s e^{*2}}\nabla \times \mathbf{j_s} = -\mathbf{B}. \tag{2.21}$$

The condition that $n_s(\mathbf{r})$ is constant can only be valid if we do not need to consider the suppression of n_s near the superconductor-normal interface. As we will see in the following sections, such interface plays an essential role in the application of superconductors.

In 1950 Ginzburg and Landau treated the superconductivity from another prospective. Instead of regarding the wave function defined in Eqn.(2.17) as the real wave function of the superconducting electron

2.4. Ginzburg-Landau Theory

which obeys the Schrödinger equation, they regard $\psi(\mathbf{r})$ as the order parameter of the superconducting state, which should satisfy the condition for thermodynamic equilibrium. Such equilibrium is achieved by taking the minimum value of the Gibbs free energy.

An order parameter, by definition, represents the degree of order in a system. In the theory of order-disorder phase transitions, it is common to define order parameter in such a way so that it is one in the order state and zero in the disordered state. The underline physical meaning of the order parameter depends on the system considered. Ginzburg and Landau postulated that if the order parameter ψ is small and varies slowly in space, the free-energy density f can be expended in a series of the form

$$f = f_{n0} + \alpha(T)|\psi|^2 + \frac{\beta}{2}|\psi|^4 + \frac{1}{2m^*}\left|\left(\frac{\hbar}{i}\nabla - e^*\mathbf{A}\right)\psi\right|^2 + \frac{h^2}{2\mu_0} \quad (2.22)$$

where $f_{n0} + h^2/2\mu_0$ is the free energy density of the normal state with local magnetic field h. $\mathbf{A}$ is the vector potential related to the local magnetic field h by $h = \nabla \times \mathbf{A}$.

There are two phenomenological parameters, α and β, in the Ginzburg-Landau free energy. As in the Landau theory of second order phase transition one takes

$$\alpha = \alpha_0(T - T_c), \quad (2.23)$$

and β is a positive constant which is independent of temperature T. The numerical value of these parameters are usually obtained by fitting the magnetic measurement data to the predictions of the Ginzburg-Landau theory.

Let us first discuss the situation where there are no external magnetic fields. From Eqn.(2.22) the free energy of superconductors can be expressed as

$$f(0) = f_{n0} + \alpha|\psi|^2 + \frac{\beta}{2}|\psi|^4. \quad (2.24)$$

By minimizing the free energy with respect to $|\psi|^2$, we obtain the density of the superconducting electrons as the system is in thermodynamic equilibrium as

$$|\psi_0|^2 = -\frac{\alpha}{\beta}.$$

Meanwhile
$$f(0) - f_{n0} = -\frac{\alpha^2}{2\beta}.$$

Comparing the above expression with the definition of the thermodynamic critical field H_c as

$$f(0) - f_{n0} = -\frac{\mu_0 H_c^2}{2}$$

we find the expression for the parameters in Eqn.(2.22) as

$$\alpha = -\frac{\mu_0 H_c^2}{|\psi_0|^2};$$

$$\beta = -\frac{\alpha}{|\psi_0|^2} = \frac{\mu_0 H_c^2}{|\psi_0|^4}.$$

If the temperature is not much lower than the critical temperature T_c we can obtain from Eqn.(2.23) and the above equations

$$H_c^2(T) = -\frac{|\psi_0|^2 \alpha}{\mu_0} = \frac{|\psi_0|^2 \alpha_0}{\mu_0}(T - T_c),$$

which, at $T \sim T_c$, agrees quite well with the empirical formula derived from experimental observations [Eqn.(2.1)].

Now let us discuss the situation with the presence of external magnetic field. In this case the free energy f is the function of both order parameter ψ^* and the vector potential $\mathbf{A}$. At thermodynamic equilibrium f has to be minimum, i.e.

$$\frac{\delta f}{\delta \psi^*} = 0,$$

$$\frac{\delta f}{\delta \mathbf{A}} = 0.$$

From the above conditions and Eqn.(2.22) we can obtain the two Ginzburg-Landau equations:

$$\frac{1}{2m^*}(-i\hbar \nabla + e^* \mathbf{A})^2 \psi + \alpha \psi + \beta |\psi|^2 \psi = 0, \qquad (2.25)$$

2.4. Ginzburg-Landau Theory

$$\mathbf{j}_s = \frac{i\hbar e^*}{2m^*}(\psi^*\nabla\psi - \psi\nabla\psi^*) - \frac{e^{*2}}{m^*}|\psi|^2\mathbf{A}. \tag{2.26}$$

These two closely-coupled equations with appropriate boundary conditions completely describe the properties of the superconductors as long as we can obtain the values of the parameters α and β.

If we ignore the spatial-variation of the order parameter ψ, Eqn.(2.26) can be simplified to London equation

$$\mathbf{j}_s = -\frac{e^{*2}}{m^*}|\psi_0|^2\mathbf{A}. \tag{2.27}$$

Combining this with the Maxwell equations and following the same procedure as the previous section, we can obtain London penetration depth as

$$\lambda_L = \left(\frac{m^*}{\mu_0|\psi_0|^2 e^{*2}}\right)^{1/2}. \tag{2.28}$$

Exactly the same as in Eqn.(2.13).

If the spacial-variation of ψ cannot be ignored, we can rewrite Eqn.(2.25) when the external magnetic field is zero as

$$-\frac{\hbar^2}{2m^*}\nabla^2\psi + \alpha\psi + \beta|\psi|^2\psi = 0. \tag{2.29}$$

Notice $\beta = -\alpha/|\psi|^2$ and assuming $\psi = \psi_0 + g$ we can obtain the differential equation with respect to g as

$$\nabla^2 g + \frac{m^*|\alpha|}{\hbar^2}g - 0. \tag{2.30}$$

For one-dimensional system, the above equation yields

$$g = \psi - \psi_0 = C\exp(-r/2\xi) \tag{2.31}$$

where C is a constant and the coherence length ξ is defined as

$$\xi^2 = \frac{\hbar^2}{2m^*|\alpha|}. \tag{2.32}$$

Therefore the coherence length defines the length-scale over which the density of superconducting electrons varies significantly.

The London penetration depth λ_L and the coherence length of the superconducting electrons ξ are the two basic parameters of the superconductors. While λ_L defines the decay length of the circulating supercurrent that is needed to cancel the external magnetic field (thus give rise to Meissner Effect), ξ tells us the decay length of the superconducting order parameters. The coherence length ξ is closely related to α thus the critical temperature T_c as shown in Eqn.(2.32). Since α is of the same order of magnitude as $k_B T_c$, we can immediately deduce that the coherence lengths of the high-T_c superconductors are much shorter than that of the conventional superconductors. In fact, while the conventional superconductors have coherence length of several thousand angstroms, the high-T_c superconductors have coherence length of only dozens of angstroms.

The short coherence length of the high-T_c superconductors has important implications to the effect of microstructural defects on its superconducting properties. We can regard the coherence length as the "wave length" of the superconducting electron wave functions. Defects with sizes that are much smaller than the coherence lengths cannot be "felt" by the superconducting electrons. Therefore in conventional superconductors, non-extended defects such as point defects and dislocations do not affect superconductivity very much. On the other hand, the strain field of a single point defects has the range comparable to the coherence length of high-T_c superconductors. As we will see in the following sections, even though the small defects may not block the path of the supercurrent completely, they play an important role on the magnetic and transport properties of high-T_c superconductors.

2.5 Two Types of Superconductors

We can classify the superconductors into two categories according to their magnetic properties. Type-I superconductors have only one critical field H_c. Its magnetization curve is shown in Fig.2.2(a). It has Meissner effect in the superconducting state and the superconducting current near its surface maintains the perfect diamagnetism of the interior. Most of the metal elements are type-I superconductors. Type-II superconductors, on the other hand, have two critical field, the lower critical field H_{c1} and upper critical field H_{c2}, as shown in Fig.2.2(b).

2.5. Two Types of Superconductors

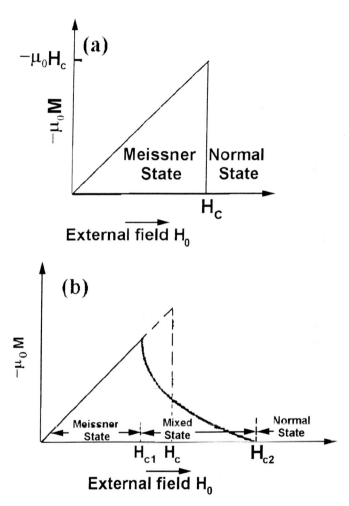

Figure 2.2 The magnetization curves of the two types of superconductors. (a) Type-I superconductor; (b) type-II superconductor.

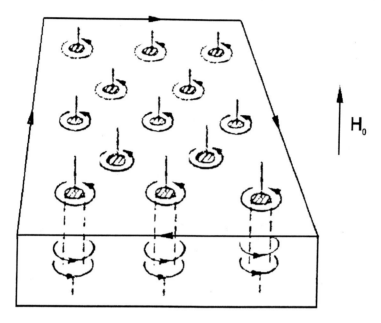

Figure 2.3 The mixed state of type-II superconductor with quantized vortices.

When the external field H_0 is smaller than the lower critical field H_{c1}, the type- II superconductor is in Meissner state with no magnetic flux in the interior, just like the type-I superconductor. When the external field H_0 is between H_{c1} and H_{c2}, the type-II superconductor is in a *mixed state* with magnetic flux in the interior. These flux lines form many small cylindrical normal regions surrounded by the superconducting region called vortices, as shown in Fig.2.3. There is still diamagnetic supercurrent near the edge of the sample. Thus the type-II superconductor in the mixed state has both diamagnetism (but $B \neq 0$) and zero resistance. the number of the normal domains increases with the external magnetic field, however the size of each normal region remains the same. When the external field reaches H_{c2} the neighboring normal regions come in contact with each other and the superconducting region can no longer form a percolating path. The sample thus has

2.5. Two Types of Superconductors

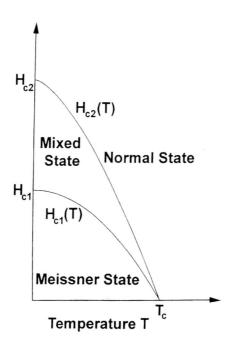

Figure 2.4 Phase diagram of a type-II superconductor.

finite resistance and becomes normal. Fig.2.4 shows the typical phase diagram of the type-II superconductor.

At temperature T and external field H_0, the energy difference between normal state and superconducting state for a type-II superconductor can be written as

$$\Delta E = -\int_0^{H_{c2}} M \, dH. \qquad (2.33)$$

The integral can be formally expressed as $\mu_0 H_c^2/2$, i.e.

$$\frac{\mu_0 H_c^2}{2} = -\int_0^{H_{c2}} M \, dH, \qquad (2.34)$$

where H_c is called *thermodynamic critical field*.

Since there are no normal/superconducting interfaces in type-I superconductor but there are many such interfaces in type-II superconductor, it is obvious that the interface energy plays an essential role in determining the types of superconductors. The interface energy is defined as the extra energy relative to that of the bulk superconducting state. There are two sources of such interface energy. Let us illustrate this by the following simple model.

Assuming the superconductor occupies the semi-infinite space in $x > 0$. The magnetic flux can then penetrate in the region $0 < x < \lambda_L$, where the magnetic induction is approximately equal to the external field, i.e. $B = \mu_0 H_0$ (see Fig.2.2). Therefore the magnetic energy at the interface is $-\lambda_L \mu_0 H_0^2/2$. On the other hand, the density of the superconducting electrons n_s decreases drastically near the interface with the typical length scale equal to the coherence length ξ. Thus the electrons in the region $0 < x < \xi$ is in normal state with energy $\xi \mu_0 H_0^2/2$ higher than the electrons in superconducting state. From these assumptions we obtain the interface energy

$$\sigma_{ns} = \xi \frac{\mu_0 H_c^2}{2} - \lambda_L \frac{\mu_0 H_0^2}{2}. \qquad (2.35)$$

Since $H_0 \leq H_c$ when the sample is in the superconducting state, we can see from the above equation that the interface energy depends sensitively on the two parameters ξ and λ_L. When $\xi > \lambda_L$ the interface energy $\sigma_{ns} > 0$. It costs energy to create a normal–superconductor interface. Therefore the superconductor shows Meissner effect (no normal regions in the interior thus total exclusion of magnetic flux), which is the case for type-I superconductor. If, however, $\xi < \lambda$ and $H_{c1} < H_0 < H_{c2}$, then $\sigma < 0$, which means that creating normal–superconductor interfaces actually *lowers* the energy. In this case the superconductor is in mixed state, which is the case of type-II superconductor.

It is clear then from the above discussion ratio of the penetration length λ_L and the coherence length ξ

$$\kappa = \frac{\lambda}{\xi} \qquad (2.36)$$

is an important parameter of the superconductor. The energy gained by dividing magnetic flux into small vortices increases with κ value. Therefore type-II superconductors with larger κ value have low H_{c1} and high H_{c2}. High-temperature superconductors are the extreme cases of type-II superconductors, with κ over 100. For some high-temperature superconductors, their H_{c2} value is so high that it is unmeasurable using the current experimental instrument.

2.6 Critical Current in Type-II Superconductors

As we discussed in the previous section, magnetic vortices penetrate into the type-II superconductors when the external field $H > H_{c1}$. The vortices consists of tiny normal regions in which $\mathbf{B} \neq 0$. If we then pass an electric current J through the sample that contains the magnetic vortices, the vortices will suffer a Lorentz force from the current

$$\mathbf{F} = \mathbf{J} \times \mathbf{B}. \tag{2.37}$$

The force will displace the vortices. Then, according to Maxwell equation, the moving vortex creates an electric field

$$\mathbf{E} = \frac{d\mathbf{\Phi}}{dt}. \tag{2.38}$$

Therefore when the external magnetic field exceeds H_{c1}, the current J dissipates energy at the rate of $\mathbf{E} \cdot \mathbf{J}$ per unit area. This energy dissipation is equivalent to electric resistivity. For type-II superconductors with no defects, the critical current at external field $H > H_{c1}$ is thus zero. In the extreme type-II superconductor such as the high-T_c oxides, H_{c1} is only several hundred Gauss. There can be no practical applications of type-II superconductors unless some mechanism exists which prevents the Lorentz force from moving the vortices. Such a mechanism is called a "pinning" force, since it "pins" the vortices to fixed locations in the material thus making $E = d\Phi/dt = 0$.

The average pinning force density F_p should be equal to or greater than the Lorentz force in order to prevent the flux line lattice from moving. From Eqn.(2.37) the maximum pinning force can be related to the critical current density J_c as $F_p = J_c B$. When $J > J_c$ the flux line lattice will move and there the system will be in a resistive state.

In the next section we will discuss the mechanism of flux-line pinning and its relationship to the microstructure of the superconductors.

2.7 Flux-line Pinning and Microstructure

As we discussed in the previous section, a magnetic flux line in type-II superconductor consists a normal region surrounded by circulating supercurrents. The size of such normal region (the so-called "vortex core") is of the same order of magnitude as the coherence length ξ of the superconductor. It costs energy ΔE_s for such normal region to reside in superconductors. This increase of energy is balanced against the reduction of magnetic energy $-\Delta E_m$. i.e.

$$\Delta E_s \leq -\Delta E_m = k_B T_c \frac{\xi^3}{V_p} \qquad (2.39)$$

where T_c is the superconducting transition temperature and V_p is the molar volume of the superconducting electrons.

The energy of the system can be further reduced if the vortex core can reside in a non-superconducting ($T_c = 0$) region since there is no cost to the superconducting energy. As we mentioned, the size of the vortex core is of the same order of magnitude as the coherence length ξ. Therefore, it is energetically most favorable for the vortex core to reside at the non-superconducting region with size similar to ξ, and it will cost energy for the vortex core to move from such none-superconducting region into the superconducting region. In other words, such non-superconducting region acts as a potential well inside which the vortex core is trapped. Such a region is generally called "flux pinning site" since it "pins" the flux line. The maximum pinning force is the external force required to move the flux line from the pinning site to the superconducting region.

The non-superconducting regions that can act as pinning sites for the flux lines are microstructual inhomogeneities and structural defects. The optimum pinning sites for each materials are different since the coherence length ξ varies greatly. For conventional superconductors the coherence length is of the order of several thousands of angstroms. The pinning sites for such materials are typically extended defects such as grain boundaries and dislocations. For high-T_c superconductors, on the other hand, the coherence lengths are of the order of tens of angstroms, thus atomic scale structural inhomogeneities such as point defects can also play roles in flux line pinning.

2.8 Good and Bad Defects

While certain types of defects with sizes close to ξ can act as pinning centers of magnetic flux lines, thus enhancing the properties of superconductors (higher critical current, better performance under high magnetic field, etc.), defects with size much bigger than the coherence length ξ can be detrimental to the superconducting properties of the materials. These extended defects act as barriers to the supercurrent flow, reducing the volume of superconducting phase in the materials.

For high temperature superconductors, the "good defects" for flux line pinning including point defects, certain dislocations, twin boundaries and certain types of low-angle grain boundaries. We will discuss the merits of each type of defects in the following chapters.

Since the high-T_c cuprates are ceramics, grain boundary structures play a very important role on the superconducting properties of the materials. Most types of grain boundaries found in high-T_c materials are bad for the superconducting properties. In high-T_c wires and tapes, these grain boundaries are the major stumbling block of the practical uses of high-T_c materials in high current, high-field applications. Much effort has been made to improve the grain boundary properties in high-T_c materials. We will discuss these efforts and the effect of different grain boundaries on the superconducting properties in later parts of this book.

Chapter 3

Tools Used to Study the Defects

3.1 Introduction

In this chapter, we describe some of the experimental as well as theoretical techniques used to study microstructure and defects in high-temperature superconductors. Since there are many fine books devoted to the details of the techniques discussed here, we will only concentrate on those techniques that are used extensively in obtaining the data discussed in the subsequent chapters of this book. Many of these techniques had been used widely in studying the ceramic structure, and have received renewed interest since the discovery of high-temperature superconductors.

3.2 Experimental Tools

3.2.1 Transmission Electron Microscopy

Electrons, x-rays and neutrons are complementary probes employed to study the structure of materials, because neutrons interact with nuclei, x-ray with electrons in the materials, and incident electrons with the electrostatic potential. The unique advantage of transmission electron microscopy (TEM) is probing an extremely small area of a sample to study local crystal structure and structural defects. With sophisticated

TEMs we can study chemical composition within nano-meter sized volumes, atomic arrangement and electronic structure of materials.

The technology of modern electron microscopes has been developed so far that nowadays commercially available instruments and the sophisticated attachments fulfill practically all requirements necessary for studying thin crystals. While the chemical resolution of a TEM in Electron Dispersive x-ray Spectroscopy (EDS) and Electron Energy-Loss Spectroscopy (EELS) experiments is mainly determined by the size of the probe, its spatial resolution is governed by the spherical aberration of the objective lens. The ultimate resolution, $r_{\min}$, can be expressed by [Buseck et al. 1988]:

$$r_{\min} = 0.66 \lambda^{0.75} C_s^{0.25} \qquad (3.1)$$

where λ is the wavelength of the incident electrons ($\lambda \approx 0.0025$nm for 200kV electrons), and Cs the constant of spherical aberration (for a high-resolution pole piece, the Cs can be as small as 0.5mm). High-resolution microscopes, usually with accelerating voltage of 200kV and above, now possess a point-to-point resolution of better than 0.19nm. Very often, ultimate resolution is not necessary for electron microscopy studies in materials science. For these studies it is more important that the specimen can be shifted and tilted over large ranges and that different signals of scattered electrons and x-rays can be detected in analytical microscopy studies.

Charged incident electron interact strongly with the transmitted specimen, the scattering cross-section for electrons is rather large compared to the cross-section of neutron or x-rays. Thin samples, range of 5nm to 1μm depending on the sample, the imaging mode and on voltage of the TEM, are required. Different kinds of electron and electromagnetic waves are emitted from a specimen which is irradiated with high-energy electrons. The different waves result from elastic or inelastic scattering processes. Different signals (see Fig.3.1) are used for different microscopy modes. Information on the crystal structure and on structural defects in the specimen can be obtained by studying the elastically scattered electrons whereas investigations of inelastically scattered electrons and of other waves at the exit surface of the specimen allow the determination of chemical composition and topology of the specimen.

3.2. Experimental Tools

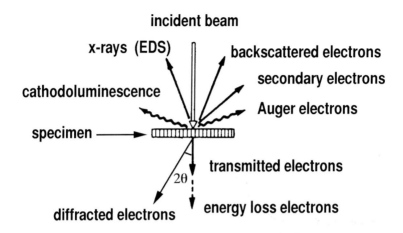

Figure 3.1 Electrons (electromagnetic waves) emitted from a transmitted specimen as a result of elastic and inelastic scattering or diffraction of the incident electron waves.

A transmission electron microscope can be operated in different modes as described briefly in the following: standard (or conventional) TEM mode with bright-field, dark-field and weak-beam imaging techniques to obtain diffraction contrast; high-resolution imaging mode to obtain structural image with phase contrast in a low-indexed Laue zone; diffraction mode including selected-area-diffraction (SAD), convergent beam electron diffraction (CBED) to obtain crystallographic structural information; nano-probe analytical mode to obtain chemical composition and electronic structural information using electron dispersive x-ray spectrometer and electron energy-loss spectrometer. Recent advancement of TEM instrumentation further enhanced its capabilities. A field-emission source, which gives much higher brightness than a W or a LaB_6 filament, can now reach a probe size smaller than 0.4nm. With a newly developed digital scanning unit, we can perform x-ray chemical mapping at resolution of 1nm, while with an annular dark-field detector we can capture composition-sensitive Z-contrast images on an atomic scale [1]. Gatan Image Filter (GIF) is also a powerful attachment to reveal structural information related to specific energy-loss electrons including chemical mapping for light elements. Digital data-recording systems with CCD (charge-coupled device) cameras are vital to any quantitative analysis of TEM observations. For more details the readers is referred to textbooks and literature on electron microscopy.

3.2.2 X-ray and Neutron Diffraction

Apart from electrons, x-ray and neutron are frequently used for structure determination and the study of crystal structures and structural defects as they provide the appropriate range of wavelengths for such investigations. For the wavelength of interest in normal diffraction work (0.05 to 2nm), corresponding photon energies are in the range of about 1-40 keV whereas neutron energies are between 0.85 and 400 meV. Excitations in condensed matter (phonons, magnons etc) are in the range of a few meV. The relative energy change of x-rays scattered inelastically (with energy loss or gain) by the sample is then very small ($< 10^{-6}$) and cannot normally be resolved (except if Mössbauer sources or synchrotron radiation combined with back-scattering techniques are used).

In contrast, neutron can suffer an appreciable relative change in energy, so that elastic (no energy change) and inelastic scattering can be distinguished. Another important difference arises form the magnetic moment of the neutron which interacts with the local magnetization density of the nuclei. This leads to magnetic scattering. This property of neutron scattering is an especially important asset in studying the oxygen related structural features in high-temperature superconductors. The electron density of oxygen atoms is too small to be "seen" by X-ray scattering. However neutrons can still interact with the nuclear magnetic moments of the oxygen nuclei. Therefore neutron scattering is an important tool to study oxygen related microstructure, as we shall see in Chapter 6.

---

[1] Test results from JEOL 3000F-BNL special, 1997

3.3 Theoretical Modeling and Computer Simulation

3.3.1 Interaction Between Atoms

The first issue we encounter when we set up models for the structural properties is how to obtain the interatomic potentials in the material. It is a especially difficult problem for high-temperature superconductors since it requires some understanding of their electronic structure.

The electronic structure of the high-temperature superconductor is a very complicated one. The part of the structure commonly considered as responsible for superconductivity, the CuO_2 plane, is highly covalent and with band structure similar to metals. Unlike simple metals, however, the electrons (and holes) in this plane are strongly correlated. The simple one-electron approximation such as tight-binding or free-electron model is not valid for these materials. On the other hand the strong antiferromagnetic coupling between copper atoms makes the exchange interaction very important in these materials. To make the matter worse, atoms in other "charge reservoir" layers act more like ionic crystal than metal, and their interaction cannot be simply modeled in band theory.

First principle calculation using local density approximations (LDA) has become the "standard theory" of solid state physics and has been very successful in predicting the properties of metals and semiconductors. However this method cannot be used in predicting the properties of high-T_c materials because of the strong exchange interaction of electrons in these materials which the LDA does not take into account.

Other "semi-empirical" method such as tight-binding model and embedded atom method (EAM), though quite effective in modeling simple metals, are not very suitable for high-T_c materials either due to the strong correlation of electrons in the materials.

The electronic structure of the high-T_c cuprate is the subject of intensive research ever since the discovery of the high-temperature superconductors, since it is related to the mechanism of the high-T_c superconductivity. Although great progress has been made, we still do not have a quantitative model of the atomic interaction from electronic structure calculations. Even if we have a complete understanding of the electronic structure, the atomic interaction model derived from it will be so complex that it would not be suited for the study of defect structures. Therefore we are forced to use some empirical models to describe the atomic interactions in these materials.

To set up empirical models for a material, one usually assumes an analytical function as the approximation of the atomic interaction potential. The parameters of the analytical function is then fitted to experimental data. To illustrate the procedure, we briefly describe two types of such models here.

The interaction potentials in the atomic scale for ionic crystals are usually established based on a Born model representation of polar solids. It consists of two-body, central potentials which represents short-range interactions as a perturbation to the Coulomb interactions between charged particles which are dominate in ionic solids. The short-range interaction potentials are commonly described by an analytical function of the Buckingham form. The interaction potential is then represented by effective pairwise potentials of the following form

$$V(r_{ij}) = \frac{Z_i Z_j e^2}{r_{ij}} + \phi_{ij}. \tag{3.2}$$

The first term is the long-range Coulombic interaction and is summed accurately by means of Ewald method. The second term ϕ_{ij} represents the short-range interactions of the Buckingham form

$$\phi_{ij} = A_{ij} \exp(-r_{ij}/\rho_{ij}) - C_{ij}/r_{ij}^6, \tag{3.3}$$

where the first term represents the Pauli repulsion that arises from overlapping electron wave functions on different atoms. It is common to associate the second term with van der Waals forces. In practice, however, this term will include contribution from other attractive forces such a small covalent terms. The interaction potential that is described by Coulomb forces and Eqn.(3.3) is called *rigid ion model*. It should be stressed that employing the rigid ion model does not necessarily imply that the electron distribution corresponds to a fully ionic system. In fact, in order to fit the experimental data, most rigid ion models assign fractional charges to the atoms. The general validity of the potential model is assessed principally by its ability to reproduce observed crystal

3.3. Theoretical Modeling and Computer Simulation

properties such as lattice constant and elastic properties. The model is then used to study the defect structure of the system.

The rigid ion model does not take into account of the effect of ionic polarizations on the atomic interactions. Therefore it cannot correctly describe the response of the crystal to the electrostatic perturbation caused by a charge defect. To simulate the polarization effects, the above model was modified using shell model [Dick and Overhauser 1958], which describes the ionic dipole in terms of the displacement of a massless shell (simulating electrons) from a core in which the mass is concentrated. The interaction between the core and shell is modeled by an isotropic harmonic spring. The overall charge on each ion is partitioned between the core and the shell, so that the free-ion polazability is given by $\alpha = Y^2/k$, where Y is the shell charge and k is the harmonic spring constant. In the shell model, the none-Coulombic interactions described in Eqn.(3.3) act between shells only. This way the shell model takes into account the vital coupling between short-range forces and polarization by the remainder of the lattice. Despite this simple mechanical representation of the ionic dipole, the shell model has been shown to correctly simulate both dielectric and elastic properties of many ionic materials, and it became an essential tool for reliable calculations of defect energies.

Sometimes, however, even such simplified treatment of the atomic interaction is too complicated for large scale simulation. In that case the model potential is further simplified by looking only at a restrict subset of the phase space. For example, when we study the oxygen ordering in $YBa_2Cu_3O_{7-\delta}$ compound (see Chapter 6), we only consider the movement of oxygen atoms on a rigid lattice. We can thus simply regard the interaction potential between oxygen atoms as screened Coulomb interaction. Such simplified models are sometimes called "lattice gas model" since atoms can only move between sites on a rigid lattice. As we will see, such model does produce results that are in reasonable agreement with experimental results and can be used as an efficient tool to study lattice defects.

3.3.2 Computer Simulation

It is well established that the microstructure of a compound is intimately related to the interaction between the atoms in that material. However even with the knowledge of the interaction potential between atoms, we still have the task of calculating the thermodynamic properties from the potential model. The number of atoms in a macroscopic system is typically of the order of 10^{23} and each atoms may carry several degrees of freedom. For such a large system statistical mechanics provides the theoretical basis of the relationship between the microscopic and macroscopic descriptions of a physical system. In statistical mechanics, a configuration of a system is described by a set of mechanical variables, Γ. Γ contains the value of all possible degrees of freedom for each particle of the system, e.g. spatial position, velocity, and magnetic moment. The *phase space*, $\{\Gamma\}$, is the parameter space spanned by all possible configurations of the system. Properties of the system are determined by a Hamiltonian function, $H(\Gamma)$, defined on the space of mechanical variables. In statistical mechanics, a probability is associated with each configuration

$$\rho(\Gamma) = \frac{e^{-H(\Gamma)/k_B T}}{Z}, \qquad (3.4)$$

where Z is a normalization factor (the partition function)

$$Z = \int_{\{\Gamma\}} e^{-H(\Gamma)/k_B T} d\Gamma. \qquad (3.5)$$

T is the absolute temperature and k_B is Boltzmann's constant. Given the probability distribution of the configurations, the thermodynamic value of a measurable physical quantity, $A(\Gamma)$, is obtained in the canonical ensemble as

$$<A(T)> = Z^{-1} \int_{\{\Gamma\}} A(\Gamma) \rho(\Gamma) d\Gamma. \qquad (3.6)$$

Equation 3.6 constitutes the formal connection between the microscopic and macroscopic physical worlds. It is obvious that this equation involves multi-dimensional integrals. Only in a very few special cases can these integrals be evaluated exactly. Approximation schemes have to be introduced for most of the physical systems of interests.

3.3. Theoretical Modeling and Computer Simulation

It is at this point that the modern computer enters as an indispensable tool to make progress. By using a fast computer, the integrals of statistical mechanics may be evaluated with an accuracy which is only limited by the computer power available. Computer is used to simulate directly the behavior of a physical system taking as its starting point the fundamental equations of statistical mechanics. It is like an experiment: it is built on as little bias as possible. Only the fundamental physical laws governing the interaction between the microscopic constituents are invoked. Contrary to a real experiment, the simulation is carried out on a well-defined system and there is full control over every "experimental parameter". However, simulation shares with the real experiment an important potential, namely that it allows for possible new discoveries which could not have been inferred from the basic physical laws of interaction. The outcome of a simulation may be thought of both as new experimental data and as a theoretical result which can be used to assess the validity of basic assumptions and predictions of analytical theories. Therefore, computer simulation interpolates between theory and experiment. By this unique ability, computer simulation may serve to illustrate and illuminate subtle and, unfortunately, not commonly recognized basic conceptual relations in scientific reasoning. There exists two major types of computer simulation techniques which have proved exceedingly successful in the study of microstructures, namely molecular dynamics method and Monte Carlo method. In a molecular dynamics simulation, the deterministic time-evolution of a classic many-body system is calculated by a numerical integration of Newtonian equations of motion. In a Monte Carlo simulation, certain stochastic elements are introduced which facilitate the evaluation of statistical mechanical averages.

Monte Carlo Method

Numerical simulation to obtain approximate solutions to statistical problems is a fairly old game well-known to the experimental mathematicians [Meyer 1953, Hammersley *et al.* 1967]. Since the numerical simulation methods are concerned with experiments on random numbers, these methods are often called *Monte Carlo* methods. To gain the full power of numerical simulation techniques, a large amount of random numbers has to be generated and processed. Therefore, it is only with the appearance of the modern computers that the use of Monte Carlo methods has really gained impetus.

There are various ways to solve problems in statistical mechanics using Monte Carlo methods (For detailed review, see [Binder 1979, Binder 1984, Mouritsen 1984]). The most commonly used technique is a certain important-sampling method which was first proposed by Metropolis et al. in 1953 [Metropolis et al. 1953]. The thermodynamic average of any mechanical variable A is defined as

$$<A> = \frac{\sum_\mu A_\mu \exp(-E_\mu/k_BT)}{\sum_\mu \exp(-e_\mu/k_BT)} \quad (3.7)$$

where E_μ represent the energy eigenvalues. Since only a few states μ which are close to the most probable values contribute significantly to the average, it is sufficient to generate only the important states by assigning enhanced probabilities to them. This is known as the "important sampling" and the averages are then defined as :

$$<A> = \frac{\sum_{\mu=1}^n A_\mu P_\mu^{-1} \exp(-E_\mu/k_BT)}{\sum_{\mu=1}^n P_\mu^{-1} \exp(-E_\mu/k_BT)} \quad (3.8)$$

where the P_μ represent the probabilities assigned and the summations are over the n configurations generated. The Metropolis method is to choose P_μ as the Boltzmann probability itself:

$$P_\mu = \frac{\exp(-E_\mu/k_BT)}{\sum_{\nu=1}^n \exp(-E_\mu/k_BT)}. \quad (3.9)$$

With this choice, the estimate for $<A>$ reduces to the simple expression

$$<A>_n = n^{-1} \sum_{\mu=1}^n A_\mu. \quad (3.10)$$

More over, the relative probability of states μ and ν is given by

$$f = \frac{P_\mu}{P_\nu} = \exp\{-(E_\mu - E_\nu)/k_BT\}. \quad (3.11)$$

This eliminates the need to calculate P_μ explicitly and only the simpler quantity f is required.

3.3. Theoretical Modeling and Computer Simulation

A possible implementation of the Monte Carlo important-sampling method is to start with some initial configuration, and then repeat again and again the following six steps which simulate the real thermal fluctuations:

1. Choose an arbitrary (e.g. random) trial configuration,

2. Compute energy difference ΔE connected to the initial and trial configurations.

3. Calculate the "transition probability" $f = \exp(-\Delta E/k_B T)$ for that change.

4. Generate a random number R between zero and unity.

5. If $f > R$, the trial configuration is accepted as the new configuration of the system, otherwise the system remains in the old state.

6. Analyze the resulting configuration as desired, store its properties to calculate the necessary averages, etc.

The trial configuration is in most cases chosen so as to make the sampling procedure as fast as possible. A popular choice for lattice models corresponds to a sequential or random visit of lattice sites and to the simplest possible single-site excitation. For a model with q discrete single-site states (Ising and Potts models), this choice may be a uniformly random choice among the q possible states. For a model with a continuous degree of freedom, e.g. characterized by a unit vector, the simplest possible excitation corresponds to choosing a new uniformly random direction on the unit hypersphere. All these types of excitation may be called Glauber-like excitation since they resemble the dynamics of the kinetic Ising model [Glauber 1972]. Use of Glauber-like excitations leads to dynamics with non-conserved order parameters. Another possible mechanism of excitation follows Kawasaki-like dynamics [Kawasaki 1972] which conserve the order parameter. A simple Kawasaki mechanism is two-site exchange of single-site properties. With these two very simple methods of excitation, it is often possible to construct, for a given problem, a mechanism which to a very high degree mimics the real physical excitations.

Molecular Dynamics Method

While Monte Carlo simulations can give us information about the equilibrium properties of the materials, molecular dynamics simulation can gives us the dynamics properties by solving the Newtonian equations of motion directly.

The earliest molecular dynamics simulation of a realistic potential were performed by Rahman [Rahman 1964, Rahman 1965] which was followed with the more systematic study of the Lennard-Jones fluid by Verlet and co-workers [Verlet 1967,1968, Levesque and Verlet 1970, Levesque *et al.* 1973]. Since then, molecular dynamics simulation method has been developed into a mature theoretical tool to study the structural as well as dynamical properties of various materials (For a more detailed review of the molecular dynamics method, see *Computer Simulation of Liquids* by All and Tildesley [Allen and Tildesley 1987]).

The basic idea of molecular dynamics simulation is quite simple. Consider a system with N identical particles. Assume also that the system is isolated from its surrounding in which case the Hamiltonian H is a constant of motion (i.e. the total energy of the system is conserved). Given the initial coordinates and momenta of the particles, the positions and velocities (momenta) at any later time can in principle be obtained as the solutions to Newton's equations of motion, i.e.

$$M\mathbf{r_i} = -\nabla_i V(\mathbf{r}^N) \tag{3.12}$$

where $\mathbf{r}^N = \{\mathbf{r_1}, \mathbf{r_2}, ..., \mathbf{r_N}\}$ and $V(\mathbf{r}^N)$ is the total potential energy of the system and M is the mass of each particle. If $\mathcal{F}(\mathbf{r}^N, \mathbf{p}^N)$ is a function of the coordinates and momenta and $\mathbf{F}$ is the associated thermodynamic property, then

$$\mathbf{F} = <\mathcal{F}(\mathbf{r}^N, \mathbf{p}^N)> \tag{3.13}$$

where the angular brackets denotes a statistical average.

Eqn.(3.13) can be viewed as a time average over the dynamical history of the system or a suitably constructed ensemble. For the case of molecular dynamics simulation the statistical averages are of the former type. The Eqn.(3.13) can, therefore, be expressed as the integral

$$\mathbf{F} = <\mathcal{F}>_t = \lim_{\tau \to \infty} \frac{1}{\tau} \int_0^\tau \mathcal{F}[r^N(t), p^N(t)]dt. \tag{3.14}$$

3.3. Theoretical Modeling and Computer Simulation

Therefore with properly constructed model potentials the molecular dynamics simulation can give us much information about the system.

A great advantage of molecular dynamics simulation over the Monte Carlo method discussed in the previous section is in its dynamics aspects. In principle, with the proper input potential energy, one may generate the entire space-time Fourier transform, $S(q,\omega)$, to compare with neutron scattering data. We can also obtain information about ion diffusion use the molecular dynamics techniques.

Chapter 4

The Structure Characteristics

4.1 Introduction

This books deals with the microstructure and defects in high- temperature superconductors. However one cannot study defects of the material without first understand its crystal structure. Although their properties vary greatly, all the high-T_c cuprates share some common structural characteristics. In this chapter we discuss several of them that are relevant to the study of the microstructure.

4.2 Crystal Structure

The crystal structures of all high-temperature superconductors are characterized by superconducting CuO_2 sheets that are separated by non-superconducting planes. All these compounds are composed of similar building blocks, which can be closely related to the classic perovskite ABO_3 (e.g. $CaTiO_3$) structure, such as vortex-sharing metal-centered octahedra, square pyramids, square planes, and chains. In the following sections, we briefly review the crystal structures of three representative cuprate families.

4.2.1 $(La_{1-x}M_x)_2CuO_4$ (La214) and Related Structures

$La_{1.85}Ba_{0.15}CuO_4$ is the first known high-temperature superconductor ($T_c \approx 30K$) discovered by Bednorz and Müller [Bednorz and Müller 1986]. It derives from the non-superconducting La_2CuO_4 by partial substitution of La by Ba. It was found that substitutions of monovalent, divalent, or tetravalent cations for lanthanum, or deviation from 4.00 oxygen, are required to obtain the superconducting state. Although $(La_{1-x}M_x)_2CuO_4$ (M=Ba, Sr, Ca) compounds do not possess high critical temperatures among the high-temperature cuprates, they will continue to provide clues to the origins of superconductivity in high-temperature superconductors.

$(La_{1-x}M_x)_2CuO_4$ (often referred to as the La214 phase) has the K_2NiF_4 structure. It is built up by flat sheets of corner-sharing CuO_6 octahedra that are separated by La(Sr, Ba) atoms, similar to those of perovskite. The $(La_{1-x}M_x)_2CuO_4$ structure differs from perovskite in that apical oxygen atoms are not linked to adjacent octahedra layers to form a three-dimensional octahedral array. The adjacent layers are offset by $(a+b)/2$ with respect to each other.

$(La_{1-x}M_x)_2CuO_4$ exhibits three distinct crystal structures, depending on the composition and dopant element M. It undergoes a complex series of perovskite-like octahedral tilt transitions from the $I4/mmm$ (No.139 in International Tables for Crystography) aristitype to lower symmetry tetragonal and orthorhombic forms. For $(La_{1-x}Ba_x)_2CuO_4$ ($0.025 < x < 0.10$), upon cooling from above room temperature, it transforms from a high temperature tetragonal (HTT) phase [Fig.4.1(a)] via a low temperature orthorhombic (LTO) phase [Fig.4.1(b)], to a low temperature tetragonal (LTT) phase [Fig.4.1(c)]. The HTT phase has a body-centered tetragonal structure ($I4/mmm$), in which the characteristic structural units are staggered planes of CuO_6 octahedral (see Table 4.1 for structure parameters). For the LTO phase, there are two variants due to the twinning. In one variant these corner shared CuO_6 octahedra rotate 3.5^o about one of the equivalent HTT [110] axes [Fig.4.1(a)], with alternate row rotating in opposite directions as dictated by the corner sharing. The resulting B-face-centered orthorhombic cell has it's a-b axes along the HTT [110] directions, and

4.2. Crystal Structure

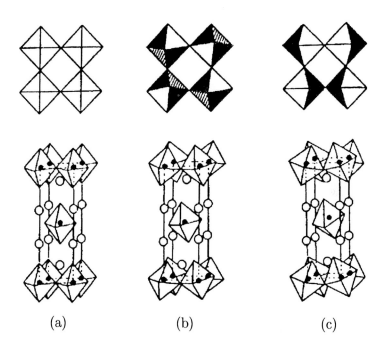

Figure 4.1 Three polymorphs of the K$_2$NiF$_4$-type compound (La$_{1-x}$A$_x$)$_2$CuO$_4$ (A=Ca, Sr, and Ba) with different space group: (a) $I4/mmm$, (b) $Bmab$, and (c) $P4_2/ncm$ (for A=Ba only).

belong to the space group $Bmab$ (=$Cmca$, No.64). Thus, the unit cell of the LTO phase is 45° rotated from that of the HTT phase and has a enlarged unit cell in a and b directions ($a_{\text{LTO}} = \sqrt{2}a_{\text{HTT}}$). The other LTO variant differs only in that the octahedra rotate about the [$\bar{1}10$] axis, and the orthorhombic $a - b$ axes are interchanged (see Table 4.2 for structure parameters). The LTT phase with space group $P4_2/ncm$ (No.138), has the same enlarged unit cell as the LTO phase but primitive rather than B-centered with space group $P4_2/ncm$. In the LTT structure, sheets of CuO$_6$ octahedra are buckled by alternate rotations of equal amplitude about both of the [110] and the [1$\bar{1}$0] axes. The resulting rotations of the octahedra are about [100] axes, adjacent layers being related by a fourfold glide operation (a 90° rotation) which

Table 4.1 System $La_{1.85}Ba_{.15}CuO_4$, temperature=296K, Tetragonal (HTT), $a = 0.3787$nm, $c = 1.3288$nm, space group: $I4/mmm$ [Jorgensen et al. 1987b].

Coordinates	x	y	z
(La, Ba)	0.0000	0.0000	0.3606
Cu	0.0000	0.0000	0.0000
O1	0.0000	0.5000	0.0000
O2	0.0000	0.0000	0.1828

Table 4.2 System: $La_{1.9}Ba_{.1}CuO_4$, temperature=91K, Orthorhombic (LTO), $a = 0.53455$nm, $b = 0.53793$nm, $c = 1.32438$nm, space group: $Bmab$ [Axe et al. 1989a].

Coordinates	x	y	z
(La, Ba)	0.0000	0.0056	0.3609
Cu	0.0000	0.0000	0.0000
O1	0.2500	0.2500	0.0055
O2	0.0000	-0.0285	0.1826

restores tetragonal symmetry (see Table 4.3 for structure parameters).

4.2.2 $YBa_2Cu_3O_{7-\delta}$ (Y123) and Related Structures

$YBa_2Cu_3O_{7-\delta}$, (often referred to as the YBCO phase), has a critical temperature above technical and psychological 77K barrier ($T_c \approx 92K$). It composed of a stack of three perovskite-like cubes as shown in Fig.4.2 (see Table 4.4 for its structure parameters). Structurally, there are oxygen deficient phase $YBa_2Cu_3O_6$ ($\delta = 1$) and fully oxygenated phase $YBa_2Cu_3O_7$ ($\delta = 0$). For $0 < \delta < 0.5$, the compound is orthorhombic and superconducting, and for $0.5 < \delta < 1$ it is tetragonal and non-metallic. The variation in oxygen concentration is easily achieved by heating under different oxygen partial pressures. For certain intermediate δ values, ordering of some of the oxygen sites may occur.

4.2. Crystal Structure

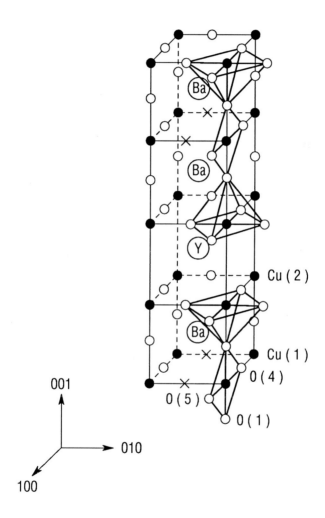

Figure 4.2 The unit cell of YBa$_2$Cu$_3$O$_7$, which consists of three cubes stacking on top of each other along the c-axis.

Table 4.3 System: $La_{1.9}Ba_{.1}CuO_4$, temperature=15K, Tetragonal (LTT), $a = 0.53559$nm, $b = 0.53665$nm, $c = 1.32381$nm, space group: $P4_2/ncm$ [Axe et al. 1989a].

Coordinates	x	y	z
(La, Ba)	0.0040	0.0040	0.3610
Cu	0.0000	0.0000	0.0000
O1	0.2500	0.2500	0.0079
O2	-0.0213	-0.0213	0.1824

Table 4.4 $YBa_2Cu_3O_{7-\delta}$, Orthorhombic, $Pmmm$ (No.47); $a = 0.38227$ nm, $b = 0.38872$ nm, $c = 1.16802$ nm (for $YBa_2Cu_3O_{6.93}$) [Jorgensen et al. 1990].

Coordinates	x	y	z
Y	0.5000	0.5000	0.5000
Ba	0.5000	0.5000	0.1843
Cu1	0.0000	0.0000	0.0000
Cu2	0.0000	0.0000	0.3556
O1	0.0000	0.5000	0.0000
O2	0.5000	0.0000	0.3779
O3	0.0000	0.5000	0.3790
O4	0.0000	0.0000	0.1590

The orthorhombic structure is described in space group $Pmmm$ with the c-axis parallel to the long cell dimension and the b-axis parallel to the longer of the two short cell edges. The structure contains an Y site with 8 nearest neighbors in nearly cubic configuration, a Ba site with 10 nearest neighbors in truncated cube-octahedra environment. The Cu-O forms CuO_4 square plane and CuO_5 square pyramids.

The tetragonal $YBa_2Cu_3O_{7-\delta}$ structure is topologically almost identical to the orthorhombic form. The most significant difference between the tetragonal and orthorhombic forms occurs at the O4 and O5 positions of the insulating plane between adjacent two BaO planes. In the tetragonal structure these sites are symmetrically equivalent and

4.2. Crystal Structure

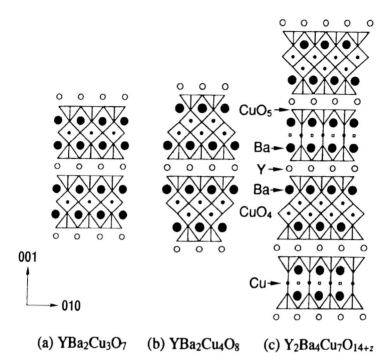

(a) YBa$_2$Cu$_3$O$_7$ (b) YBa$_2$Cu$_4$O$_8$ (c) Y$_2$Ba$_4$Cu$_7$O$_{14+z}$

Figure 4.3 Projections of the structures for (a) YBa$_2$Cu$_3$O$_7$, (b) YBa$_2$Cu$_4$O$_8$, and (c) Y$_2$Ba$_4$Cu$_7$O$_{14+\delta}$ along the [100] direction. Open square: partially pccupied O site.

must be equally occupied. In the orthorhombic structure, however, oxygen concentrates in O4, halfway along the b-axis, so b is slightly longer than a. In ideal YBa$_2$Cu$_3$O$_7$, therefore, O4 is fully occupied while O5 is empty, leading to distinctive CuO$_3$ chain oriented parallel to the b-axis. The a and b axes generally differ by less than 2%, giving a pseudotetragonal unit cell.

There are structures nearly identical to orthorhombic YBa$_2$Cu$_3$O$_{7-\delta}$, but with extra copper and oxygen in the Cu-O strip. They are often refereed as YBa$_2$Cu$_4$O$_8$ (often referred to as the Y124 phase) and Y$_2$Ba$_4$Cu$_7$O$_{14+\delta}$ (often referred to as the Y247 phase) structures (Fig.4.3) (see Tables 4.5 and 4.6 for their structure parameters). YBa$_2$Cu$_4$O$_8$ deviates from YBa$_2$Cu$_3$O$_{7-\delta}$ by intercalation of

Table 4.5 YBa$_2$Cu$_4$O$_8$: orthorhombic, $Ammm$ ($=Cmmm$, No.65) $a = 0.38411$nm, $b = 0.38718$nm, and $c = 2.7240$nm;

Coordinates	x	y	z
Y	0.5000	0.5000	0.0000
Ba	0.5000	0.5000	0.1356
Cu1	0.0000	0.0000	0.2127
Cu2	0.0000	0.0000	0.0614
O1	0.0000	0.0000	0.1454
O2	0.5000	0.0000	0.0524
O3	0.0000	0.5000	0.0528
O4	0.0000	0.5000	0.2187

a second Cu-O chain adjacent to the original Cu-O chains that extend along the b-axis. The intercalation (-Y-CuO$_2$-Ba-CuO-CuO-BaO-CuO$_2$-Y-) leads to a translation period perpendicular to the layers that is more than twice as long as in YBa$_2$Cu$_3$O$_{7-\delta}$. The structure of Y$_2$Ba$_4$Cu$_7$O$_{14+\delta}$ can be described as an intergrowth between tetragonal YBa$_2$Cu$_3$O$_6$ and orthorhombic YBa$_2$Cu$_4$O$_8$. The difference between Y247 and YBa$_2$Cu$_3$O$_{7-\delta}$ is that in Y247 every other layer of single corner-shared chains is replaced by a layer of double edge-sheared chains. The unit cell dimension of Y$_2$Ba$_4$Cu$_7$O$_{14+\delta}$ along the c-axis is equivalent to the sum of two YBa$_2$Cu$_3$O$_6$ and one YBa$_2$Cu$_4$O$_8$. Both Y124 and Y247 are orthorhombic and belong to space group $Ammm$.

4.2.3 Bi$_2$Sr$_2$Ca$_{n-1}$Cu$_n$O$_{2n+6}$ (Bi2201, Bi2212, Bi2223) and Related Structures

In the Bi-Sr-Ca-Cu-O system a homologous series of phases exists with idealized composition of Bi$_2$Sr$_2$Ca$_{n-1}$Cu$_n$O$_{2n+6}$. They consist of a sequence of perovskite-type units and BiO double layers with rock salt coordination as shown in Fig.4.4. Depending on the number of CuO layers in the unit-cell, the phases show superconducting transitions at temperatures of T$_c$ = 10K (n = 1, often referred to as the Bi2201, phase), T$_c$ = 80K (n = 2, often refer to as the Bi2212 phase) and

4.2. Crystal Structure

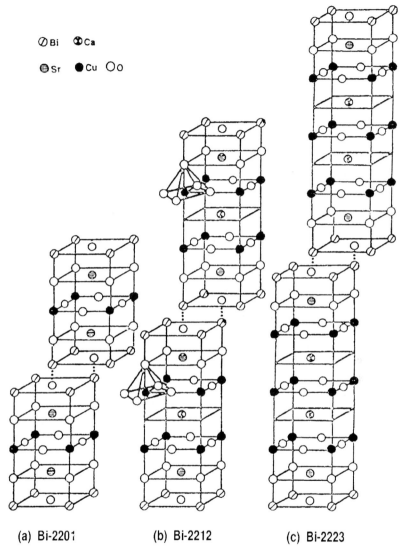

(a) Bi-2201   (b) Bi-2212   (c) Bi-2223

Figure 4.4 Illustration of the crystal structures of (a) $Bi_2Sr_2CuO_{6+y}$, $a = 3.81$Å, $c = 24.61$Å. Similar structure for Tl compound with $a = 5.45$Å, $c = 23.17$Å. (b) $Bi_2Sr_2CaCu_2O_{8+y}$, $a = 3.81$Å, $c = 30.66$Å. Similar structure for Tl compound with: $a = 3.86$Å, $c = 29.39$Å. (c) $Bi_2Sr_2Ca_2Cu_3O_{10+y}$, $a = 3.81$Å, $c = 37.10$Å. Similar structure for Tl compound with $a = 3.82$Å, $c = 36.26$Å.

Table 4.6 $Y_2Ba_4Cu_7O_{15-\delta}$: orthorhombic, $Ammm$ (=$Cmmm$, No.65) $a = 0.3851$nm, $b = 0.3869$nm, and $c = 5.029$nm (for $\delta = 0.4$) [Bordet et al. 1988b].

Coordinates	x	y	z
Y	0.5000	0.5000	0.1155
Ba1	0.5000	0.5000	0.0431
Ba2	0.5000	0.5000	0.1880
Cu1	0.0000	0.0000	0.0000
Cu2	0.0000	0.0000	0.0829
Cu3	0.0000	0.0000	0.1483
Cu4	0.0000	0.0000	0.2301
O1	0.0000	0.5000	0.0353
O2	0.5000	0.0000	0.1430
O3	0.0000	0.5000	0.1432
O4	0.0000	0.0000	0.1937
O5	0.5000	0.0000	0.0871
O6	0.0000	0.5000	0.0865
O7	0.0000	0.5000	0.2328
O8	0.0000	0.5000	0.0000
O9	0.5000	0.0000	0.0000

$T_c = 110$K ($n = 3$, often referred to as the Bi2223 phase), respectively. For the $n = 1$ phase, the structure contains two Sr_2O_2 layers that are separated by Bi_2O_2 double layers, and CuO_2 single layers and no Ca layers. The $(a + b)/2$ lattice translation between the Bi_2O_2 double layers leads to the periodicity perpendicular to the layers be doubled (10 layers per unit cell) to about 2.45nm (see Table 4.7 for its structure parameters). The $n = 2$ phase contains 14 layers per unit cell. The CuO sheets in Bi2201 are replaced by CuO_2-Ca-CuO_2 sandwiches in Bi2212 with a layer sequence -Bi_2O_2-Sr_2O_2-CuO_2-Ca-CuO_2-Sr_2O_2-Bi_2O_2-. Ca atoms adopts eight coordination, similar to the Y environment in $YBa_2Cu_3O_{7-\delta}$. There are no oxygen at this level, so Cu atoms have only five nearest neighbors in square pyramidal coordination, rather than the elongated octahedra coordination of Bi2201. Its

4.2. Crystal Structure

Table 4.7 System: $Bi_2Sr_2CuO_6$, orthorhombic, $a = 0.5361$nm, $b = 0.5370$nm, and $c = 24.369$nm, Space group $Cmmm$ [Torardi et al. 1988].

Coordinates	x	y	z
Bi	0.0000	0.2758	0.0660
Sr	0.5000	0.2479	0.1790
Cu	0.5000	0.7500	0.2500
O1	0.7500	0.5000	0.2460
O2	0.0000	0.2260	0.1450
O3	0.5000	0.3340	0.0640

translation period along the c-axis is 3.09nm (see Table 4.8 for its structure parameters). The structure of Bi2223 phase is similar to that of the Bi2212 phase, but additional CuO_2 and Ca layers are inserted within the CuO_2-Ca-CuO_2 sandwich of Bi2212, yielding a CuO_2-Ca-CuO_2-Ca-CuO_2 sandwich. The unit cell consists of 18 layers with $c = 3.7$nm. Although the topology of the structure are very similar among Bi2201, Bi2212 and Bi2223, due to the Ca layer the CuO bonding in the crystal structure of the different phases differs significantly. $Bi_2Sr_2CuO_6$ consists of CuO_6 octahedra arrays, $Bi_2Sr_2CaCu_2O_8$ consists of Cu_5 square pyramids arrays, while $Bi_2Sr_2Ca_2Cu_3O_{10}$ consists of CuO_4 sheets and CuO_5 square pyramids (see Table 4.9 for its structure parameters). The homologous series of bismuth compounds also suggests the possible existence of other structures with more copper layers. Structure of isolated slabs or entire grain of $Bi_2Sr_2Ca_3Cu_4O_{12}$ and $Bi_4Sr_4Ca_3Cu_5O_{18}$ have been observed by high-resolution electron microscopy and electron diffraction.

$Bi_2Sr_2Ca_{n-1}Cu_nO_{2n+6}$ ($n = 1, 2, 3$) posses incommensurate modulations. It is generally assumed that the average structures of these phases have a B-centered orthorhombic Bravais lattice. The basic building blocks may be easy to understand, however, the structures of these real materials are much complicated by incommensurate modulation, oxygen nonstochiometry, cation disorder, layer stacking faults, and other nonperiodic behavior. A good example is the Bi2223 phase. To date, high quality single phase $Bi_2Sr_2Ca_2Cu_3O_{10}$ still has not been synthe-

Table 4.8 System: $Bi_2Sr_2CaCu_2O_8$, orthorhombic, $a = 0.5408$nm, $b = 0.5413$nm, $c = 3.0871$nm, space group: $Amaa$ (No.66) [Gao et al. 1988].

Coordinates	x	y	z
Bi	0.5000	0.2285	0.0522
Sr	0.0000	0.2537	0.1409
Ca	0.5000	0.2500	0.2500
Cu	0.5000	0.2498	0.1967
O1	0.7500	0.0000	0.1950
O2	0.2500	0.5000	0.2020
O3	0.5000	0.2800	0.1220
O4	0.0000	0.1500	0.0530

sized, and a complete structure refinement has not been reported.

Related structures of $Bi_2Sr_2Ca_{n-1}Cu_nO_{2n+6}$ can be found in Tl-based cuprates. $Tl_2Ba_2CuO_6$, $Tl_2Ba_2CaCu_2O_8$, and $Tl_2Ba_2Ca_2Cu_3O_{10}$ are regarded as basically isomorphous with Bi2201, Bi2212, and Bi2223, respectively [Hazen 1990]. On the other hand, Hg-containing compounds $HgBa_2CuO_4$, $HgBa_2CaCu_2O_6$ and $HgBa_2Ca_2Cu_3O_8$ are isostructural with $TlBa_2CuO_5$, $TlBa_2CaCu_2O_7$ and $TlBa_2Ca_2Cu_3O_9$. For their structural details, see [Radaelli et al. 1993, Huang et al. 1993, Wagner et al. 1993].

4.3 Bonding and Structural Features

It is obvious from the previous section that a common structural feature of the high-temperature superconductors is the CuO_2 planes separated by rock-salt type rare-earth metal-oxide blocks. It is well-established by now that the superconductivity happens at the CuO_2 plane, while the other layers act as charge reservoirs. Because of the non-stoichiometry of the rare-earth metal-oxide layers (for La214, Bi- and Tl- compounds) or the CuO chain planes (for YBCO compounds), mobile holes (deficiency of valence electrons) are doped into the CuO_2 plane, which acts as charge carriers of the superconductors. Such hole-doping is achieved

4.3. Bonding and Structural Features

Table 4.9 System: $Bi_2Sr_2Ca_2Cu_3O_{10}$, orthorhombic, $a = 0.54029$nm $b = 0.54154$nm, $c = 3.7074$nm, space group $A2aa$ [Miehe et al. 1990].

Coordinates	x	y	z
Bi	-0.0360	0.2330	0.0411
Sr	-0.0100	0.7450	0.1148
Ca	0.4680	0.2570	0.2072
Cu1	0.0000	0.2460	0.1619
Cu2	0.0000	0.2500	0.2500
O1	0.2520	-0.0060	0.1650
O2	0.2740	0.4980	0.1637
O3	0.0470	0.2530	0.0956
O4	0.2300	0.0020	0.2486
O5	0.4250	0.1390	0.0434
O6	0.2300	0.5440	0.0372

by creating many types of point defects, which include cation substitution or vacancy (e.g. in La214), as well as excess oxygen (e.g. in YBCO, Bi- or Tl compounds).

The layered structure of the high-temperature superconductor affect the physical properties of these compounds greatly. In the more anisotropic compounds such as Bi2212 and Bi2223, it is very easy to shear the material along the ab-plane without damaging the basic structure of the compound. These properties leads to new techniques to produce high quality tapes and wires by applying strong shear stress to these materials. It is also make Bi-2212 and Bi-2223 have clean grain boundaries as shown in chapter 12. In addition, the anisotropy in thermal conductance along ab plane and c-axis affect the shape and size of the columnar defects created by heavy-ion irradiation (see chapter 13).

As a general structural model, we can view the high temperature superconductors as intergrowths of peroskite and rock-salt blocks. The rock-salt block is composed of rare-earth metal oxide AO, where A is the rare-earth metal element such as Bi or Tl. The perovskite block is composed of CuO_2 planes and metal layers M such as Ca in Bi- 2212 and Y in YBCO. The intergrowth regions in common BO such as LaO,

SrO, or BaO are structurally related to both the rock-salt and pervskite blocks, and act as buffers between these two structures. The CuO_2 layers are built from corner-sharing square-planar CuO_4 units, and Cu can be bonded to another oxygen atom, creating a square-pyramidal environment, or to two additional oxygen atoms, giving an axially elongated octahedra surrounding. Compositionally, these compounds have the formula $(AO)_m(BO)_2(M)_{n-1}CuO_2)_n$ and are often referred to as $m2(n-1)n$ phase. All the compounds discussed in previous section can be regarded as speciual cases of the $m2(n-1)n$ structure [Whangbo and Torardi 1991].

The CuO_2 plane form a rigid, two-dimensional network with short Cu-O bonds. In an ideal square-planar perovskite layer, in in-plane Cu- O distance would be one-half the perovskite a axis, i.e. ≈ 1.9Å. This imposes the restriction that if the atoms of the AO and BO layers are coplanar, the A-O and B-O intralayer distances be close to $\approx 1.9 \times \sqrt{2} \approx 2.7$Å for the ideal structures [Whangbo and Torardi 1991]. On the other hand, the ideal A-O and B-O distances are quite different. According to the ionic radii for the 9-coordinate B cations (i.e. La^{3+}, Sr^{2+}, Ba^{2+}), the 6-coordinate A cations (i.e. Ti^{3+}, Bi^{3+}), and the 6-coordinate O^{2-} anion, the ideal A-O and B-O distances are estimated as Ba-O=2.87Å, Sr-O=2.71Å, La-O=2.62Å, Bi-O=2.3Å, and Ti-O=2.29Å. Therefore there is severe mismatch for the CuO_2 plane and the rare-earth metal oxide plane as well as the intergrowth region. The interlayer strain for these material is quite large. To relieve the strain, metal atoms are displaced from their ideal position to form shorter and stronger bonds with oxygen. This causes the CuO_2 plane to buckle, creating either periodic modulations or arrays of dislocations in the materials. This in turn gives rise to charge modulations. We will discuss the effect of the lattice mismatch on the microstructures of the oxide-superconductors in more detail in Chapter 8 and Chapter 10.

Chapter 5
Types of Defects

5.1 Intragranular Defects

5.1.1 Point Defects

Point defects are lattice defects of zero dimensionality, i.e., they do not possess lattice structure in any dimension. Typical point defects are vacancies, interstitials, and impurity atoms (substitutional and interstitial). Although isolated, or randomly distributed point defects are very difficult to detect experimentally (clusters of such point defects were observed at grain boundaries, especially at dislocation cores), ordered point defects can be easily observed.

Oxygen/holes are the most important point defects in determining superconductivity in the high T_c oxides. Oxygen ordering in the CuO basal plane (or chain plane), containing the -O-Cu-O- chains along the b-direction, is responsible for the tetragonal-to-orthorhombic phase transition, [Jorgensen *et al.* 1987c, Cai and Mahanti 1988, Cai and Mahanti 1989, de Fontaine *et al.* 1989, de Fontaine *et al.* 1990] and hence, for the formation of twins in $YBa_2Cu_3O_{7-\delta}$. Oxygen ordering also gives rise to ordered superstructures (oxygen-vacancy ordering) in oxygen-deficient samples [van Tendeloo *et al.* 1987a, Reyes-Gasga *et al.* 1989, Beyers *et al.* 1989, Chen *et al.* 1988, Zhu *et al.* 1991d]. The characteristic feature of the oxygen/vacancy ordering is that the oxygen vacancies form long vacancy chains (-V-Cu-V-) along the a-direction (V-represents a vacant oxygen site in the O-Cu-O- chain) rather than

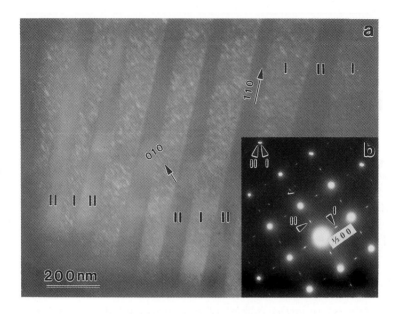

Figure 5.1 "$2a$" oxygen ordering in a nominal $YBa_2Cu_3O_{6.7}$ sample: (a) A dark-field image formed by the $1/2, 0, 0$ superlattice spot (marked by arrow I in (b)). The white-island contrast in the twin domain I corresponds to the regions of the "$2a$" ordering. (b) $(001)^*$ zone diffraction pattern. The $1/2, 0, 0$ superlattice spot corresponding to the "$2a$" ordering in domain II is marked by arrow II.

adopting a random distribution to form a superstructure. For example, alternating chains of filled and vacant oxygen sites give a superstructure with a wave vector $\mathbf{q} = [1/2, 0, 0]$ at $YBa_2Cu_3O_{6.5}$, while the ordered removal of every fourth oxygen atom (or oxygen vacancy) from every other Cu-O chain gives a superstructure with a wave vector $\mathbf{q} = [\pm 1/4, \pm 1/4, 0]$ for $O_{6.85}$ or $O_{6.15}$ [Alario-Franco et al. 1987]. In the latter case, the ordering scheme is the same in $O_{6.85}$ and $O_{6.15}$, but the roles of the oxygen and vacancies are reversed. Other ordered structures such as, $\mathbf{q} = [2/5, 0, 0]$ and $\mathbf{q} = [1/3, 0, 0]$, are also observed [Mitchell et al. 1988, Werder et al. 1988a]. Although oxygen itself is usually difficult to detect, oxygen-ordering on a micro-scale can be seen in the diffraction pattern as well as in a dark-field image by TEM. Fig.5.1 shows an example of a $2a$ ordering. Fig.5.1(a) is a dark-field im-

5.1. Intragranular Defects

age using the superlattice reflection [1/2, 0, 0] marked by arrow I. The islands of white contrast in twin lamella I, corresponding to the superlattice reflection, suggest the existence of a $2a$ (-O-V-O-V-) ordering along the [100] direction. The superlattice reflection marked by arrow II corresponds to the $2a$ ordering in twin lamella II. To observe the ordering in twin lamella II, the superlattice reflection II (see Fig.5.1(b)) must be used.

Analysis of electron diffuse scattering from the oxygen short range ordering in $YBa_2Cu_3O_{7-\delta}$ ($0.1 < \delta < 0.8$) provides strong evidence that the material contains small domains (Magneli-like units [Khachaturyan and Morris 1988]) with alternating planes of occupied and vacant chain sites along the [100] direction. The extent of ordering in the (100) plane is > 5nm, while the width of the ordering domain in the [100] direction is a function of δ, reaching the maximum ($\sim$ 10 nm) near $\delta \approx$ 0.4. The dependence of the domain size with oxygen substoichiometry can be qualitatively explained using a simple model involving only the local rearrangement of oxygen, and agrees with the theoretical work on this system [Cai and Mahanti 1988, Khachaturyan and Morris 1987, Khachaturyan and Morris 1988, de Fontaine et al. 1989, de Fontaine et al. 1990].

For samples with a near-stoichiometric oxygen content ($\delta < 0.1$), oxygen or vacancy ordering do not usually generate a superstructure, although local ordering or clustering may still exist. Nano-scale oxygen fluctuation was observed in nominal $YBa_2Cu_3O_{7-\delta}$ ($\delta \approx 0$) using high-resolution electron energy-loss spectroscopy (EELS). The observation confirmed the suggestion that, based on the observed anomalous double-peaked magnetization curve and intragrain weak-links, the possible presence of nano-scale oxygen deficient regions can act as significant pinning centers in stoichiometric single crystals [Daeumling et al. 1990].

Local oxygen ordering also occurs when a small fraction of the Cu atoms in the basal plane is replaced by certain trivalent cations in $YBa_2(Cu_{1-x}M_x)_3O_{7-\delta}$ (M=Fe [Wördenweber et al. 1989, Xu et al. 1989], Co [Maeno et al. 1987, Schmahl et al. 1990], or Al [Siegrist et al. 1988, Tarascon et al. 1988b]). Although substitution did not enhance flux pinning as originally expected, the defect structure associated with such point defects has attracted much attention [Bordet et al. 1988c,

Yang et al. 1989, Banngärtel and Bennemann 1989, Dunlap et al. 1989, Jiang et al. 1991, Kreckels et al. 1991]. We will discuss such defect structure in more details in later chapters.

5.1.2 Line Defects-Dislocations

Dislocation is a line defect with strain field around its core. Dislocation's motion produces plastic deformation of crystals at stress well below the theoretical shear strength of a perfect crystal. For high-temperature superconductors, dislocations can play an important role in determining their mechanical properties. On the other hand, since the coherence length is very small (0.4nm in [001] direction and 3.1nm in the a-b plane for $YBa_2Cu_3O_{7-\delta}$) [Gallagher et al. 1987b], any defect which is associated to a disturbed part of crystal larger than the coherence length can be efficient in pinning the flux lines. These small coherence length and the strong coupling between elastic and superconducting behaviors could lead to the existence of strained regions where T_c can be significantly altered [Guinea 1988].

The commonly observed dislocations in high-temperature superconductors are perfect dislocations with a [100], [010] or [110] Burgers vector, partial dislocations with stacking fault in between were also observed [Nakahara et al. 1989, Rabier and Denanot 1990, Yoshida et al. 1990]. An example of [110] dislocations in $YBa_2Cu_3O_{7-\delta}$ is shown in Fig.5.2 with different reflections to determine the Bergers vector of the dislocations. Fig.5.3 shows a set of bow-out dislocations piling up at the a-b plane. Although high-T_c cuprates are highly anisotropic, the characteristics of the dislocations can be determined unambiguously using **g·b** criterion, especially when a high-order reflection is used [Cockayne 1978]. When $\mathbf{g} \cdot \mathbf{b} = 0$, i.e., the Burgers vector **b** of a dislocation is in the reflected lattice planes, the dislocation is out of contrast (Fig.5.2(b) and Fig.5.3(b)). By characterizing the line direction of the dislocations ξ (note, not the line projection **l**), we can learn whether the dislocation has a character of edge, or screw, or a mixture of the two. For deformation-induced dislocations pile-up, the slip plane normal **n** can be assessed by adding an additional reference direction **m**, and recording the dislocations under two different beam directions $\mathbf{B}_1$ and $\mathbf{B}_2$.

5.1. Intragranular Defects

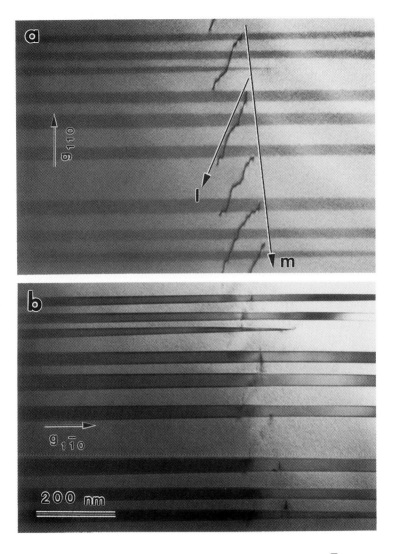

Figure 5.2 [110] dislocations imaged with (a) **g** = 100, (b) **g** = 1$\bar{1}$0. Note when **g** · **b** = 0, the dislocations are out of contrast, however, the twinning contrast is still visible. The dislocation slip plane can be determined using the projected dislocation line direction **l** and the reference direction **m** under two different beam directions **B**$_1$ and **B**$_2$.

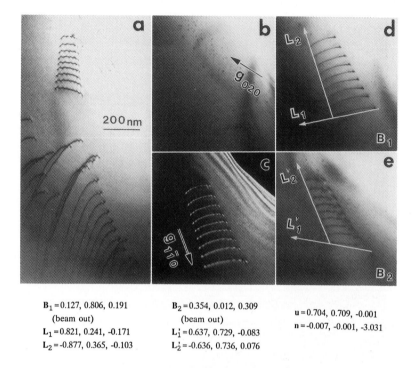

B_1 = 0.127, 0.806, 0.191
(beam out)
L_1 = 0.821, 0.241, -0.171
L_2 = -0.877, 0.365, -0.103

B_2 = 0.354, 0.012, 0.309
(beam out)
L_1^i = 0.637, 0.729, -0.083
L_2^i = -0.636, 0.736, 0.076

u = 0.704, 0.709, -0.001
n = -0.007, -0.001, -3.031

Figure 5.3 [100] perfect dislocations. (a) Specimen normal is nearly parallel to the beam. (b) $\mathbf{g} \cdot \mathbf{b} = 0$; dislocations are out of contrast. (c) Dark-field image. (d) and (e): Specimen was tilted about 60° to determine the dislocation line direction and plane normal. Note that here the real direction of the incident electron beam is considered to have negative sign (beam out).

This gives:

$$\xi = (\mathbf{B}_2 \times \mathbf{l}_1) \times (\mathbf{B}_2 \times \mathbf{l}_2), \tag{5.1}$$

$$\mathbf{n} = [(\mathbf{B}_1 \times \mathbf{l}_1) \times (\mathbf{B}_2 \times \mathbf{l}_2)] \times [(\mathbf{B}_1 \times \mathbf{m}_1) \times (\mathbf{B}_2 \times \mathbf{m}_2)], \tag{5.2}$$

where the subscripts denote the corresponding beam directions, $\mathbf{B}_1$ or $\mathbf{B}_2$. Using Eqn.(5.1) and Eqn.(5.2), we can accurately determine the crystallographic orientations of a line and a plane. The slip plane in $YBa_2Cu_3O_{7-\delta}$ is always found to be (001), i.e. the a-b plane.

5.1.3 Planar Defects

Planar defects are two-dimensional defects, which are often referred to stacking faults and planar interfaces. In this section, we only discuss stacking faults, and leave interfaces, i.e., twin boundaries and grain boundaries and in section Chapter 7 and 12, respectively.

The perovskite cuprates have layered structures, they form a very large family with difference composition and layered sequence. This provide a probability to form stacking faults in crystals. As a matter of fact, a stacking fault, or an intercalation of CuO/Ca bi-layer in Bi-2212 forms a local Bi-2223 structure. The intercalation mechanism is crucial in understanding the Bi/2212-Bi/2223 phase transition (see Chapter 9).

Stacking faults lying in the (001) plane and bounded by a pair of partial dislocations are also frequently encountered in $YBa_2Cu_3O_{7-\delta}$ [Zandbergen et al. 1988, Tafto et al. 1988, Kramer et al. 1990]. These faults are usually wide, about hundred nm, indicating low fault energy. In $YBa_2Cu_3O_{7-\delta}$, most of the stacking faults have a plane normal of [001] with a displacement vector $\mathbf{R} = 1/6[301]$. The dislocations which bond the fault are often screw type with Burger vector ($\mathbf{b} = 1/2[100]$. Using $\mathbf{g} \cdot \mathbf{b} = 0$ and $\mathbf{g} \cdot \mathbf{R} = 0$ extinction rules, we can determine the Burgers vector $\mathbf{b}$ of the dislocations and the displacement vector $\mathbf{R}$ of the fault, respectively. Fig.5.4 is an example of dislocations associated with stacking faults. In Fig.5.4(a), where $\mathbf{g} \cdot \mathbf{R} = 0$ ($\mathbf{g} = 10\bar{3}$), the faults are invisible; in Fig.5.4(b), where $\mathbf{g} \cdot \mathbf{b} = 0$ ($\mathbf{g} = 0\bar{1}3$), the dislocations are invisible, and in Fig.5.4(c), where $\mathbf{g} \cdot \mathbf{b} = \mathbf{g} \cdot \mathbf{R} = 0$ ($\mathbf{g} = 020$), both the dislocations and faults are invisible. Fig.5.4(d) shows the faults edge-on from near the [010] zone axis. The 1/6[301] displacement can result from an edge-shear arrangement of the Cu-O_5 truncated octahedron, as sketched in Fig.5.5. Such faults are non-conservative, i.e., cannot be generated by the motion of dislocations. They are chemical faults, mainly due to one or two extra CuO planes (extrinsic fault, $YBa_2Cu_4O_8$ or $YBa_2Cu_5O_9$, also see Fig.13.29 in Chapter 13), although faults due to the lack of a CuO plane (intrinsic fault) also were observed [Zhu et al. 1990f]. The intrinsic or extrinsic character of the faults can be determined by the characteristics of the fault fringe contrast [Amelinckx and van Landuyt 1976].

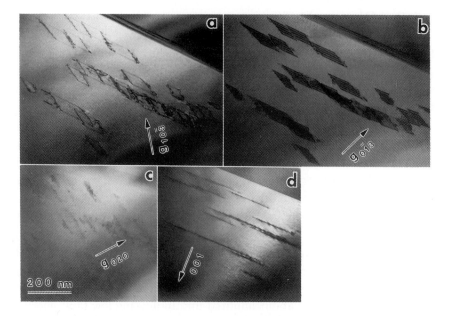

Figure 5.4 1/2[100] partial dislocations associated with stacking faults ($\mathbf{R} = 1/6[301]$). (a) $\mathbf{g} \cdot \mathbf{R} = 0$; faults are invisible. (b) $\mathbf{g} \cdot \mathbf{b} = 0$; dislocations are invisible. (c) $\mathbf{g} \cdot \mathbf{R} = \mathbf{g} \cdot \mathbf{b} = 0$; both dislocations and faults are invisible. (d) Fault planes are seen edge-on.

5.1.4 Volume Defects - Inclusions

Volume defects are three-dimensional defects, such as second phase particles inclusions. Depending on the sample fabrication procedures, especially for industrial tapes and wires, there are abundant identified and unidentified minor phases.

Most second phases are not superconducting. Such non-superconducting phases may not always be detrimental to superconductivity. It has been reported that Y_2BaCuO_5 inclusions, which are due to an incomplete peritectic reaction during crystal growth, randomly distributed in melt-textured $YBa_2Cu_3O_{7-\delta}$, are beneficial to the mechanical properties as well as the current-carrying capability of the superconductors. Both magnetic and transport measurements have shown that the J_c increases with increasing Y_2BaCuO_5 content and decreasing particle size [Murakami *et al.* 1991, Lee *et al.* 1992]. A controversy had

5.1. Intragranular Defects

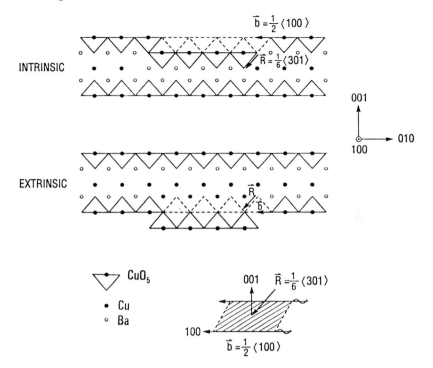

Figure 5.5 Sketches of the intrinsic and extrinsic stacking faults formed through the shear arrangement of CuO_5. The triangles represent the CuO_5 octahedron, viewing along the [100] direction. Both chemical faults have a displacement vector $\mathbf{R} = 1/6[301]$, plane normal [001], and are bounded by two sessile dislocations with Burgers vector of $1/2[100]$.

arisen on the interpretation of the observations. Now, it is generally accepted that because the size of these Y_2BaCuO_5 particles is at least an order of magnitude larger than the coherent length of $YBa_2Cu_3O_{7-\delta}$, the particles themselves would not be effective in flux pinning. Rather, the inclusion related defects, such as $Y_2BaCuO_5/YBa_2Cu_3O_{7-\delta}$ interfaces, dislocations, and stacking faults, are responsible for the increase in J_c [Murakami et al. 1991, Mironva et al. 1993, Wang et al. 1993b]. The interface can act as a source of dislocations, as shown in Fig.5.6, apparently due to the lattice mismatch at the interface. However, not all the interfaces are likely to generate dislocations, due to the high

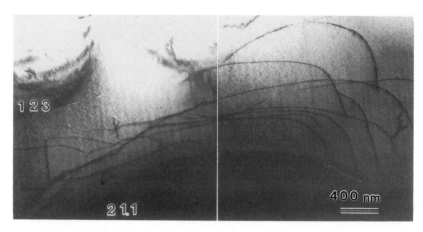

Figure 5.6 [100] dislocations generated by a 211 particle. Here 123 and 211 denote the YBa$_2$Cu$_3$O$_{7-\delta}$ matrix and Y$_2$BaCuO$_5$ inclusion, respectively.

anisotropy of the elastic coefficient both of the inclusion and the matrix. Furthermore, like the deformation-induced dislocations and stacking faults, their contribution in pinning may not be significant because these defects are confined in the a-b planes, and they only can be beneficial to pinning when the magnetic field is parallel to the a-b planes.

5.1.5 Structural Modulations

There are two classes of structural modulations in high temperature superconductors: periodic modulation and quasi-periodic modulation. Incommensurate periodic modulation occurs in Bi-based superconductor family, while quasi-periodic modulation presents mainly in YBa$_2$(Cu$_{1-x}$M$_x$)$_3$O$_{7-\delta}$ (M= Fe [Xu *et al.* 1989, Wördenweber *et al.* 1989], Co [Maeno *et al.* 1987, Schmahl *et al.* 1990], or Al [Siegrist *et al.* 1988, Tarascon *et al.* 1988b]).

For incommensurate modulations the periodicity of the modulation wave is an irrational multiple of at least one of the basic lattice periods of the crystal. The modulation in Bi-2212 is one dimensional and incommensurate along the *b*-axis. Using the superspace-group approach, one- and two-dimensionally incommensurately modulated crystal can be described in four- and five-dimensional periodic space, respectively

5.1. Intragranular Defects

[de Wolff et al. 1981].

Fig.5.7 shows an atomic image of the modulation in Bi-2212 with a body-centered supercell of 2.69nm × 3.09nm. Systematic buckling of the (010) planes along the [001] direction is clearly visible. Based on high-resolution electron microscopy observations, early researches only focused on lattice displacement [Horiuchi et al. 1988, Matsui et al. 1988, Eibl 1991, Budin et al. 1993]. However, recently studies showed that in addition to the lattice displacement, charge modulation is also play an important role in the structure [Zhu and Tafto 1996a]. The dark cage-like contrast with a body-centered symmetry is attributed to the pile-up of electrons in the BiO layer. The embedded image in Fig.5.7 is a simulated image, which was calculated based on a model superimposing periodic lattice displacement and charge transfer (for details see Chapter 10). The model suggests the modulation in Bi-2212 involves more than the metal atoms. The modulation of the O atoms in the Bi-O plane is pertinent as it provides the mechanism for the incorporation of extra O atoms in the layer and thereby contributes to the hole concentration in the Cu-O plane. A study using combined x-ray single-crystal and neutron powder refinement of the modulated structure revealed the extra oxygen content y in $Bi_2Sr_2CaCu_2O_{8+y}$ corresponds to 0.14 [Gao et al. 1993a].

In $YBa_2(Cu_{1-x}M_x)_3O_{7-\delta}$ (M= Fe [Xu et al. 1989, Wördenweber et al. 1989], Co [Maeno et al. 1987, Schmahl et al. 1990], or Al [Siegrist et al. 1988, Tarascon et al. 1988b]), when $x < 0.02$, the substitution of Cu with M causes the reduction of twin spacing (Fig.5.8 regions A and B). For $x > 0.025$, it generates a homogeneous structural modulation with an overall tetragonal symmetry of the Bravais lattice (Fig.5.8, region C). The modulated structures, sometimes called tweed because of their appearance in TEM images, have a quasi-periodic domain contrast associated with streaks of diffuse scattering as seen in electron diffraction patterns (Fig.5.8(c)). We will discuss such quasi-periodic modulations in more detail in Chapter 6.

5.1.6 Artificially Created Defects

Although numerous efforts have been made to create randomly distributed structural defects through synthesis processes, the effectiveness

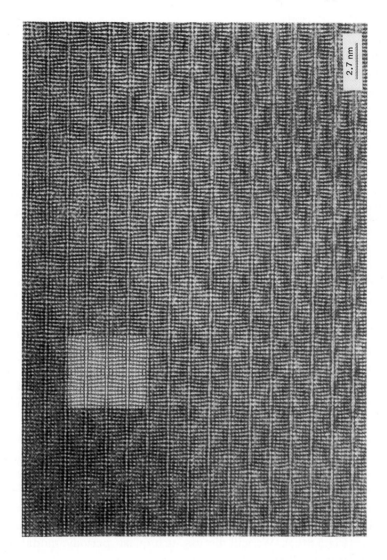

Figure 5.7 High-resolution image of Bi-2212 viewing along the [100] direction. Incommosurate structural modulation with a supercell of 26.9 × 30.9Å^2 is clearly visible. The embedded image is a calculated one.

5.1. Intragranular Defects

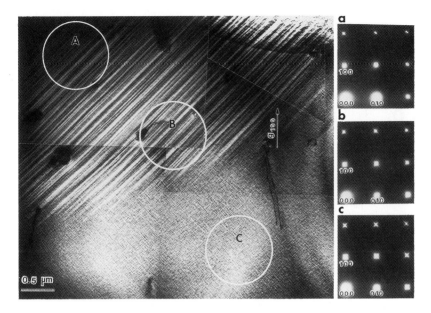

Figure 5.8 Coexistence of twin and tweed due to the inhomogeneous concentration of dopant Fe in a nominal YBa$_2$(Cu$_{0.98}$Fe$_{0.02}$)$_3$O$_{7-\delta}$ sample. (a), (b), and (c) show selected area diffraction of $(001)^*$ zone from the area of A, B, C, respectively. Typical tweed image and the associated diffuse scattering are shown in region C and diffraction (c), respectively.

of such defects in pinning flux lines has been disappointing. One of the reasons is because the cuprates are highly anisotropic with a layered structure, while deformation induced defects such as dislocations and stacking faults are generally distributed only in the a-b plane. As an alternative, the creation of defects by heavy-ion irradiation with several hundred MeV heavy-ions, has became popular [Bourgault et al. 1989, Hensel et al. 1990, Civale et al. 1991]. The resulting linear ion-tracks were found to provide strong flux pinning in temperature and field regimes where the other types of defects were ineffective [Konczykowski et al. 1091, Gerhauser et al. 1992, Hardy et al. 1991 1992, Budhani et al. 1992b, Budhani et al. 1992a, Budhani and Suenaga 1992]. The investigations of such defects on flux pinning and critical current density were widely reported [Budhani et al. 1992b, Budhani et al. 1992a,

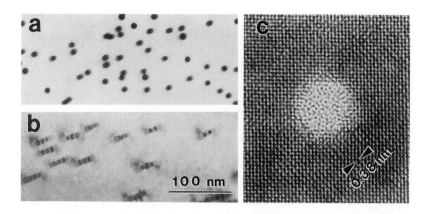

Figure 5.9 Typical morphology of a bulk $YBa_2Cu_3O_{7-\delta}$ sample irradiated with 300 MeV gold ions viewing along the direction of the incident ion-beam.

Budhani and Suenaga 1992].

Figure 5.9 shows typical morphology of a bulk $YBa_2Cu_3O_{7-\delta}$ sample irradiated with 300 MeV gold ions viewing along the direction of the incident ion-beam. (Fig.5.9(a)) and 18 degrees away from the incident beam (Fig.5.9(b)). In the latter case, the ion-tracks, running from the top to the bottom of the sample, give rise to thickness fringes as well as to contrast due to their intersection with the specimen surface and crystal matrix. Nano-diffraction and HREM of the damaged regions, viewed from the cross section of the ion track (Fig.5.9(c)), shows that they are continuous columns of amorphous material (also see: [Raveau 1992]). This is expected because the samples are extremely thin ($<$ 0.5μm) compared to the range of these ions in $YBa_2Cu_3O_{7-\delta}$ ($>$ 14μm) [Zhu et al. 1993b].

A comparative study of the defect formation in Bi-cuprate and oxygen-reduced and ozone-treated $YBa_2Cu_3O_{7-\delta}$, shows that the degree of the radiation damage by the heavy ions depends on: (a) the rate at which ions lose their energy in the target; (b) the crystallographic orientations with respect to the incident ion-beam; (c) thermal conductivity and chemical state (oxygen concentration for $YBa_2Cu_3O_{7-\delta}$) of the sample, and (d) the extent of preexisting defects in the crystal. A theoretical model based on ion-induced localized melting and anisotropic thermal conductivity of these materials provided a basis for

5.2. Intergranular Defects – Grain Boundaries

the size and shape of the amorphous tracks. Measurements of the superconducting properties of Au^{+24}- and Ag^{+21}-irradiated $YBa_2Cu_3O_{7-\delta}$ thin-film show a universal linear scaling between the fractional areal damage vs. the superconducting transition temperature and the normal state resistivity. A marginal enhancement in the critical current density, J_c, occurs when the density of the defects is $\leq 5 \times 10^{10}/cm^2$ and their diameter is about the coherence length. These criteria are satisfied by the silver ions. The damage due to the gold ions is much too severe to improve J_c. For more details, readers are referred to Chapter 13.

5.2 Intergranular Defects – Grain Boundaries

Grain boundaries are interfaces between identical crystals (or grains) which have different orientations. Grain boundaries are a subset of interphase boundaries, which are interfaces between unlike crystals. All such interfaces are commonly regarded as defects in crystals, reflecting a tradition of referring all ideal structures to the perfect crystal. Grain boundaries are characterized in terms of the relative orientation of the crystals that they separate, and the orientation of the grain boundary plane. A grain boundary has five degrees of freedom with respect to crystallographic orientation. Three for the misorientation (among them two for the rotation axis and one for the angle), and two for the grain boundary plane.

Grain boundaries can be divided into three category in terms of their crystallographic orientations. Twist boundary, tilt boundary, and arbitary boundary. As illustrated in Fig.5.10, a tilt boundary has its boundary plan normal perpendicular to its common rotation axis, while twist boundary has its boundary plan-normal parallel to the rotation axis. An arbitary boundary is usually referred to as a boundary in which the axis of rotation and the boundary plane have no special relative orientation. The boundary can be considered to be with a mixed character of both twist and tilt components.

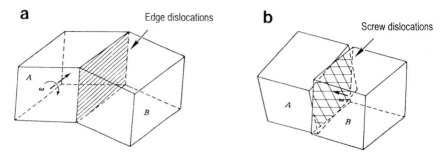

Figure 5.10 (a) A tilt boundary, where the rotation axis ω is in the boundary plane. (b) A twist boundary, where ω is normal to the boundary plane.

5.2.1 The Concept of Coincidence Site Lattice

In general, the microstructure of a grain boundary is much more complicated than a local perturbation in an otherwise perfect crystal. Therefore, it is often appropriate to consider a boundary as a thin region with its own structure such as periodicity and local chemistry. To understand the structures and properties of grain boundaries, two major theoretical approaches were developed: (1) the Coincidence Site Lattice (CSL) model [Bollmann 1982], and (2) the Structural Unit Model (SUM) [Sutton and Vitek 1983]. Although SUM can provide detailed atomic positions of a grain boundary, it requires a precise knowledge of the interatomic potentials, thus limiting its application to simple crystals. The CSL model, based on geometric considerations between two crystals, has successfully explained the structural features observed at grain boundaries in many cubic systems with metallic, ionic, and co-valent bonding [Smith and Pond 1976, Balluffi 1979, Sun and Balluffi 1982, Balluffi and Bristowe 1984, Forwood and Clarebrough 1991]. An important reason for the popularity of the CSL stems from the fact that the CSL description of a boundary structure can be verified by the experimentally determined grain boundary geometry and the configuration of grain boundary dislocations (GBDs) using TEM. A CSL can be produced by rotating two crystals relative to one another about a common crystallographic axis, using a lattice site as the origin. The ratio of the unit-cell volume of the CSL to that of the crystal is defined as Σ. The physical significance of the CSL is that if the CSL formed by two abutting grains is very dense (i.e., Σ is small), the two grains

5.2. Intergranular Defects – Grain Boundaries

will share many lattice sites at the boundary and the boundary should thus have low energy. Low Σ boundaries are, therefore, expected to form preferentially in polycrystalline materials compared with high Σ or non-CSL boundaries.

Based on the CSL model, a grain boundary has a localized structure in which the atoms occupy sites that are displaced from normal lattice sites in two adjoining crystals. These lattice displacements accommodate the different orientations of the neighboring crystals, i.e., deviations from a perfect crystal or from a coincidence orientation, and they generate regions in the boundary where there is a good lattice fit, separated by localized regions of bad fit. The localized discordant areas are the cores of grain boundary dislocations. The CSL model was developed to describe the structure of large-angle boundaries. In the model, the misorientation between the grains is accommodated in two parts. The first part of the misorientation corresponds to an exact low- energy CSL orientation which is accommodated by arrays of closely space "primary" grain boundary dislocations (PGBD). The reminder of the misorientation, corresponding to the deviation from the CSL orientation, is accommodated by arrays of more coarsely spaced "secondary" grain boundary dislocations (SGBD). The density of the SGBD increases with the increase of the deviation. In general, pure twist boundary may have two or three sets of screw dislocations depending on the type of boundary plane, while asymmetrical tilt boundaries contain two sets of edge dislocations, and symmetrical tilt boundaries contain only one set (Fig.5.10). Technically, the structure of any grain boundary can be described formally in terms of arrays of dislocation.

CSL-related grain boundaries occur at unique special orientations, which may exhibit distinct properties of "special" grain boundaries extend over some range of misorientation close to certain exact coincidence values. The most commonly used criteria of specialness to describe such "special boundaries" is the Brandon criterion [Brandon 1966]:

$$\Delta\theta_{\max} = 15^o(\Sigma)^{-1/2}. \quad (5.3)$$

This is the permissible rotation deviation from the coincidence misorientation for which a grain boundary is considered to be "coincidence related". The structure of "coincidence related" or "near-coincidence" grain boundaries is described by introduction of dislocation arrays, whose spacing changes with misorientation.

5.2.2 The Concept of O-lattice

An important development in the geometrical approach of grain boundary structure has been Bollmann's O-lattice theory [Bollmann 1970, Bollmann 1982]. The O-lattice is a useful geometrical approach to characterizing interfacial structure, and describes the matching and mismatching of the misoriented lattices at an interface. If two misoriented crystal lattices are allowed to inter-penetrate, there will be a periodic set of points in crystal space (not necessarily the lattice points of either lattice) where the two lattices shear the coincidence sites, known as the O-elements. The term O is used in O-lattice because these pairs may serve as an origin for the transformation that transforms one crystal into another. These sites are the locations of the best match, or equivalently, sites of minimum strain. The O- elements can be separated from one another by the so-called O-cell- wall, i.e., planes bisecting the connection between two O-elements. In this way, a cell structure representing the area of the maximum disregistry between the two lattices is introduced into the crystal space. Consequently, the intersection of a grain boundary with the cell walls is the dislocation network of the boundary.

The position of the O-elements (defined by vector $\mathbf{X}^{(O)}$) and the Burgers vector of the dislocations (defined by vector $\mathbf{b}^{(L)}$) can be determined by the O-lattice equation [Bollmann 1970]:

$$\mathbf{X}^{(O)} = (\mathbf{I} - \mathbf{A}^{-1})^{-1}\mathbf{b}^{(L)} \qquad (5.4)$$

where $\mathbf{A}$ represents a homogeneous, linear transformation of lattice 1 into lattice 2, $\mathbf{I}$ is identity, and $\mathbf{b}^{(L)}$ are Burgers vectors of dislocation arrays. The O-elements may be a point, line, or plane lattice, depending on whether the rank of $(\mathbf{I} - \mathbf{A}^{-1})$ is 3, 2, or 1, respectively. When transformation $\mathbf{A}$ is a pure rotation, the O-element forms a line lattice parallel to the rotation axis. The projection of the lattice along the rotation axis is the O-lattice. The basis of the theory is the identification of locations of minimum strain between the two crystals, which are identified as those points of identical structure that occupy coincident positions, or O- lattice. If the coincidences between two crystals occur

5.2. Intergranular Defects – Grain Boundaries

at lattice points, it creates coincidence site lattice, or CSL. CSL is a superlattice of the O-lattice (Fig.5.11).

5.2.3 Frank Formula

Frank was the first to develop a theory which provided a dislocation description for the case of arbitrary boundary [Frank 1950]. The Frank formula gives the net Burgers vector **b** intersected by any vector **V** lying in the plane of the grain boundary:

$$\mathbf{b} = 2\sin(\theta/2)(\mathbf{V} \times \mathbf{a}). \tag{5.5}$$

For an arbitrary grain boundary, three sets of independent boundary dislocations are necessary [Amelinckx and Dekeyser 1959], with non co-planar Burgers vectors $\mathbf{b}_i$, $\mathbf{b}_j$, and $\mathbf{b}_k$, with line directions of ξ_i, ξ_j, and ξ_k, and with spacings of d_i, d_j, and d_k. ξ_i and d_i can be expressed as:

$$\xi_i = [\mathbf{a} \times (\mathbf{b}_j \times \mathbf{b}_k)] \times \mathbf{n} \tag{5.6}$$

$$d_i = \left\{ 2\sin\left(\frac{\theta}{2}\right) \left| \frac{[\mathbf{a} \times (\mathbf{b}_i \times \mathbf{b}_j)] \times \mathbf{n}}{\mathbf{b}_i \cdot (\mathbf{b}_j \times \mathbf{b}_k)} \right| \right\}^{-1} \tag{5.7}$$

where **n** is the boundary plane normal, θ is the deviation angle from the coincidence orientation, **a** is the rotation axis, and i, j, and k are the permutations of the ith array of grain boundary dislocations. For any given boundary misorientation and boundary planar normal, we can calculate the line direction and line spacing of the boundary dislocations for the corresponding Burger vector of the dislocation.

Frank's dislocation model is essentially equivalent to Bollmann's O-lattice approach [Christian 1976]. With Frank's formula, calculating the line directions and line spacings of the dislocations is more straightforward; however, the Burgers vectors of *all* dislocation arrays which accommodate the misorientation at the boundary must be known. Experimentally, this is not always possible. In contrast, the O-lattice approach allows us to determine the character of an individual array of dislocations. Based on a knowledge of the b-net derived from the O-lattice construction, then, using Frank's dislocation model makes the analysis of a grain boundary structure much easier and faster, at least on a purely geometrical basis.

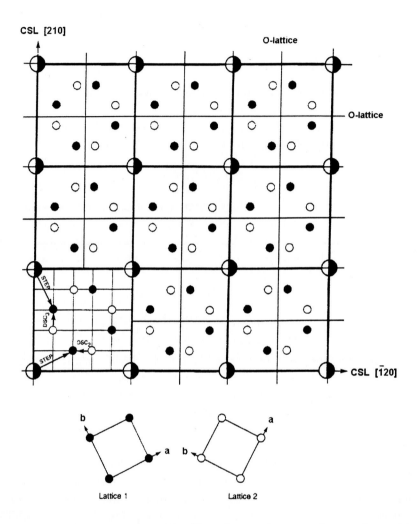

Figure 5.11 Schematic diagram of a $\Sigma = 5$ coincidence site lattice, denoted by large half-black and half-white circles, formed by two square lattices (Lattice 1 and Lattice 2) in a tetragonal crystal for a rotation of $36.87°$ about the [001] axis. The CSL vectors are [210], [$\bar{1}$20], and [001] and the DSC vectors are $1/5[\bar{1}20]$, $1/5[\overline{21}0]$, and [001]. The O-lattice is also shown as a sub-lattice of the CSL.

Chapter 6

Oxygen Ordering Related Defects in YBa$_2$Cu$_3$O$_7$

6.1 Introduction

We will concentrate our discussions in this chapter on the defects in YBa$_2$Cu$_3$O$_{7-\delta}$ materials that are related to the oxygen ordering in the Cu-O plane as well as the effect of trivalent impurity substitutions on the microstructure of YBa$_2$Cu$_3$O$_{7-\delta}$. While there are some studies of other type of defects, the defects related to oxygen-ordering is unique to YBa$_2$Cu$_3$O$_{7-\delta}$ systems. Extensive studies in YBa$_2$Cu$_3$CuO$_{7-\delta}$ microstructure have greatly increased our understanding towards many important aspect of materials science, as we will illustrate in the following sections.

The microstructure YBa$_2$Cu$_3$O$_7$ is among the most extensively studied partly because it is the first superconductor which has critical temperature higher than the boiling point of the liquid nitrogen. Therefore there is great hope for its application. Another reason is that it is relatively easy (compared to other high-T$_c$ materials) to make high-quality samples. The basic structure of YBa$_2$Cu$_3$O$_7$ is also quite stable in high-temperature, with the exception of the oxygen order-disorder phase transition. Since order-disorder transition was a hot research topic in the early 80's, we have many theoretical tools which can be readily applied to the YBa$_2$Cu$_3$O$_7$ systems.

Soon after the discovery of high-temperature superconductivity in the Y-Ba-Cu-O system, the phase responsible for the 90 K transition was found to have a cation stoichiometry of 1Y:2Ba:3Cu. The unit cell dimensions determined by electron and x-ray diffraction identified the structure as being related to a cubic perovskite with one of the cube axes tripled (see the excellent review by R.M. Hazen [Hazen 1990] for references) and there are three copper-oxygen planes in one unit cell. The neutron diffraction studies using Rietveld refinement technique showed unambiguously that the oxygen atoms in the oxygen deficient copper plane (basal plane with layer stoichiometry CuO) of the unit cell were ordered in the orthorhombic structure as shown in Fig.4.2 and the 90 K material contained nearly seven oxygen atoms per unit cell. The plane containing Cu1 in this figure is the oxygen deficient basal plane. In the perfectly ordered situation, this plane consists of Cu-O chains parallel to the crystallographic b-axis.

Superconducting properties of this system is closely related with the oxygen content and oxygen ordering. It has been found that in the thermally quenched samples, the superconducting transition temperature changes from 92 to 0K as the oxygen content is varied from 7.0 to 6.5 [Kwok et al. 1988]. On the other hand, transition temperature measurements on samples obtained by the gettered annealing method at low temperatures show a clear plateau in T_c versus the oxygen concentration curve [Cava et al. 1987a]. The width of the transition was found to narrow in the "plateau" region. This plateau at $T_c \approx 60$ K is a strong evidence for the existence of two superconducting phases in the $YBa_2Cu_3O_{7-\delta}$ system and the two transition temperature 92K and 60K have been related to the hole concentration in the CuO_2 plane which in turn depends on the oxygen structure in the basal plan [Warren et al. 1989]. The plateau is thought to be related to the existence of a double-cell orthorhombic structure, which has been seen in several experiments [Chaillout et al. 1988, Kubo et al. 1988, Fleming et al. 1988, Chen et al. 1988]. On the other hand, experiments where samples have been annealed for a long time with fixed oxygen content shows an increase in superconducting transition temperature with increase in the orthorhombicity of the system [Veal et al. 1990]. It has been shown that the increased orthorhombicity is caused by increased oxygen ordering. These experiments clearly indicate that there is a direct link between oxygen ordering and oxygen content in the CuO basal plane and the superconducting transition temperature.

The observed structure and properties of $YBa_2Cu_3O_{7-\delta}$ (we refer to this as YBCO compound) below room temperature depend critically on how the material is processed at higher temperatures. This situation arises because the superconducting properties are largely controlled by the oxygen content and ordering in $YBa_2Cu_3O_{7-\delta}$, which in turn, are controlled by the annealing times and temperatures, the oxygen partial pressures, and the quench rates used in preparing the material.

In the following sections we will see how a relatively simple structural model for $YBa_2Cu_3O_7$ was established using experimental findings, and how such model was extended to study the defect structure of the system. This illustrate the fruitfulness of the close collaboration between experimentalists and theorists to study a complex system.

6.2 Models for Oxygen Ordering in $YBa_2Cu_3O_{7-\delta}$

In this section we will establish a rather simple theoretical model to be used to calculate defect structures in later part of this chapter [Cai and Mahanti 1988, Cai and Mahanti 1989, Cai and Mahanti 1990]. We base this model on the experimental findings of orthorhombic to tetragonal phase transitions in YBCO compounds under various oxygen partial pressures. The parameters in the model were chosen so that the calculated structural properties fits well with the experimental data. This model can then be modified to study the defect structure of the system.

This phenomenological approach is widely used in materials science to study the defect structure of a complex system when the model for its electronic structure is unavailable or to complex to be used in large-scale computer simulations (see Sec.3.3 for more detailed discussions). We will see a good example of such an approach in this chapter.

The orthorhombic to tetragonal phase transition in YBCO compounds was initially identified by electron beam heating in a transmission electron microscope (TEM) and was subsequently studied *in situ* by numerous techniques [Beyers *et al.* 1987a]. Hot-stage X-ray diffraction showed a sudden change of lattice parameters above 500°C and the

a and b axes became equal [Beyers et al. 1987a, Schuller et al. 1987]. The orthorhombic YBCO compound turns tetragonal at a temperature which depends on the oxygen partial pressure [Schuller et al. 1987, Beyers et al. 1987b]. Thermogravimetric studies have shown that O_7 starting materials begins to lose oxygen reversibly above ~ 350–$400°C$ ([Beyers et al. 1987b, Gallagher et al. 1987a, Kishio et al. 1987, Yamaguchi et al. 1988] and [Specht et al. 1987]). The orthorhombic to tetragonal transition is assumed to occur when there is a discontinuity in the curve of weight loss versus temperature plots [Gallagher et al. 1987a].

In situ neutron diffraction showed that the observed temperature dependent changes both in the lattice parameters are caused by changes in oxygen content and oxygen order in the basal copper plane [Jorgensen et al. 1987a]. The oxygen that is lost above $\sim 400°C$ comes primarily from the $O(4)[=(0,1/2,0)]$ site and goes to the normally vacant $(1/2,0,0)$ site (see Fig.4.2), resulting in the contraction of the b-axis and expansion of the a-axis just below the transition. These results show that the oxygen structure changes from fully ordered at low temperature, to partially ordered at higher temperatures in the orthorhombic phase, to completely disordered at still higher temperatures in the tetragonal phase. Jorgensen et al. also find that although the orthorhombic to tetragonal phase transition temperature depends on the oxygen partial pressure, the oxygen concentration at the transition temperature is relatively insensitive to the oxygen partial pressure and the transition always occurs near the oxygen concentration of 6.5 [Jorgensen et al. 1987a].

These experimental findings clearly indicate that the orthorhombic to tetragonal structural transition in $YBa_2Cu_3O_{7-\delta}$ is an oxygen order-disorder transition occurring in the CuO basal plane. Thus a two-dimensional lattice gas model is a reasonably good starting point to study this phenomena.

To develop a lattice gas model for the order-disorder transition, the structure of CuO basal plane can be described by a 2-dimensional square lattice consisting of Cu atoms, oxygen atoms, and oxygen vacancies (V). In the ordered (orthorhombic) phase oxygen atoms and vacancies are ordered so that there are linear Cu-O chains parallel to the b-axis and Cu-V chains parallel to the a-axis, as show in Fig.6.1.

6.2. Models for Oxygen Ordering in YBa$_2$Cu$_3$O$_{7-\delta}$

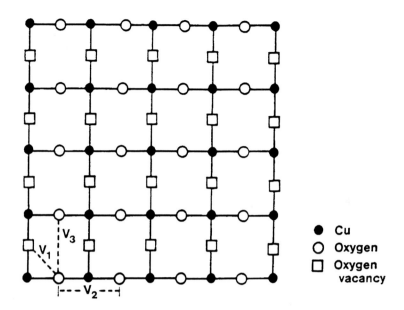

Figure 6.1 The ground state structure of YBa$_2$Cu$_3$O$_7$ basal plane. The back dots indicate the positions of Cu ions, the open circles indicate the positions of the oxygen atoms and the open square indicate oxygen vacancy positions.

The stoichiometry of the arrangement is YBa$_2$Cu$_3$O$_7$. If the oxygen atoms are absent from this plane, then one has YBa$_2$Cu$_3$O$_6$ and the structure of the compound is tetragonal. The relationship between oxygen concentration x in a lattice shown in Fig.6.1 and the value of oxygen deficiency in YBa$_2$Cu$_3$O$_{7-\delta}$ can be written as

$$x = \frac{1-\delta}{2}. \tag{6.1}$$

To write the order parameter of the order-disorder transition of oxygen atoms in the basal plane, we denote the fractional site occupancy numbers x_1 and x_2 as the concentration of oxygen atoms at O$_4$ and O$_5$ site, respectively (see Fig.4.2), where

$$x = \frac{x_1 + x_2}{2}. \tag{6.2}$$

The order parameter of the transition Φ can then be written as

$$\Phi = \frac{|x_1 - x_2|}{x_1 + x_2}. \tag{6.3}$$

Since we only allow oxygen atoms to occupy sites on a rigid lattice defined in Fig.6.1, this model is called lattice gas model.

To obtain an approximate interaction potential of this lattice gas model, let us assume that the interaction between oxygen ions in this system is predominately Coulomb interaction with metallic screening for $6.5 < 7 - \delta < 7$. The metallic screening is due to the holes in the oxygen band. For $\delta > 0.5$, it is believed that the oxygen holes are localized and do not take part in metallic screening. The lattice gas Hamiltonian has the form:

$$H_{LG} = V_1 \sum_{<ij>} n_i n_j + V_2 \sum_{(ij)Cu} n_i n_j + V_3 \sum_{(ij)} n_i n_j + (\epsilon(x) - \mu) \sum_i n_i, \tag{6.4}$$

where the single site energy $\epsilon(x)$ can be defined as

$$\epsilon(x) = E_b + E_d/2 + \alpha x, \tag{6.5}$$

where E_b is the binding energy of the oxygen atom in the basal plane and its origin is the Madelung energy; E_d is the disassociation energy of an oxygen molecule and x is the average oxygen site occupancy which is given by:

$$x = \frac{1}{N} \sum_i n_i. \tag{6.6}$$

The physical origin of the αx term in Eqn.(6.5) can be ascribed either to a change in the copper valency with increased oxygen concentration x [Iwazumi et al. 1988] or to a mean field representation of the longer-range part of the oxygen-oxygen repulsion [Cai and Mahanti 1988].

The chemical potential μ can be calculated using the equation of state of the diatomic ideal gas [Landau and Lifshitz 1965]:

$$\mu = k_B T \ln \left\{ \left(\frac{P}{P_0} \right) \left[1 - \exp\left(-\frac{T_E}{T} \right) \right]^{1/2} \right\}, \tag{6.7}$$

where P is the oxygen partial pressure, T_E is the characteristic temperature defining the vibrational energy of the oxygen molecule. The

6.2. Models for Oxygen Ordering in $YBa_2Cu_3O_{7-\delta}$

characteristic pressure P_0 is given by

$$P_0 = \left(\frac{2\pi M k_B}{h^2}\right)^{3/2} \left(\frac{4\pi^2 I k_B}{h^2}\right) k_B T^{7/2} \qquad (6.8)$$

where the moment of initia I of the O_2 molecule is given by the relation $(4\pi^2 I k_B)/h^2 = 2.1 K$; M is the mass of the O_2 molecule.

In Eqn.(6.4), V_1 is the interaction between two nearest neighbor oxygen atoms. The interaction between two next nearest neighbor oxygen atoms is either V_2 or V_3 depending on whether a copper atom bridges these two or not, respectively. de Fontaine et al. were the first to realize that V_2 and V_3 could be different [de Fontaine et al. 1987]. The physical meaning of these parameter will be discussed later.

The first lattice gas model calculation was performed by Bakker et al. who assumed that the interaction between oxygen atoms in the basal plane of $YBa_2Cu_3O_{7-\delta}$ can be modeled by an isotropic lattice gas Hamiltonian ($V_2 = V_3 = 0$) which is equivalent to a 2-dimensional antiferromagnetic Ising model [Bakker et al. 1987]. Using quasi-chemical approximation, they found that for $P = 1 atm.$, the system showed an order-disorder transition which agreed with the experimental observation. However later Monte Carlo simulation results showed that the oxygen concentration at the critical temperature was always near $x = 0.37$ and was relatively insensitive to the oxygen partial pressure [Cai and Mahanti 1988] . This result is in disagreement with the experimental results of Jorgensen et al. [Jorgensen et al. 1987a], which shows that the concentration of oxygen at the critical temperature is near $x - 0.25$.

The value of the critical concentration ($x_c = 0.37$) for the nearest neighbor repulsion model was first discussed by Binder and Landau who argued that it is the percolation threshold of a system consisting of a filled ($\sqrt{2} \times \sqrt{2}$) lattice (the open circles in Fig.6.1. This gives $x_c \sim 0.25 + 0.25 \times 0.59 = 0.37$, where 0.59 is the site percolation threshold for a square lattice [Binder and Landau 1980]. Therefore Bakker et al.'s model with isotropic repulsive interaction is not adequate to describe the orthorhombic to tetragonal phase transition of the $YBa_2Cu_3O_{7-\delta}$ system. In addition to the problem of x_c, this model also gives a much stronger temperature dependence of x near the critical temperature T^* which is not seen experimentally. This latter problem was addressed by Salomons et al. who introduced a concentration dependent binding energy term ($E_b + \alpha x$ of Eqn.(6.5)) as discussed earlier in this section [Salomons et al. 1987] .

The isotropic repulsive model has another drawback, namely it does not favor chain formation for oxygen deficient systems as seen in experiments. The fact that one finds Cu-O chains in oxygen deficient systems strongly suggest that $V_2 < V_3$ or that V_2 be negative. Bell introduced a model with isotropic attractive interaction for next nearest neighbor oxygen atoms [Bell 1988]. Although Bell's model gives the temperature dependence of the oxygen fractional site occupancy in good agreement with experiments, there is a phase separation at low temperature and Cu-O chain formation is not energetically favored. On the other hand, de Fontaine and Wille et al.'s model [de Fontaine et al. 1987, Wille et al. 1988] which has $V_2 < 0$ and $V_3 > 0$ removes this difficulty. They have obtained the temperature-concentration phase diagram using a cluster variational method which agrees very well with Cai's Monte Carlo simulation results [Cai and Mahanti 1988]. However the cluster variation method does not work well at low temperature.

We will now discuss the Monte Carlo simulations of the lattice gas model with Hamiltonian defined in Eqn.(6.4). These simulations were carried out on a 40×40 square lattice as shown in Fig.6.1 using periodic boundary condition and $10{,}000 \sim 20{,}000$ Monte Carlo sweeps per site. We have done both heating and cooling runs and have used different shapes of the boundaries to perform the simulation. These studies give results which agree very well with each other and the studies of the size dependence of the critical behavior gives evidence that for the parameters we have chosen, the transition is continuous.

First, the Monte Carlo simulations were carried out for a number of different sets of interaction parameters. We find that the oxygen concentration (x_c) at the transition temperature T^* depends on the relative strengths of the next nearest neighbor interaction parameter V_2 and V_3. For $V_3 \geq |V_2|$, $x_c \sim 0.37$. However for $V_3 < |V_2|$, x_c can have values ranging from $0.25 \sim 0.33$ depending on their relative strengths.

To fit the general features of the experimental data shown in Fig.6.1, we have made the following choice for the values of the parameters in Eqn.(6.4): $V_1 = 0.25eV$, $V_2 = -0.2eV$, $V_3 = 0.1eV$, $E(x) = -1.45 + 1.5x$.

6.2. Models for Oxygen Ordering in $YBa_2Cu_3O_{7-\delta}$

In Fig.6.2 we show the temperature dependence of the order parameter Φ (defined in Eqn.(6.3)) and the susceptibility χ, which is defined as

$$\chi = \frac{N}{k_B T}[<\Phi^2> - <\Phi>^2]. \qquad (6.9)$$

The susceptibility χ measures the thermal fluctuation of the order parameter. Since lattice gas model is mathematically equivalent to Ising model which represents a ferromagnet [Binder 1979], the χ is mathematically equivalent to the susceptibility of a magnetic system.

The transition temperature T^* is identified with the peak position of the $\chi \sim T$ curve. Its values are 700°C for $P = 1$bar, 660°C for $P = 0.2$bar and 620°C for $P = 0.02$bar; These values agree very well with the experimental data (700, 670, and 620°C for the three values of pressure respectively). With the above values of the parameters and for the pressure given above, the low-T phase has $x = 0.5$ which corresponds to the stoichiometry of $YBa_2Cu_3O_7$.

In Fig.6.3, we give the oxygen concentration x and the fractional site occupancies x_1, x_2 as a function of T for different values of P. It can be seen that the occupation of the "wrong" sublattice (x_2) decreases drastically below T^*. The value of the oxygen concentration at T^* is close to 0.25 in good agreement with the experimental results. Also the T-dependence of x for $T > T^*$ is very well reproduced by our Monte Carlo simulation. To see if the model and the parameter values we choose accurately represent the physical properties of the system we study, we show in Fig.6.4 the pressure dependence of oxygen concentration isotherm compared with the experimental data obtained by Salomons et al. [Salomons et al. 1987]. The simulation results agree very well with the experimental data.

To further quantify the oxygen ordering, we have calculated the average and maximum chain lengths along a and b axes respectively. These are shown in Fig.6.5. We can see that even at temperature well above T^*, the average and the maximum Cu-O chain length still have very high value, indicating strong short range order in the tetragonal phase.

Since Monte Carlo simulations of this lattice gas model with the interaction parameter chosen as above give results which are in very good agreement with experimental data, it is important the discuss in

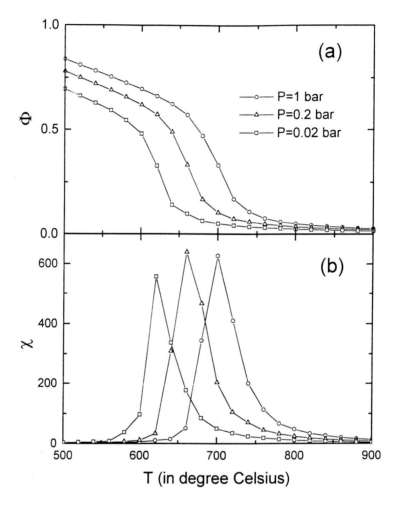

Figure 6.2 Temperature dependence of (a) the order parameter Φ; and (b) the susceptibility χ for different oxygen partial pressure.

6.2. Models for Oxygen Ordering in $YBa_2Cu_3O_{7-\delta}$

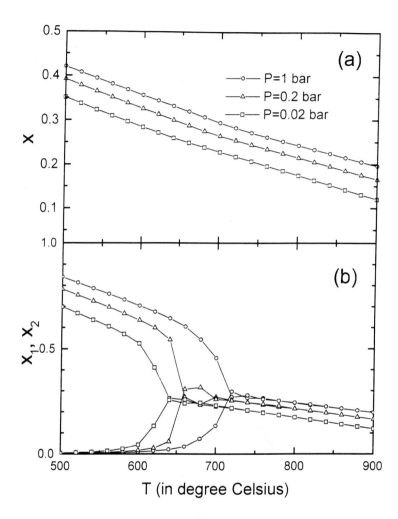

Figure 6.3 Temperature dependence of (a) the total oxygen concentration x; and (b) the fractional site occupancies at O4 and O5 site (x_1, x_2), respectively, for different oxygen partial pressure.

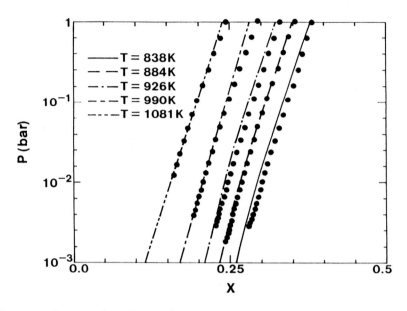

Figure 6.4 Pressure dependence of oxygen concentration isotherm compared with the experimental data of Salomons *et al.* [Salomons *et al.* 1987]. The dots are experimental data.

detail the physical meaning and the assumption made in establishing the values of these parameters. In particular, we would like to discuss why a model with only short range interaction is a good approximation to the system $YBa_2Cu_3O_{7-\delta}$ which is apparently dominated by ionic bonding.

First, let us consider the system $YBa_2Cu_3O_6$. This is an antiferromagnetic insulator with no oxygen atoms in the basal plane. The charge state of Cu(1) is 1+. When oxygen is intercalated into the system, the number of Cu^{1+} ions per unit cell ($\rho_{Cu^{1+}}$) decreases linearly as the oxygen concentration x increases, as shown by the X-ray absorption measurements of the Cu K-edge by Tranquada *et al.* [Tranquada *et al.* 1988]. The slope of the $\rho_{Cu^{1+}}$ versus x curve is between 1 and 2 depending on whether the sample is in the tetragonal or the orthorhombic phase. Thus our assumption of the binding energy of oxygen as a linear function of x with the slope of 1.5 is quite reasonable for the whole

6.2. Models for Oxygen Ordering in $YBa_2Cu_3O_{7-\delta}$

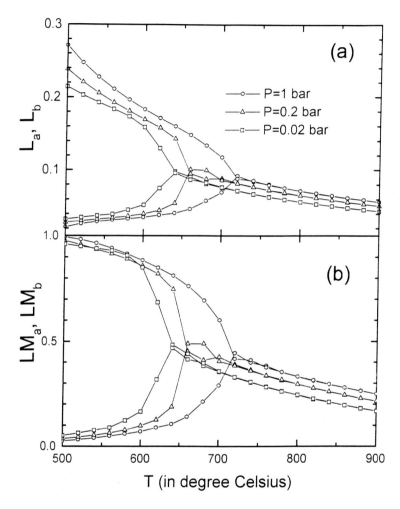

Figure 6.5 The temperature dependence of (a) average Cu-O "chain" length; and (b) maximum Cu-O "chain" length for different values of oxygen partial pressure.

range of oxygen concentration of $YBa_2Cu_3O_{6+2x}$ system ($0 < x < 0.5$). For the $YBa_2Cu_3O_7$ system, the nominal charge state of the basal plane Cu is 3+. However, due to the strong on site Coulomb repulsion, d^8 state (two holes in the d-shell) is highly improbable theoretically and also has not been seen experimentally. XPS experiments indicate that the extra "hole" goes to the oxygen site and is mobile. Thus when the concentration of oxygen in the basal plane $x \geq 0.25$, the system goes through an metal-insulator transition and becomes metallic. We suggest that it is the metallic screening by these mobile holes which makes the interaction between oxygen ions in the basal plane short-ranged so that the lattice gas model with short range interaction is a good approximation to the real system. We can reasonably assume that there is one mobile hole per unit cell and the distribution of the "hole gas" is such that there is no mobile hole near the Y layer and that holes are uniformly distributed between two nearest CuO_2 layers (see Fig.4.2). Including both Thomas-Fermi and the ion dielectric screening ($\epsilon = 7$), we estimate that $V_1 \approx 0.3eV$ and $V_3 \approx 0.1eV$, which are consistent with the choice of the parameters given in Eqn.(6.4).

The interaction between two oxygen atoms bridged by a Cu atom, V_2, is negative primarily due to the indirect interaction caused by the transfer of oxygen holes via Cu. Unfortunately there is no satisfactory method to calculate V_2 because it depends not only on the oxygen hole density but also on the interaction with the localized spins at the Cu sites. To estimate the order of magnitude of V_2, we consider the CuO chains only and construct an one dimensional tight binding model to describe the motion of the oxygen hole. We further assume that the localized spins on the Cu sites are either parallel or antiparallel to each other. Using the values of the parameters derived from the Emery's model [Emery 1987], we find that V_2 is negative and is of the order of $0.1eV$. Thus we believe that this lattice gas model is physically reasonable and the oxygen-oxygen interaction is basically anisotropic and of short range nature in the "metallic" phase. It is known that the Thomas-Fermi screening length increases with the decreasing of oxygen content which decreases the hole density, and the lattice gas model with only short range interaction fails to be a good approximation for $x < 0.25$ when the system becomes insulating and the interaction becomes long ranged.

The phase diagram in the $T \sim x$ plane has been studied by various groups using cluster variation method, transfer matrix method, and by Monte Carlo simulation [Aukrust et al. 1990]. For similar sets of parameters, these calculations show that in addition to the orthorhombic phase of YBa$_2$Cu$_3$O$_7$ (OI) and the tetragonal phase of YBa$_2$Cu$_3$O$_6$, there is a thermodynamically stable phase of YBa$_2$Cu$_3$O$_{6.5}$ with alternate full and empty Cu-O chains. The nature of the transition between the tetragonal and orthorhombic phases is found to be second order. In the next section, we will discuss the ordered oxygen arrangements under different processing conditions.

6.3 Effect of Thermal Quenching on Oxygen Ordering

Soon after the orthorhombic YBa$_2$Cu$_3$O$_7$ and tetragonal YBa$_2$Cu$_3$O$_6$ phases were identified, researchers began to investigate the structures and properties of YBa$_2$Cu$_3$O$_{7-\delta}$ samples with intermediate oxygen contents ($0 < \delta < 1$). In a large number of such studies [Jorgensen et al. 1987c, Tokumoto et al. 1987, Nakanishi et al. 1988, Farneth et al. 1988] the samples were prepared by rapidly quenching the high-temperature, low oxygen content tetragonal phase from 600–1000°C to room temperature or liquid nitrogen temperature. These studies usually found a smooth variation of T_c with oxygen content with the samples becoming insulators near O$_{6.5}$, corresponding to an orthorhombic-to-tetragonal phase transition [Jorgensen et al. 1987a]. Alternatively, intermediate oxygen content samples were prepared by removing oxygen at lower temperatures (usually 400–500°C) from an O$_7$ starting material. These studies employed a variety of methods to remove the oxygen, including annealing in reducing atmospheres [Farneth et al. 1988, Monod et al. 1987], equilibrating O$_6$ and O$_7$ YBCO mixtures in sealed quartz tubes [Cava et al. 1987a], gettering annealing with zirconium in a sealed tube [Cava et al. 1987b], and titrating electrochemically in a solid-state cell. The variation of T_c with oxygen content in samples prepared at lower temperatures is different from that observed in samples quenched from the tetragonal phase. The superconducting transition temperature does not vary smoothly with oxygen content in these samples. Instead, T_c remains at 91K between O_7 and $O_{6.9}$, then falls to a $\sim$ 60K "plateau" between $O_{6.7}$ and $O_{6.6}$, and finally drops to zero (i.e. not superconducting) between $O_{6.5}$ and $O_{6.4}$. In their attempt to explain the plateau in the oxygen concentration dependence of T_c, Zaanen et al. [Zaanen et al. 1988] developed a simple model by assuming that the superconducting transition temperature is directly linked to the hole concentration in the CuO_2 plane. The hole concentration in turn depends on the oxygen concentration and oxygen ordering in the CuO plane through the electronic structure of the CuO chains formed in the CuO basal plane. To apply this model to the real system, however, it is important to know the CuO chain length distribution under various procession methods.

Because of the close correlation between the oxygen ordering and the superconducting properties of $YBa_2Cu_3O_{7-\delta}$ systems, the structure of the oxygen deficient $YBa_2Cu_3O_{7-\delta}$ produced under various processing conditions had been the subject of both theoretical [Wille et al. 1988, Khachaturyan and Morris 1987, Khachaturyan and Morris 1988, Cai and Mahanti 1988, Cai and Mahanti 1989, de Fontaine et al. 1989, Cai and Mahanti 1990] and experimental work [Cava et al. 1987a, Specht et al. 1988, van Tendeloo et al. 1987b, Chaillout et al. 1987, Werder et al. 1988b, Chen et al. 1988]. Analysis of diffuse scattering from the early stage of oxygen ordering provides strong evidence that the material contains small domains with alternating planes of occupied and vacant chain sites along the [100] direction (see Fig.5.1), which seem to support the model we discussed in previous section. The width of these domains depends strongly on the oxygen content.

Figure 6.6 shows narrow streaks of diffuse scattering for quenched samples of $YBa_2Cu_3O_{7-\delta}$ ($0.2 < \delta < 0.7$). The streaks point in the [100] direction and the intensity maxima are often at $g + (1/2, 0, 0)$ (as shown in Fig.6.6(a)), $(1/3,0,0)$ $(2/3,0,0)$ and $(2/5,0,0)$ $(3/5,0,0)$. For $0.33 < \delta < 0.5$ the streaks are short, approximate superstructure reflections. For lower and higher oxygen content the streaks become longer. The streaks are very narrow, suggesting coherence over more than 10nm in the [010] direction. Also, along the [001] direction there is coherence of up to 10 nm as seen in Fig.6.6(c). Almost no diffuse scattering is observed for $\delta > 0.7$ or $\delta < 0.2$.

The observed diffuse scattering can be analyzed using a simple

6.3. Effect of Thermal Quenching on Oxygen Ordering

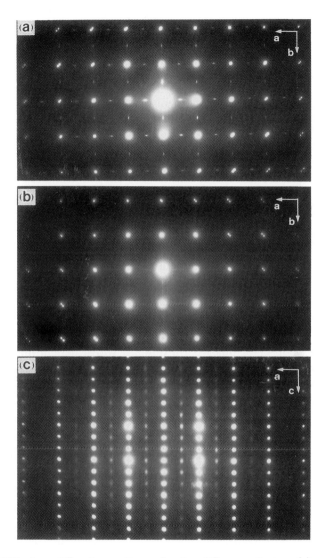

Figure 6.6 Electron diffraction patterns showing diffuse scattering. (a) and (b) are the $(001)^*$ projection for compositions $\delta = 0.33$ and $\delta = 0.19$, respectively. Contributions from the two twin orientations give two superimposed patterns rotated $90°$ relative to each other. (c) is the $(010)^*$-projection for $\delta = 0.33$.

model involving only the local rearrangement of oxygen [Zhu et al. 1990a]. Fig.6.7(a) shows the calculated diffuse scattering from randomly distributed oxygen (O) and vacancy (V) planes along the [100] direction. In terms of one-dimensional short-range order parameters this means that $\alpha_o = 1$ and the other α_1, α_2, etc. are zero. Here α_r is related to the probability P_{OVr} of finding a V plane at a distance r from an O plane by $\alpha_r = 1 - P_{OVr}/\delta$. Figures 6.7(b) - 6.7(d) show calculations based on models with alternating (100) planes where the chain sites are fully oxygen occupied (O) and vacant (V). The *deviation* from the average structure, which gives the diffuse scattering is δ oxygen atoms on the O sites, and $(1-\delta)$ oxygen atoms on the V sites. Thus, the scattering amplitudes in the expression for diffuse scattering are $f_o(\delta)$ and $f_o(\delta-1)$ for the O and the V sites, respectively. Here f_o is the scattering amplitude for oxygen. The width of the OVOV domain along the [100] direction in number of unit cells, N, is related to the observed width of the diffuse maxima Δ by $N = \Delta^{-1}$. Figures 6.7(e) and 6.7(f) demonstrate that diffuse scattering near the fundamental reflections would be a signature of oxygen diffusion and tendency towards phase separation.

The calculated diffuse scattering of Figs.6.7(b)–6.7(d) agrees with experimental observations with a domain width depending upon oxygen composition. The variation of the width of the OV domains with composition may be understood by considering the initial stage of their growth. A fluctuation that initiates an OV ordering sequence for $\delta = 0.5$ is likely to expand over a large distance along the [100] direction because the transport of oxygen to neighboring planes is sufficient. On the other hand, for $\delta = 1/3$ only [OVO] is possible before transport of oxygen over larger distances is necessary and, similarly, for $\delta = 2/3$ with [VOV]. With the transport of oxygen only to neighboring planes no formation of VO sequences with fully occupied and fully vacant planes is possible for $\delta > 2/3$ and $\delta < 1/3$. The composition of the OV sequence will from this consideration be the same as the average composition:

$$\delta = 1 - \frac{n_o}{n_v + n_o} = 1 - \frac{n_v + 1}{2n_v + 1}; \qquad \delta \leq 0.5 \qquad (6.10)$$

where n_o and n_v are the numbers of oxygen and vacancy planes. A symmetry related expression exists for $\delta > 0.5$. The width of the

6.3. Effect of Thermal Quenching on Oxygen Ordering

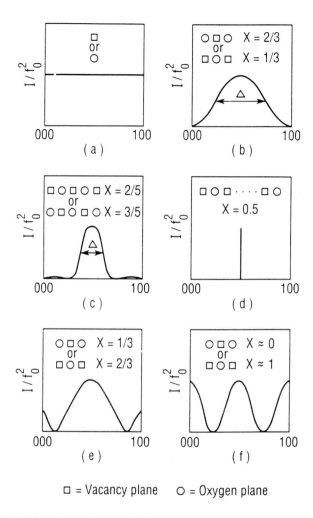

□ = Vacancy plane   O = Oxygen plane

Figure 6.7 Calculated streak profiles for the indicated sequences of oxygen and vacancy planes ($x = 1-\delta$). (a) Randomly distributed oxygen vacancy. (b) through (d) are based on models with fully occupied and vacant (100) planes with sequences and relative occupancies indicated. (e) and (f) have a sequence composition different from the average composition, and are included to explore the possibility of phase separation.

domain along the [100] direction is thus $[(n_o - 1) + n_v]a = 2n_v a = Na$, where N is the number of unit cells. The calculated N shows a qualitatively similar behavior with the maximum occurring at $\delta = 0.5$.

The composition dependent OV sequences along the [100] direction suggests YBa$_2$Cu$_3$O$_{7-\delta}$ may form a homologus series of Magneli phases for $1/3 < \delta < 2/3$ [Khachaturyan and Morris 1988]. The next step in the ordering process may be the ordering of these Magneli-like units to form long range ordered phases. This requires stacking faults in the OV sequence. Neighboring OO or VV planes have been predicted to be energetically unfavorable [Wille et al. 1988], but the alternative of phase separation requires transport of oxygen over larger distances. A possible explanation for the tendency towards two maxima near (1/3,0,0) and (2/3,0,0) would be the initiation of the further ordering into Magneli-like phases. However, preliminary calculations where correlation between [OVO] domains are considered, suggest that domains shifted by one unit cell along the [100] direction (antiphase due to e.g. limited correlation along the [001] direction) also may account for this additional modulation.

In order to see how the CuO chain length distribution depends on the oxygen concentration and quench rates, we have carried out a series of Monte Carlo simulations for constant oxygen concentration (Kawasaki dynamics) using the lattice gas model we developed in the previous section. We explored the low temperature structure of the CuO plane for different oxygen concentrations and thermal cooling rates. A 40 × 40 lattice was used for the simulation. The high temperature configuration was obtained by using Monte Carlo simulation at 2300K (well above the order-disorder phase transition temperature for O$_7$) and the system was equilibrated at this temperature with 10,000MCS/site (MCS =Monte Carlo steps). The low temperature data were obtained by either slowly cooling (annealing) the system with a temperature interval of 100K or rapid cooling (quenching) the system to 300K.

Figure 6.8 shows the oxygen configurations of the annealed and the quenched systems at 300K for the YBa$_2$Cu$_3$O$_{6.5}$ system. We find that the structure for the annealed samples is a double-cell orthorhombic (OII) structure with half of the CuO chains intact and half empty. The quenched system, on the other hand, shows very small domain of

6.3. Effect of Thermal Quenching on Oxygen Ordering

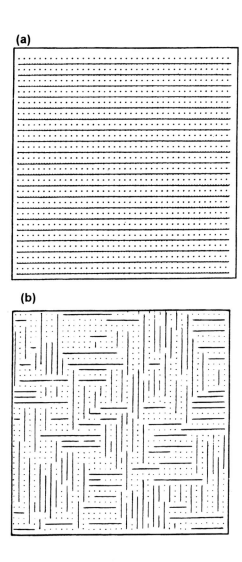

Figure 6.8 Oxygen configuration of (a) the annealed $YBa_2Cu_3O_{6.5}$; (b) the quenched $YBa_2Cu_3O_{6.5}$.

orthorhombic structure, with CuO chains oriented along the a and b directions. The overall symmetry is, however, tetragonal as indicated by the structure factor along the a and b directions.

The difference between the structure of quenched and annealed systems has also been studied for the oxygen concentration $O_{6.7}$. Since this oxygen concentration is not appropriate for either the OI structure or the OII structure, new structural features arise. Typical oxygen configuration for the annealed and quenched $YBa_2Cu_3O_{6.7}$ compound at 300K are shown in Fig.6.9. We find that the low-temperature structure for the annealed $YBa_2Cu_3O_{6.7}$ (Fig.6.9(a)) is a highly degenerate one. Instead of a single domain of ordered structure or a two-domain structure which is phase separated into OI and OII domains, we find long Cu-O chains "intercalated" into each other so that one sees a mixture of structure with OI and OII symmetry as indicated by the "snapshots" of the oxygen configurations (Fig.6.9). We estimate the typical OII domain size along the a axis to be about 5 – 6 unit cells corresponding to about 20–25Å, which agrees very well with the experimental results [Fleming et al. 1988, Chen et al. 1988]. It is also shown, both by the oxygen structure factor calculation and the "snapshots" of the oxygen configuration, that the average Cu-O chain length depends sensitively on the thermal history of the sample. This may in turn, as indicated by Zaanen et al., affect the hole density in the CuO_2 plane, and the superconducting transition temperature T_c [Zaanen et al. 1988]. The quenched $YBa_2Cu_3O_{6.7}$ system shows short chains in both a and b directions and the overall symmetry is tetragonal.

6.4 [110] Tweedy Modulation

In this section we explore the lattice distortion due to the short-range oxygen ordering discussed in the previous section. Below about 750°C, $YBa_2Cu_3O_{7-\delta}$ undergoes a tetragonal to orthorhombic structural phase transition resulting in twinning on the (110) planes. When a small fraction of the Cu atoms (Cu(1) atoms) is replaced by certain trivalent cations, e.g. Fe, Co, or Al, a structural modulation occurs, generating an overall tetragonal symmetry of the Bravais lattice [Xu et al. 1989, Wördenweber et al. 1989, Schmahl et al. 1990, Zhu et al. 1991b]. Similar modulated structures, sometimes called tweed because of the

6.4. [110] Tweedy Modulation

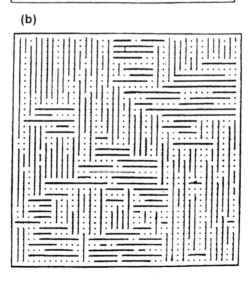

Figure 6.9 Oxygen configuration of (a) the annealed $YBa_2Cu_3O_{6.7}$; (b) the quenched $YBa_2Cu_3O_{6.7}$.

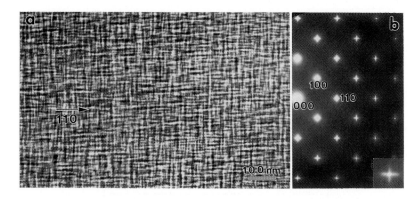

Figure 6.10 (a) Tweed contrast in $YBa_2(Cu_{0.97}Fe_{0.03})_3O_{7-\delta}$, imaged under $\mathbf{g} = 200$ two-beam condition. (b) $(001)^*$ selected-area diffraction pattern from (a).

appearance of transmission electron microscopy (TEM) images, have been encountered in some binary alloy systems which exhibit statistical fluctuations in their composition (or in an order parameter), or undergo a phase transformation with the reduction of the crystal symmetry [Khachaturyan 1983, Shapiro et al. 1986]. The modulated structure usually shows a roughly periodic domain image associated with the streaks of diffuse scattering seen in electron diffraction patterns [Tanner 1966, Robertson and Wayman 1983].

Tweed was observed in $YBa_2(Cu_{1-x}M_x)_3O_{7-\delta}$, tweed was observed for M=Co, Fe, or Al, $x > 0.025$ and M=Cu, $\delta \approx 0.6 - 0.8$, and its image contrast was most visible when $x = 0.03 - 0.04$ (Fig.6.10(a)) [Xu et al. 1989, Zhu et al. 1990e, Zhu et al. 1991c, Schmahl et al. 1990, Kreckels et al. 1991]. In diffraction pattern, streaks are seen along the $[110]^*$ and $[\bar{1}10]^*$ direction, resulting in a cross, through the fundamental reflections (Fig.6.10(b)).

The morphologies of tweed near the (001) projection in for M=Fe, $x = 0.015, 0.02, 0.025, 0.03, 0.05, 0.1$, and 0 are shown in Fig.6.11. The period of the modulation is inversely proportion to the dopant concentration x, being approximately $p = 5.6 + x^{-1}$ (Fig.6.12). For $x > 0.1$, tweed image is hardly visible, however, the structural modulations can be detected by observing the diffuse scattering, especially for $x > 0.22$, when satellite spots due to the modulations become

6.4. [110] Tweedy Modulation

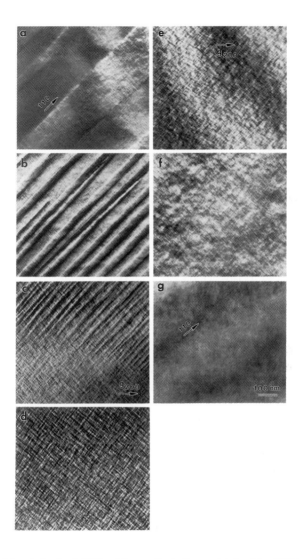

Figure 6.11 Morphologies in $YBa_2(Cu_{1-x}Fe_x)_3O_{7-\delta}$ obtained by TEM (under $\mathbf{g} = 200$ two-beam condition) at different dopant content x: (a) 0.045, (b) 0.06, (c) 0.075, (d) 0.09, (e) 0.15, (f) 0.3, and (g) 0.45. All images have the same scale.

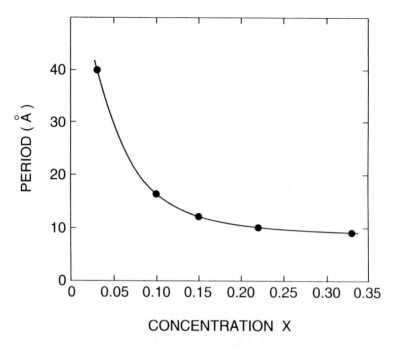

Figure 6.12 Observations of tweed period as a function of the dopant concentration x in $YBa_2(Cu_{1-x}M_x)_3O_{7-\delta}$ (M=Fe, $0.025 < x < 0.15$), (M=Co, $0.025 < x < 0.33$), (M=Al, $0.025 < x < 0.06$). Note that the period is inversely proportional to x ($p = 5.6 + 1/x$). The $x = 0.03$ and $x = 0.33$ are shown in inset (a) and inset (b), respectively.

evident [Zhu et al. 1991b, Zhu et al. 1993d, Skjerpe et al. 1992, Zhu et al. 1993f]. For a typical dopant concentration of $x = 0.03$, the average period of tweed is about 4nm, while for $x = 0.33$, the average period is 0.8-0.9nm. Having noted that the structural modulation caused by Co, Fe, or Al doping is almost identical (although they have different solubility: $x_{Co} = 0.33$, $x_{Fe} = 0.15$, $x_{Al} = 0.06$, in $YBa_2(Cu_{1-x}M_x)_3O_{7+\delta}$), we hereafter consider the relation of the modulation to the concentration of x without distinguishing between these cations unless specified otherwise. Tweed contrast was also observed for undoped but oxygen reduced $YBa_2Cu_3O_{7-\delta}$. Fig.6.13 is an example for $\delta = 0.65$.

6.4. [110] Tweedy Modulation

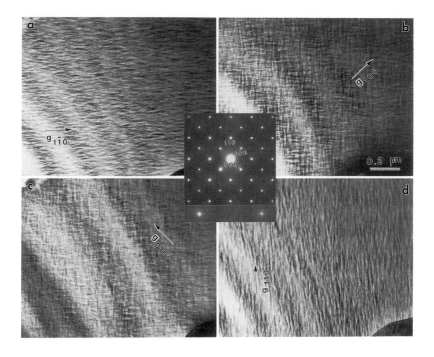

Figure 6.13 Tweed image in $YBa_2Cu_3O_{6.23}$, recorded from the same area but with different diffraction conditions: $\mathbf{g} = 1\bar{1}0$ (a), $\mathbf{g} = 200$ (b), $\mathbf{g} = 020$ (c) and $\mathbf{g} = 110$ (d). The tweed contrast obeys the $\mathbf{g} \cdot \mathbf{R} = 0$ extinction rule. When $\mathbf{g} = 1\bar{1}0$ or $\mathbf{g} = 110$, $[110]$ or $[\bar{1}10]$ tweed perpendicular to $\mathbf{g}$ is out of contrast. The selected-area diffraction pattern is also shown in the inset.

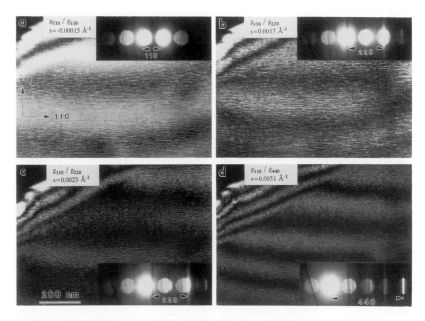

Figure 6.14 $\mathbf{g} = 110$ dark-field images of $YBa_2(Cu_{0.97}Fe_{0.03})_3O_{7-\delta}$ for different deviations s, from the Bragg position: (a) $s = -0.00015\text{Å}^{-1}$; (b) $s = 0.0017\text{Å}^{-1}$; (c) $s = 0.0023\text{Å}^{-1}$; (d) $s = 0.0051\text{Å}^{-1}$.

In diffraction imaging, tweed appears as a linear variation of contrast, or striations, which lies nearly parallel to the traces of the (110) and ($\bar{1}10$) planes. For a ($h00$) (Fig.6.13(b)) or a ($0k0$) (Fig.6.13(c)) reflection both two sets of striations are visible, while one set of striations perpendicular to the diffraction vector is out of contrast for a ($hh0$) reflection (Figs.6.13(a) and 6.13(d)). This suggests that tweed contrast obeys the $\mathbf{g} \cdot \mathbf{R} = 0$ extinction rule, and tweed contrast is strain contrast. Figures 6.14(a)-(d) are dark-field images from the same area under the diffraction conditions of $\mathbf{g}_{110}/\mathbf{g}_{hh0}$ (imaged by $\mathbf{g}_{110}$ while $\mathbf{g}_{hh0}$ is excited, where $hh0 = 110, 220, 440$, (see inset, Fig.6.14(a)-(d)). By increasing the deviation s ($|s| = 0.00015 \sim 0.0051\text{Å}^{-1}$, was accurately measured by the distance of the corresponding Kikuchi line from the Bragg spot), from the Bragg position for a (110) reflection, the thickness fringes come closer together indicating that there is a decrease in the effective extinction distance, ξ_g. However, the tweed spacing, which is

6.4. [110] Tweedy Modulation

on a finer scale, is relatively independent of diffraction conditions. The appearance of the wider striation of tweed contrast in Fig.6.14(a) and (b) is due to poorer resolution close to the Bragg position where the effective extinction distance, ξ_g, is larger because the two-beam image resolution of a defect is approximately $\xi_g/3$ [Thomas and Goringe 1979]. To confirm this finding, the optical diffractograms from Fig.6.14(a)-(d) were analyzed, and the results showed that the typical dimension of the domain is about 4 nm across and 35 nm along the tweed.

In HREM images, tweed appears as mosaic assembly of nonuniformly distorted lattice, which is most clearly seen by observing the image at a grazing incidence along [100] and [010] directions (Fig.6.15), giving rise to a quasi-periodic change in contrast. However, a similar observation along the [110] and [$\bar{1}$10] directions reveals no lattice distortion. This confirms our two beam observation that the displacement is along $<110>$ direction, and the $<110>\{\bar{1}10\}$ shear strain may be responsible for the displacement. Fig.6.16(a) is a digitized micrograph of a high-resolution tweed image. By digitally placing a set of small soft-edge windows on one of the two sets of the $<110>$ streak of the diffractogram (inset of Fig.6.16(a)), a subsequent inverse Fourier transformation showed an improved visibility of the displacive structure modulation (Fig.6.16(b)).

The displacive nature of the tweedy modulation can also revealed from electron diffraction. However, the cross-shaped diffuse scattering, which superimposes on every Bragg spot in the $(001)^*$ SAD pattern (Fig.6.10(b)) does not reflect the intrinsic features of the structure due to the strong multiple scattering in a high-symmetry Laue, whereby a diffracted electron beam acts as an incident beam. Such multiple scattering can be minimized by using a thinner sample, or tilting the sample away from the zone axis. After eliminating the multiple scattering effect, the shape of the diffuse scattering corresponding to the tweed consists of two separate sets of streaks rather than a cross pattern, as sketched in Fig.6.17(a) and (b). The length of the streaks increases with increasing distance from the origin, while their intensity falls off. For an $(h00)$ reflection, there are two sets of equivalent streaks, while for an $(hh0)$ spot, there is only one set of streaks. For an $(hk0)$ ($h \neq k$) spot, there are two sets of asymmetric streaks and their relative lengths depend on their indices (see Fig.6.17). Such characteristics of the diffuse

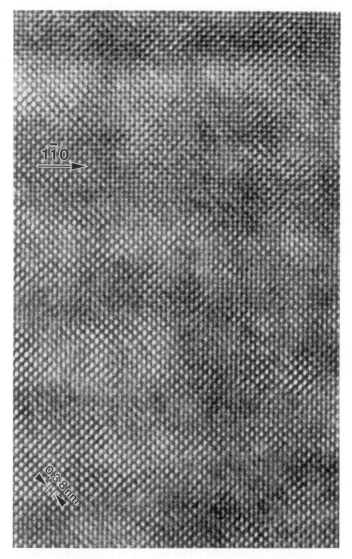

Figure 6.15 HREM image of tweed structure from $YBa_2(Cu_{0.97}Fe_{0.03})_3O_{7+\delta}$ showing the modulations on an atomic scale. Note the modulation in contrast and the lattice displacement along $[110]$ and $[1\bar{1}0]$ directions.

6.4. [110] Tweedy Modulation

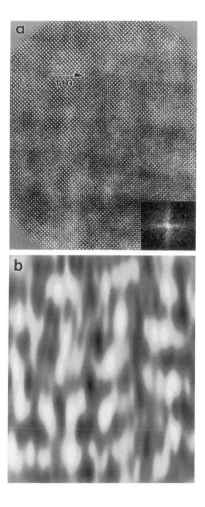

Figure 6.16 A digitized tweed image from a HREM micrograph such as shown in Fig.6.15. A Bragg spot of the diffractogram Fourier transformed from (a) is shown in the inset. (b) An image of wave contrast image in $[\bar{1}10]$ direction formed by using only $[110]^*$ streaks [as shown in the inset in (a)] represents the quasi-periodic displacement wave of the lattice in tweed structure.

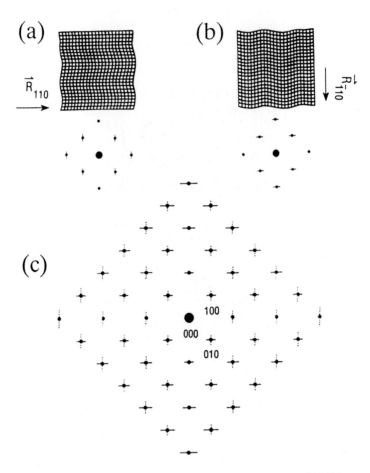

Figure 6.17 A schematic drawing of the tweed domain and its $(001)^*$ diffraction pattern. The modulated lattices formed by [110] shear (a) and $[\bar{1}10]$ shear (b) are shown, together with the corresponding diffraction patterns. Note, there are no radial streaks; all the streaks are perpendicular to the invariant plane. Superimposition of the diffraction patterns of (a) (dashed-line) and (b) (solid-line) is depicted in (c). The tweed can be considered as two separate overlapping displacement waves, each having a set of streaks of diffuse scattering.

scattering suggests that the modulation involves a shear-displacement on {110} plane in $<\bar{1}10>$ direction.

6.5 Theoretical Model for Fe Doped $YBa_2Cu_3O_{7-\delta}$ and Tweed

In this section we describe a theoretical model of oxygen ordering and tweed modulation in Fe doped YBCO compound, to shed light on the effect of oxygen ordering in Fe doped YBCO system on the structural modulation we discussed in the previous section.

6.5.1 Model for Oxygen Ordering in Fe Doped $YBa_2Cu_3O_{7-\delta}$

X-ray and neutron diffraction studies indicate that the structure of $YBa_2(Cu_{1-x}M_x)_3O_{7-\delta}$ (where M=Fe, Co ,Ga, Al) becomes tetragonal for $x > 0.18$. Unlike the tetragonal phase of $YBa_2Cu_3O_{7-\delta}$ ($\delta > 0.5$), the tetragonal $YBa_2(Cu_{1-x}M_x)_3O_{7-\delta}$ system is a superconductor with T_c as high as 80K[Xiao et al. 1988]. The Cu(1) site can also be preferentially substituted by magnetic elements such as Fe and Co [Oda et al. 1987, Micheli et al. 1988]. We will briefly discuss how to model these systems to study the oxygen ordering problem by developing a generalized lattice gas model with the assumption of metallic screening of oxygen interaction similar to the case of the undoped system as we discussed in section 6.2.

As we mentioned in section 6.2, the lattice gas model for $YBa_2Cu_3O_7$ system is based on the assumption that the interaction between oxygen atoms can be approximated by a metallically-screened Coulomb interaction on a rigid lattice. From experimental observations we can assume that the impurity doping and oxygen ordering happens only at the CuO basal plane for trivalent impurities such as Fe, Co, Al, or Ga. Since the system is metallic for $0 < \delta < 0.5$, this assumption seems reasonable for oxygen concentrations within that range.

When a di- or monovalent Cu ion is replaced by an Fe ion, an extra hole will tend to localize near the Fe sites, i.e., the Fe ion becomes effectively trivalent. This will have two major effects: (a) the binding

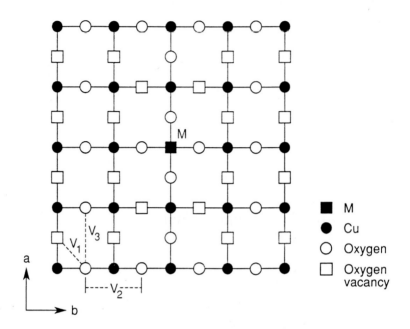

Figure 6.18 The structure of the basal plane of the Fe-doped $YBa_2Cu_3O_7$ system with a single Fe impurity.

energy of an oxygen ion near the Fe site increases; (b) the repulsion between oxygen atoms near the Fe site increases either due to the increase of negative charge of oxygen ions or due to the decrease in local screening because of the decrease in the mobile hole density. This causes the formation of a "cross-link" of twin -O-M-O- chains at the impurity site, as shown in Fig.6.18.

Based on this picture, we modify the lattice-gas model of Eqn.(6.4) to incorporate the effects of the impurities. The Hamiltonian representing the Fe doped YBCO system is

$$H = \sum_{<ij>} V_{1ij} n_i n_j + V_2 \sum_{(ij)b} n_i n_j + \sum_{(ij)a} V_{3ij} n_i n_j + \sum_i E_i n_i, \qquad (6.11)$$

where the following interaction parameters were used in the calcula-

6.5. Theoretical Model for Fe Doped $YBa_2Cu_3O_{7-\delta}$ and Tweed

tions:

$$V_2 = -0.2eV, \tag{6.12}$$
$$V_{1ij} = 0.25eV, V_{3ij} = 0.1eV, \tag{6.13}$$

if neither site i nor site j are the nearest neighbors of Fe;

$$V_{1ij} = 0.5eV, V_{3ij} = 0.2eV, \tag{6.14}$$

if either site i or site j is the nearest neighbor of Fe and the other site is not; and

$$V_{1ij} = 1.0eV, V_{3ij} = 0.4eV, \tag{6.15}$$

if both sites i and j are the nearest neighbors of Fe. In addition, the single-site binding energy $E_i = -4eV$ if site i is the nearest neighbor of Fe and $E_i = 0$ otherwise. n_i is the oxygen occupational operator ($n_i = 1$ if site i is occupied by an oxygen atom and $n_i = 0$ if site i is an oxygen vacancy site).

Since the oxygen concentration in $YBa_2(Cu_{1-x}M_x)_3O_{7-\delta}$ is very close to 7, we assume that the oxygen concentration is independent of x and remains to be 7. We also assume that the impurity M is randomly distributed in the CuO plane. We have performed Monte Carlo simulation with constant oxygen concentration and for several values of x on a 60×60 square lattice with periodic boundary conditions. For a given configuration of impurity M, 10,000 MCS/site are used to calculate the order parameter Φ, which is given by the difference in the oxygen fractional site occupancy x_1 and x_2 defined as:

$$\Phi = \frac{|x_1 - x_2|}{x_1 + x_2}. \tag{6.16}$$

For each value of x, Φ is averaged over 20 configurations of the impurities. The calculated order parameter Φ at $T = 300K$ is shown as squares in Fig.6.19. The order parameter deduced from the experiment of Xiao et al. [Xiao et al. 1988] [assuming that $\Phi = (b-a)/(b+a)$] is also shown in Fig.6.19 as crosses. The general feature agree reasonably well, thus providing a justification of the basic assumptions behind our modified lattice gas model. The discrepancy between theory and experiment can be either due to the long-range effect of the impurity

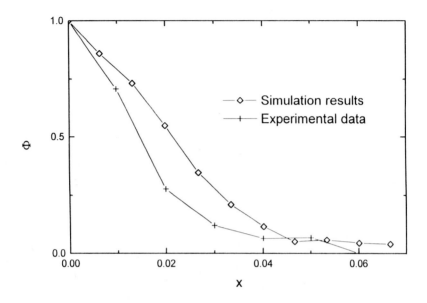

Figure 6.19 Oxygen order parameters Φ of the $YBa_2(Cu_{1-x}M_x)_3O_7$ system as a function of the impurity concentration x at 300K from Monte Carlo simulations and from data of Xiao *et al.*[Xiao *et al.* 1988].

on local oxygen ordering or the fact that the distribution of impurity atoms is not completely random.

From a "snapshot" of the oxygen configuration of the M-substituted system we find that local "cross-links" are formed around the M sites at the low impurity limit (see Fig.6.20). These cross-links destroy the Cu-O chains along the b axis and form short Cu-O chains along the a axis. As the concentration of the impurity increases, the number of the short Cu-O chains along the a axis also increases. At high impurity density, there are long Cu-O chains in both directions thus overall the system shows tetragonal symmetry (see Fig.6.21). However, unlike the tetragonal phase of the quenched $YBa_2Cu_3O_{7-\delta}$ systems, the length of the Cu-O chains are rather long (due to large oxygen concentration).

The chain lengths maybe sufficiently large to produce a large enough density of conducting holes in the CuO_2 plane. This may be the reason why the superconducting transition temperature T_c is still quite

6.5. Theoretical Model for Fe Doped $YBa_2Cu_3O_{7-\delta}$ and Tweed

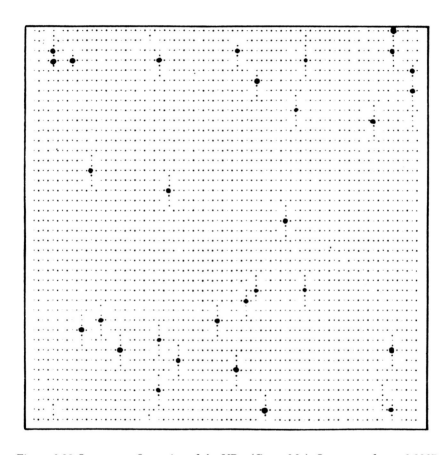

Figure 6.20 Oxygen configuration of the $YBa_2(Cu_{1-x}M_x)_3O_7$ system for x=0.0067 obtained from Monte Carlo simulation at 300K.

Figure 6.21 Oxygen configuration of the YBa$_2$(Cu$_{1-x}$M$_x$)$_3$O$_7$ system for x=0.04 obtained from Monte Carlo simulation at 300K.

6.5. Theoretical Model for Fe Doped $YBa_2Cu_3O_{7-\delta}$ and Tweed

high in these Fe substituted compounds even though the structure is tetragonal.

In summary, the effect of substituting Cu by trivalent impurities on the oxygen ordering has been investigated using a modified lattice gas model established under the assumption of screened Coulomb interaction. The orthorhombic to tetragonal transition in Fe-doped system is found to be due to the formation of short cross-linked CuO chains near these trivalent impurities. As we will see the next section, such cross-link gives rise to the tweedy modulation discussed in the previous section.

It is important to realize though, that the lattice gas model we have established is only a very crude model of the effective interaction in $YBa_2Cu_3O_{7-\delta}$ system. In this model we have completely ignored the long-range interaction due to the lattice distortion and treated the long range part of the Coulomb interaction in a mean field fashion which clearly is not applicable for the case of $\delta > 0.5$. Khachaturyan et al. [Khachaturyan and Morris 1987, Khachaturyan and Morris 1988] have argued that, to correctly describe the ordering process, the full three-dimensional nature of the materials must used. Although we argue that this lattice gas model gives a good description of the oxygen ordering under various processing conditions, it obviously cannot explain the stability of the twin boundary and some of the superlattice features found in oxygen deficient compounds. More realistic Hamiltonian is needed to describe these phenomena, which we will discuss next.

6.5.2 Modification of the Lattice Gas Model to Include Elastic Strain

Extensive Monte Carlo simulations using an anisotropic lattice gas model potential have been performed [Cai and Mahanti 1988, Cai and Mahanti 1989, Cai and Mahanti 1990, de Fontaine et al. 1987, de Fontaine et al. 1989] to describe oxygen ordering in $YBa_2Cu_3O_7$ and related systems and seem to describe reasonably well the experimental data on the structural phase diagram [de Fontaine et al. 1989], the orthorhombic-to-tetragonal phase transition in the Fe-doped system, and the metastable structure of quenched samples [Cai and Mahanti 1989, Cai and Mahanti 1990]. However, these lattice gas models cannot explain without modification the microstructures of the $YBa_2Cu_3O_7$ system caused by elastic distortion of the lattice, such as the formation of "tweed" in $YBa_2(Cu_{1-x}Fe_x)_3O_7$ system [Zhu et al. 1990e, Zhu et al. 1991b].

Here We treat the elastic energy as a perturbation of the lattice gas Hamiltonian in order to calculate the diffuse scattering spectrum associated with the tweed formation [Cai et al. 1992a, Cai et al. 1993a, Cai et al. 1993b]. The oxygen atom configuration obtained from our Monte Carlo simulation of the lattice gas model on a rigid square lattice is used to obtain the input concentration wave amplitude for continuum elasticity theory calculations [Semenovskaya and Khachaturyan 1991, Jiang et al. 1991]. In essence, we assume that the oxygen distribution over lattice sites is not affected by the orthorhombic distortion of the lattice. In light of the excellent agreement between previous results calculated based on the lattice gas model and experimental results, this is a reasonable assumption.

The oxygen configurations obtained from Monte Carlo simulation show that the local "cross-links" formed around M atoms disrupt the orthorhombic structure, and overall the system shows tetragonal symmetry. Since the oxygen atoms can be regarded as interstitial in CuO plane causing expansion of the lattice parameter in the direction of the Cu-O chain and contraction perpendicular to the Cu-O chain, the resulting lattice distortion caused by the "cross-links" is along [110] and [$\bar{1}$10] direction. High-resolution electron-microscopy and TEM image analysis show that this shear displacement are responsible for the tweed contrast [Zhu et al. 1990e].

Since the displacements of oxygen atoms in the orthorhombic domains, relative to the square lattice, are small, a linear elastic approximation can be used to study the lattice distortion due to oxygen ordering in the $YBa_2(Cu_{1-x}M_x)_3O_7$ system. In the long wave length limit, the displacement of the lattice site s can be written to first order as follows: [Jiang et al. 1991, Krivoglaz 1969]

$$\delta \mathbf{R}_s = \sum_\mathbf{q} \mathbf{A}_\mathbf{q} c_\mathbf{q} \exp(-i\mathbf{q} \cdot \mathbf{R}_s) \qquad (6.17)$$

where $c_\mathbf{q}$ is the amplitude of the planar oxygen concentration wave with wave vector $\mathbf{q}$. $\mathbf{A}_\mathbf{q}$ is determined by three inhomogeneous linear

6.5. Theoretical Model for Fe Doped $YBa_2Cu_3O_{7-\delta}$ and Tweed

equations:

$$q\lambda_{ijlm}n_jn_lA_{\mathbf{q}m} = \lambda_{ijlm}n_jL_{lm} \tag{6.18}$$

where summation is made over repeated indices; λ_{ijlm} is a component of the elastic modulus tensor; n_i is a component of the unit vector $\mathbf{n} = \mathbf{q}/q$, and $L_{lm} = \delta u_{lm}/\delta c_0$ where u_{lm} is a component of the strain tensor; c_0 is the average oxygen concentration of the CuO basal plane. For tetragonal crystals, the tensor components L_{lm} are given by

$$L_{lm} = \frac{1}{d_l}\frac{\partial d_l}{\partial c_0}\delta_{lm} \tag{6.19}$$

where d_l is the lattice constant. In this section we treat the oxygen atoms in the basal plane of $YBa_2(Cu_{1-x}Fe_x)_3O_7$ as interstitial in a square Cu lattice where the interstices are mid-points of the edges (x and y) of the unit cell. The tensor components are

$$L^x_{11} = \frac{1}{b}\frac{\partial b}{\partial c_0};\ L^x_{22} = \frac{1}{a}\frac{\partial a}{\partial c_0}; \tag{6.20}$$

$$L^y_{11} = L^x_{22},\ L^y_{22} = L^x_{11}, \tag{6.21}$$

$$L^x_{22} = L^y_{33} = \frac{1}{c}\frac{\partial c}{\partial c_0} \tag{6.22}$$

where a, b, c are the lattice constants of the $YBa_2Cu_3O_7$ system with oxygen concentration in Cu-O basal plane to be c_0. If we treat the oxygen atoms in Cu-O plane as interstitial in a square lattice, Eqn.(6.18) may then be modified as

$$q\lambda_{ijlm}n_jn_lA^\gamma_{\mathbf{q}m} = \lambda_{ijlm}n_jL^\gamma_{lm},\ \gamma = x, y. \tag{6.23}$$

Once $\mathbf{A_q}$ is obtained from the above equation, the intensity of the diffuse scattering due to lattice distortion for small displacement $\mathbf{q}$ of the scattering vector away from a reciprocal lattice point (Bragg peak) $\mathbf{Q}$ can be derived from the following expression: [Krivoglaz 1969]

$$I_H(\mathbf{Q}+\mathbf{q}) = |F(\mathbf{Q})|^2|\mathbf{Q}\cdot(\mathbf{A}^x_\mathbf{q}c^x_\mathbf{q} + \mathbf{A}^y_\mathbf{q}c^y_\mathbf{q})|^2 \tag{6.24}$$

where $c^x_\mathbf{q}$ and $c^y_\mathbf{q}$ are the amplitudes of the oxygen concentration wave of two different kinds of interstitial oxygen atoms (along b and a axis, respectively) and $F(\mathbf{Q})$ is the atomic structure factor of the full unit cell.

The thermal average of the diffuse scattering intensity can then be obtained as:

$$\begin{aligned}
<I>_H &= <I(\mathbf{Q}+\mathbf{q})>/|F(\mathbf{Q})|^2 \\
&= |\mathbf{Q}\cdot\mathbf{A}^x_\mathbf{q}|^2 <|c^x_\mathbf{q}|^2> + |\mathbf{Q}\cdot\mathbf{A}^y_\mathbf{q}|^2 <|c^y_\mathbf{q}|^2> \\
&+ |(\mathbf{Q}\cdot\mathbf{A}^{*x}_\mathbf{q})(\mathbf{Q}\cdot\mathbf{A}^y_\mathbf{q})<c^{*x}_\mathbf{q}c^y_\mathbf{q}> \\
&+ |(\mathbf{Q}\cdot\mathbf{A}^{*y}_\mathbf{q})(\mathbf{Q}\cdot\mathbf{A}^x_\mathbf{q})<c^{*y}_\mathbf{q}c^x_\mathbf{q}>
\end{aligned} \quad (6.25)$$

where the correlation functions of $c^x_\mathbf{q}$ and $c^y_\mathbf{q}$ are obtained by the Fourier transform of $<|c^x(\mathbf{r})|^2>$, $<|c^y(\mathbf{r})|^2>$, $<|c^{*x}(\mathbf{r})c^y(\mathbf{r})|>$, which are obtained from the Monte Carlo simulation. The subscript H denotes the fact that this scattering intensity comes from a generalized Huang scattering [Krivoglaz 1969].

Following Jiang et al. [Jiang et al. 1991] we chose $c_0 = 0.5$, $L^x_{11} = -L^x_{22} = 0.011$, $L^x_{33} = 0.010$ and the values of λ_{ijlm} to be those given by Reichardt et al. [Reichard et al. 1988] for the $YBa_2Cu_3O_7$ compound.

Since most diffuse scattering experiments are performed at room temperature, the thermal diffuse scattering (TDS) can also make a major contribution to the total diffuse scattering intensity. To the first order approximation [Wooster 1961]

$$I_{TDS} = |F(\mathbf{Q})|^2 \frac{kT}{\tau}\frac{(\mathbf{Q}+\mathbf{q})^2}{q^2}K(f)_g \quad (6.26)$$

where T is the temperature, τ is the volume of the unit cell,

$$K(f)_g = g_i g_m (\lambda_{ijlm}f_j f_l)^{-1}, \quad (6.27)$$

and $\{g\}$, $\{f\}$ are the wave normals of the $\mathbf{Q}$ and $\mathbf{q}$, respectively. The total diffuse scattering intensity is then the sum of the thermal and Huang diffuse scatterings, $I = I_H + I_{TDS}$.

Figure 6.22(a)-(i) shows the diffuse scattering intensity, calculated as described above, compared with the diffuse scattering data obtained by TEM, of $YBa_2(Cu_{1-x}Fe_x)_3O_7$ for $x = 0.03$. Fig.6.22(a)-(c) shows the calculated diffuse scattering intensity I_H due to lattice distortion, and Fig.6.22(d)-(f) shows the calculated total diffuse scattering intensity I due to both lattice distortion and thermal diffuse scattering.

6.5. Theoretical Model for Fe Doped YBa$_2$Cu$_3$O$_{7-\delta}$ and Tweed

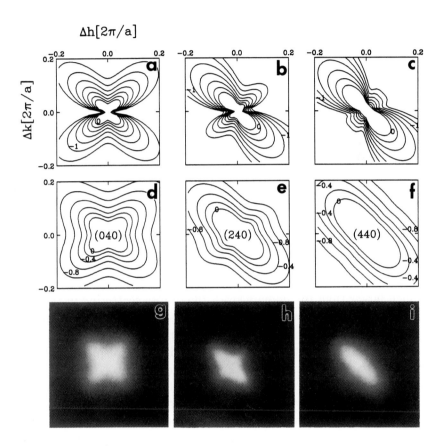

Figure 6.22 (a)-(c) Contour plots in logarithmic intervals of the diffuse scattering intensity of YBa$_2$(Cu$_{1-x}$M$_x$)$_3$O$_7$ due to lattice distortion for $x = 0.033$, around (a) (040), (b) (240) and (c) (440) reciprocal-lattice points. (d)-(f) Contour plots in logarithmic intervals of the total diffuse scattering intensity of YBa$_2$(Cu$_{1-x}$M$_x$)$_3$O$_7$ for $x = 0.033$, around (d) (040), (e) (240) and (f) (440) reciprocal-lattice points. (g)-(i) Diffuse scattering measured by TEM of YBa$_2$(Cu$_{1-x}$Fe$_x$)$_3$O$_7$ for $x = 0.03$, around (g) (040), (h) (240) and (i) (440) reciprocal-lattice points.

Fig.6.22(g)-(i) shows the diffuse scattering data obtained by TEM. The theoretical diffuse scattering intensity profile, as shown in Fig.6.22(d)-(f), agrees very well with the TEM data, as shown in Fig.6.22(g)-(i). This indicates that the lattice distortion due to oxygen ordering is of the same order of the magnitude as the thermal vibration of the lattice. The diffuse patterns are streaked along the two equivalent $<110>$ directions in reciprocal space for $\mathbf{Q} = (h00)$ or $(0h0)$, while for $(hh0)$, one set of the $\{110\}$ streaks (the radial streaks) vanishes. Around $(hk0)$, $h \neq k \neq 0$, the intensity and the length of the streaks depend on the distance from the origin [Zhu et al. 1991b]. The characteristic features of the simulated diffuse scattering intensity, which arises from $[110]/(\bar{1}10)$ shear displacements of the crystal due to oxygen distribution, are consistent with the electron diffuse scattering, tweed image [Zhu et al. 1990e, Zhu and Cai 1993] and the X-ray diffuse scattering observations [Jiang et al. 1991].

Figure 6.23 shows the calculated diffuse scattering intensity due to lattice distortion for various M concentrations, compared with the X-ray diffuse scattering intensity for $YBa_2(Cu_{1-x}Al_x)_3O_7$ where $x = 0.046$ [Jiang et al. 1991]. In the long wave length limit (where the wave length is much larger than the domain size), the diffuse scattering in the long wave length limit shows the characteristics of Huang scattering, i.e. $I \propto 1/q^2$. The region of the Huang scattering, however, is restricted to small q, where the scattering is governed by the displacement field far away from the defect centers. For larger q-values (i.e. shorter wavelength), c_q is no longer a constant. In fact, the results suggest that $<|c_q^2|> \propto 1/q^2$, which results in a decrease of diffuse scattering intensity proportional to $1/q^4$. The minimum q-value where the $1/q^4$ scattering is observed gives a critical radius $R_{crit} = 1/q_c$, which at low Miller indices, is approximately the average size of the orthorhombic microdomain. We find that when x increases from 0.017 to 0.067, the average size of the microdomains is estimated to decrease from 80Å to 10Å.

In conclusion, we have shown that based on the assumption that the interaction between oxygen atoms in $YBa_2(Cu_{1-x}M_x)_3O_7$ is predominantly a screened Coulomb interaction, the elastic distortion of the lattice can be treated as a perturbation of the two dimensional lattice gas model. The streaks of the diffuse scattering for $x = 0.03$ were

6.5. Theoretical Model for Fe Doped YBa$_2$Cu$_3$O$_{7-\delta}$ and Tweed

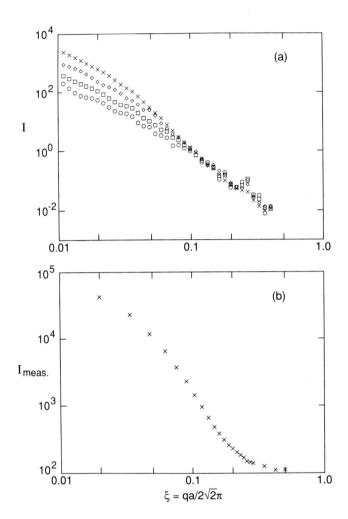

Figure 6.23 (a) Calculated diffuse scattering intensity of YBa$_2$(Cu$_{1-x}$M$_x$)$_3$O$_7$ for $x = 0.017$ (solid line), 0.033 (dashes), 0.05 (dot-dash), 0.067 (dot), vs. reduced wave vector ξ in the $[\bar{1}10]$ direction at the (660) point in reciprocal lattice. (b) Measured X-ray diffuse scattering intensity of YBa$_2$(Cu$_{1-x}$Al$_x$)$_3$O$_7$ for $x = 0.046$, vs. reduced wave vector ξ in the $[\bar{1}10]$ direction at the (660) point [Jiang et al. 1991].

simulated using a concentration wave / displacement wave approach. As discussed before, the "cross-links" and the local oxygen ordering in the CuO plane are essential in determining the lattice displacement of the crystal; therefore, we first generated low-energy configurations of the CuO plane with a random distribution of dopant using Monte Carlo simulations. The detailed lattice distortion associated with that configuration then was derived using a linear elastic approximation. From a knowledge of the displacement, the intensity of the diffuse scattering (Huang scattering) finally was calculated and compared to the experimental data. The diffuse scattering data obtained by this technique are in good agreement with the results of TEM and X-ray scattering experiments. From quantitative measurements of X-ray diffuse scattering profiles, the average size of the orthorhombic microdomains can be obtained.

6.6 Reduction and Reoxideation in $YBa_2(Cu_{1-x}Fe_x)_3O_{7-\delta}$

The effect of systematic substitution of Fe into the $YBa_2Cu_3O_7$ system has been the subject of intensive experimental and theoretical study. It is now generally agreed that Fe substitutes into Cu sites, although opinions vary on which of the two Cu site Fe goes into [Xu et al. 1989, Dunlap et al. 1989]. One of the interesting effects of Fe substitution is that it induces an orthorhombic to tetragonal structural phase transition without a severe depression of the superconducting transition temperature T_c. This tetragonal phase differs in nature from the tetragonal phase found in oxygen deficient $YBa_2Cu_3O_{7-\delta}$ in the sense that the oxygen content of the former system is close to or above 7.

Theoretical calculations based on a lattice gas model with interactions arising from screened-Coulomb repulsion and short-range covalent bonding bridged by Cu between oxygen atoms [Cai and Mahanti 1989, Cai and Mahanti 1990, Baumgärtel and Bennemann 1989] show that there is local "cross-link" formation around trivalent impurity ions. As the impurity concentration increases, these "cross-links" destroy the orthorhombic symmetry, and overall the system shows tetragonal symmetry. However local orthorhombic domains and long Cu-O chains still

6.6. Reduction and Reoxidation in $YBa_2(Cu_{1-x}Fe_x)_3O_{7-\delta}$

exist in these "tetragonal" systems thus providing an explanation of why the superconducting properties were not changed by Fe additions as the oxygen deficiencies in $YBa_2Cu_3O_{7-\delta}$. The results of Monte Carlo simulation shows that this model represents very well the experimental situation (see Sec.6.5).

Unfortunately, this earlier work did not resolve the controversy over the Fe location. Since Fe has a higher valency than Cu, Coulomb effects increase the local binding energy of oxygen isotropically and crosslink formation will appear as a consequence of Fe doping into either Cu site [Cai and Mahanti 1989, Cai and Mahanti 1990, Baumgärtel and Bennemann 1989]. In this section we try to shed light into this question by studying the effect of reduction and subsequent low temperature reoxidation on the oxygen ordering and Fe distribution in $YBa_2(Cu_{1-x}Fe_x)_3O_{7-\delta}$ systems [Cai et al. 1992b].

Bulk samples of $YBa_2(Cu_{1-x}Fe_x)_3O_{7-\delta}$ ($x = 0.03$ and $x = 0.12$) were produced by a conventional sintering method (standard samples). The oxygen-reduced samples were prepared by heating the standard Fe-doped samples in argon at 830^oC for 20 hours, then furnace cooling to below 100^oC. The reoxidized samples were then prepared by heating the oxygen-reduced samples in flowing oxygen at 400^oC for 10 hours.

So called "tweed" structure has been observed by transmission electron microscopy in $YBa_2(Cu_{1-x}Fe_x)_3O_{7-\delta}$ systems. It usually exhibits a roughly periodic domain image which is associated with streaks of diffuse electron scattering in the diffraction pattern. Fig.6.24 is an example from a standard $YBa_2(Cu_{0.97}Fe_{0.03})_3O_{7-\delta}$ sample. Fig.6.24(a) is a typical selected-area diffraction pattern of the (001) zone. The overall pattern has a 4-fold symmetry with diffraction intensity which falls off monotonically from one Brillouin zone to the next. Two cross-streaks of diffuse intensity in two equivalent $< 110 >$ directions (see the insert)are superimposed on the Bragg diffraction spots. Fig.6.24(b) shows the tweed contrast observed near (001) orientation in real space. It consists of two sets of mutually perpendicular microdomains elongated in [110] and [$\bar{1}$10] directions, corresponding to the streaking observed in the diffraction pattern. Fig.6.24(c) shows the tweed contrast and diffraction (inset) from an oxygen-reduced sample. Both diffraction and morphology show no visible modulated structure after oxygen is moved from the standard $YBa_2(Cu_{0.97}Fe_{0.03})_3O_{7-\delta}$. However, after reoxida-

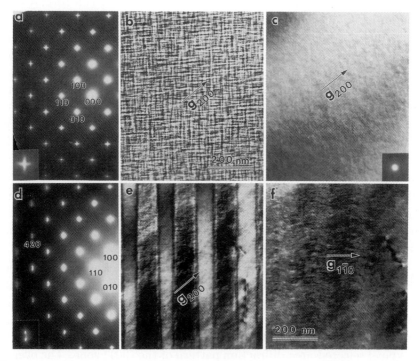

Figure 6.24 Microstructure of $YBa_2(Cu_{0.97}Fe_{0.03})_3O_{7-\delta}$: (A) (001) zone diffraction pattern and (b) tweed contrast of the sintered sample; (c) morphology and a diffraction spot [insert of (c)] of the oxygen reduced sample; (d) (001) zone diffraction pattern of the reoxidized sample, with the (420) spot is shown in insert of (d); weak tweed and strong twin contrast of reoxidized sample under (e) $\mathbf{g} = 200$ diffraction condition and (f) $\mathbf{g} = 1\bar{1}0$ diffraction condition.

6.6. Reduction and Reoxidation in $YBa_2(Cu_{1-x}Fe_x)_3O_{7-\delta}$

tion at $400°C$, modulated structure (twin and tweed) forms again. The diffraction pattern (Fig.6.24(d)) exhibits splitting of the Bragg spots, which indicates an orthorhombic symmetry, and very weak streaks superimposed on the Bragg spots. The splitting arises from the twinning, the streaks from the tweed, as shown in Fig.6.24(e). Fig.6.24(e) and Fig.6.24(f) were imaged in the same area. The twin and tweed contrast obey the conventional $\mathbf{g} \cdot \mathbf{R} = 0$ extinction rule. Under the $\mathbf{g} = 200$ diffraction condition (Fig.6.24(e)), both twins and two sets of tweed are visible, while under $\mathbf{g} = 1\bar{1}0$ (Fig.6.24(f)) twin and one of the two sets of tweed, that perpendicular to $\mathbf{g}$, are out of contrast (also see Fig.6.25(e,f)).

Similar results were also observed in a reduced/reoxidized $YBa_2(Cu_{0.88}Fe_{0.12})_3O_{7-\delta}$ samples. For standard $YBa_2(Cu_{1-x}Fe_x)_3O_{7-\delta}$ ($x > 0.33$) samples, structure modulation no longer can be observed in the image morphology (Fig.6.25(b)), but can be deduced from electron diffraction, as shown in Fig.6.25(a). The [110] diffuse streaks tend to form an intensity maximum indicating enhanced periodicity of the modulated structure. However, after oxygen reduction, the streaks of the diffuse scattering disappear (Fig.6.25(c)). When oxygen is added back into the sample, twin and tweed structure appears (Fig.6.25(d,e,f)). This can be seen clearly from both the diffraction pattern (split and streak in Fig.6.25(d)) and the morphology (Fig.6.25(e,f)). Fig.6.25(e) and (f) show the twin and tweed image from the same area but under different diffraction conditions. For reoxidized samples, both $x = 0.03$ and $x = 0.12$, the microstructure is much different from the standard ones. The former shows weak tweed modulation but strong twinning with large orthorhombicity, whereas the latter shows strong tweedy modulation but weak twinning with small orthorhombicity. The orthorhombicity for $x = 0.03$ sample increases approximately by a factor of 3 compared with that for $x = 0.12$.

To study the changes in oxygen ordering during the reduction and reoxidation process, as manifested in the TEM results described in the previous section, we performed Monte Carlo simulations of a lattice-gas model as we described in the previous section. The oxygen concentration of the $YBa_2(Cu_{1-x}Fe_x)_3O_{7-\delta}$ system is found to be roughly $y = 7 + 1.5x$ where x is the concentration of Fe impurities in this system. The distribution of Fe ions is assumed to be uniform in the system

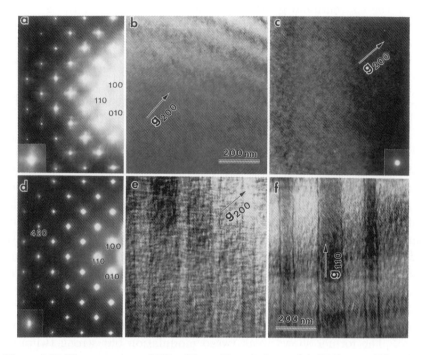

Figure 6.25 Microstructure of $YBa_2(Cu_{0.88}Fe_{0.12})_3O_{7-\delta}$: (a) (001) zone diffraction pattern (b) morphology of the sintered sample; (c) morphology and a diffraction spot [insert of (c)] of the oxygen reduced sample; (d) $(00\bar{1})$ zone diffraction pattern of the reoxidized sample, with the (420) spot is shown in insert of (d); strong tweed and weak twin contrast of reoxidized sample under (e) $\mathbf{g} = 200$ diffraction condition and (f) $\mathbf{g} = 1\bar{1}0$ diffraction condition.

6.6. Reduction and Reoxidation in $YBa_2(Cu_{1-x}Fe_x)_3O_{7-\delta}$

prepared using the sintering procedure. During the high-temperature reduction process, all of the oxygen atoms are removed from the system except the those near the Fe site, which are tightly bound. This gives rise to a residual oxygen concentration of 6.4 for systems doped with 10% of Fe compared with a residual oxygen concentration of 6 for the system without Fe. It is energetically favorable for nearby Fe atoms to cluster and share the few oxygen atoms left. We investigated this clustering by means of simulations utilizing a strong effective Fe–Fe attraction for nearest neighbor only, i.e. simulation of the nucleation process of a dilute ferromagnetic Ising model. During the low temperature reoxidation process, the Fe configuration is fixed because of the low Fe mobility and acts as pinning centers for oxygen atoms. The ability of the clustered Fe to disrupt the orthorhombic structure by forming cross-links is much smaller than that of uniformly distributed Fe impurities, therefore the orhorhombicity of the reduced and re- oxidized sample increases drastically, as seen by the results shown below.

Monte Carlo simulations of the reduction and reoxidation process were performed with Kawasaki dynamics using the model discussed in the previous section. A 32×32 lattice with periodic boundary condition was used. The high temperature configuration was obtained by doing MC simulation at $T = 2300K$ (above the order- disorder phase transition temperature for all the systems concerned), and the system was equilibrated at this temperature with 10,000 MCS/site (MCS = Monte Carlo steps). The low temperature configurations were obtained by slowly cooling the system to 300K in steps of 100K. 10,000 MCS/site were used to obtain thermodynamic averages for each temperature.

Figure 6.26 shows the oxygen and Fe configuration of $YBa_2(Cu_{0.967}Fe_{0.033})_3O_{7.05}$ with a random Fe distribution. The local "cross-links" formed around Fe atoms disrupt the orthorhombic structure, and overall the system shows tetragonal symmetry. Fig.6.27 shows the Fe configuration of the oxygen-reduced sample (oxygen atom positions are not shown). We can see that the Fe atoms clustered into "droplets" by the effective attraction between them due to the "starvation" of oxygen. Fig.6.28 shows the oxygen configuration of $YBa_2(Cu_{0.967}Fe_{0.033})_3O_{7.05}$ with the Fe configuration obtained in the reduction process (same as in Fig.6.27). For this system each "droplet" behaves not very differently from a single Fe atom, thus the effective

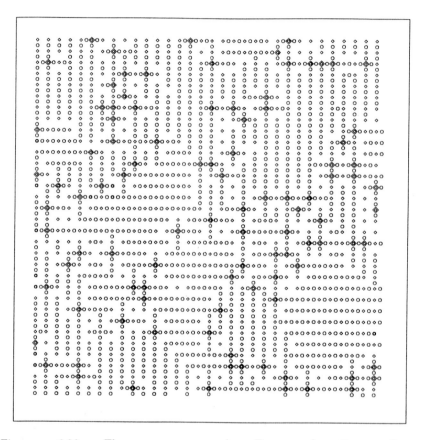

Figure 6.26 The oxygen and Fe configuration of $YBa_2(Cu_{0.967}Fe_{0.033})_3O_{7.05}$ with a random Fe distribution. The big diamond represents Fe, the small diamond represents Cu, and the circle represents oxygen atoms.

6.6. Reduction and Reoxidation in $YBa_2(Cu_{1-x}Fe_x)_3O_{7-\delta}$

Figure 6.27 The Fe configuration of the oxygen-reduced sample. The big diamond represents Fe and the small diamond represents Cu.

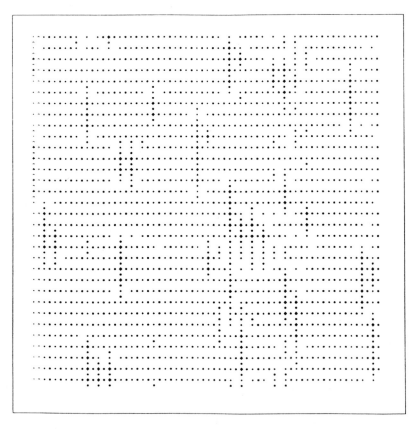

Figure 6.28 The oxygen and Fe configuration of $YBa_2(Cu_{0.967}Fe_{0.033})_3O_{7.05}$ with the Fe configuration obtained in the reduction process (re-oxidized sample). The big diamond represents Fe, the small diamond represents Cu, and the circle represents oxygen atoms.

6.6. Reduction and Reoxidation in YBa$_2$(Cu$_{1-x}$Fe$_x$)$_3$O$_{7-\delta}$

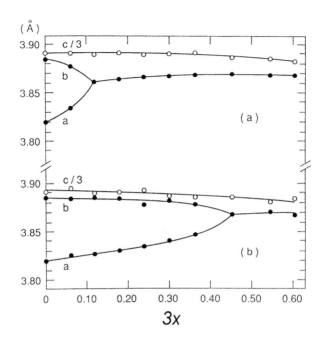

Figure 6.29 Lattice parameters vs. $3x$ curve for YBa$_2$(Cu$_{1-x}$Fe$_x$)$_3$O$_y$: (a) sample prepared via oxygen annealing, (b) sample prepared via the reduction and reoxidation process.

impurity level is lower compared to that of samples prepared using a sintering procedure (i.e. with a random Fe configuration). We see that the reduced-reoxidized (Fe clustered) system this system shows overall orthorhombic symmetry even though its Fe and oxygen concentration is the same as the sintered system.

In Fig.6.29 experimental data of lattice constant a and b are plotted as a function of for Fe concentration for systems prepared under either sintered or reduction–reoxidation condition [Katsuyama et al. 1990]. The reason we plot it as a function of $3x$ instead of x is that we assume all the Fe is in the CuO basal plane. Previous neutron diffraction studies [Jorgensen et al. 1987a] have indicated that the orthorhombic to tetragonal transition is due to the disordering of the oxygen atoms

130   Chapter 6. Oxygen Ordering Related Defects in $YBa_2Cu_3O_7$

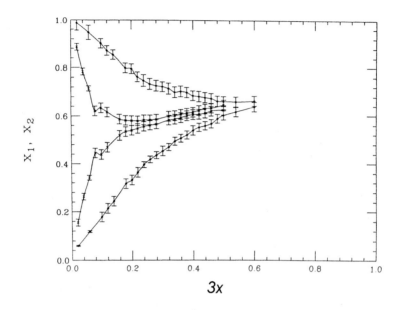

Figure 6.30 Oxygen fractional site occupancy vs. $3x$ curve for $YBa_2(Cu_{1-x}Fe_x)_3O_y$, (a) system with a random Fe distribution, (b) system prepared via the reduction and reoxidation process.

and the changes in lattice constant a and b are due to the change of fractional site occupancy of oxygen atoms in the two sublattices (shown as open circles and squares in Fig.6.18) defined as follows:

$$x_1 = \frac{\sum_{i_1} n_{i_1}}{N} \qquad (6.28)$$

$$x_2 = \frac{\sum_{i_2} n_{i_2}}{N} \qquad (6.29)$$

where $N = L \times L$, i_1 and i_2 are the sites of the two sublattices shown in Fig.6.18 as open circles and square. Fig.6.30 shows the simulation results for the oxygen fractional site occupancy as a function of Fe concentration for systems with (a) uniformly distributed Fe atoms and (b) "droplets" of Fe obtained by the clustering of Fe atoms in the former sample. It is clear that the Fe concentration at which the system become tetragonal obtained from simulation, agrees qualitatively with the experimental data.

In summary, we have performed experimental as well as simulation studies of the reduction and reoxidation process of Fe doped $YBa_2Cu_3O_y$. Using an anisotropic lattice-gas model with interactions between atoms based on the screened Coulomb interactions, we show that the experimental data of Katsuyama et al. can be explained by Fe clustering in CuO basal plane [Katsuyama et al. 1990]. While it is possible that some of the Fe may migrate into CuO_2 planes in the reduction process, it is very improbable that the majority of the Fe atoms, with such low interplanar mobility, would do so. Also if the Fe moved into CuO_2 plane, it is hard to understand why the superconducting transition temperature T_c remains essentially the same, and in some experiments, actually increases after the reduction and reoxidation procedure.

6.7 Three Dimensional Structural Modulation

6.7.1 $0.02 < x < 0.10$

Figures 6.31(a)-(c) shows typical zone-axis selected-area diffraction patterns of the $(001)^*$, $(101)^*$ and $(100)^*$ orientations of $YBa_2(Cu_{0.96}Fe_{0.04})_3O_{7-\delta}$, respectively. By analyzing the geometry of the intersection of the Ewald sphere and the reciprocal lattice [Fig.6.31(d)-(f)], we concluded that the Bragg spots consist of paddle-shaped wings of diffuse scattering (Fig.6.31(g) for the diffuse scattering around an ($h00$) reflection), rather than streaks as we treated in previous sections. The edges of the wings do not point exactly in the $[110]^*$ direction, but also have a component along the c^*-axis. Measurement of the expansion of the wing along the c-axis suggests that the correlation length perpendicular to the a-b plane is about 40-50nm [Zhu et al. 1993d].

Since the period of the structural modulation is so small (about 4nm in the [110] direction for $x = 0.03$), nano-diffraction with a field emission source is very suitable for studying tweed. Fig.6.32(a) and (b)

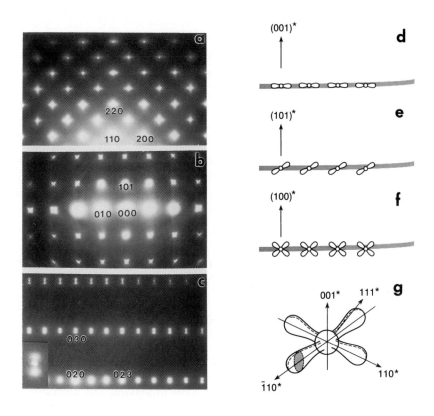

Figure 6.31 (a)-(c) Selected-area diffraction showing diffuse streaks superimposed on the fundamental reflections for $x = 0.04$ in (a) the $(001)^*$ zone (note that the radial streaks are due to the double diffraction and the non-radial streaks reflect the structural modulation), (b) the $(\bar{1}10)^*$ zone. (d)-(f) Sketches of the intersection of the Ewald sphere (shaded curve) and the reciprocal lattice of (d) the $(001)^*$ zone, (e) the $(101)^*$ zone and (f) the $(100)^*$ zone. (g) The shape of an $(h00)$ reciprocal spot for $x=0.03$-0.05.

6.7. Three Dimensional Structural Modulation

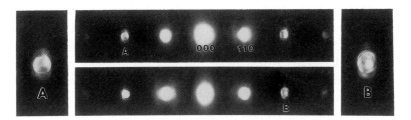

Figure 6.32 Examples of nanodiffraction patterns near the (001)* zone for $x = 0.03$ observed using a 20Å probe from a thick (about 1000Å) sample, showing the split of the fundamental reflections along the $[\bar{1}10]^*$ direction. Two enlarged 220 spots, labeled A and B, show the disc with maximum intensity at the ideal 220 position (A) and off the ideal 220 position (B) respectively.

show examples of a 2nm beam probing an $(hh0)$ systematic row near the $(001)^*$ projection. The fundamental reflections split into several reflections rather than giving the streaking we observed when a much larger probe is used (Fig.6.10(b) and Fig.6.31). The amplitude of the splitting increases with the distance from the origin. Such splitting can be attributed to a small lattice rotation about the c- axis, such as micro-twinning, and/or to the local change of orthorhombicity (rotation angle: $\theta = 2\tan^{-1}(b/a) - \pi/2$). In addition, we observed intensity at the ideal $(hh0)$ position (see 220 spots, marked as A and B in Fig.6.32) suggesting the presence of regions with a tetragonal symmetry. Because the period of the structural modulation along the [110] direction is close to the probe size, it is likely that the splitting of the spot is due to several domains with different orientational variants (including differences in orthorhombicity) stacking along the projection ($\sim 50 - 100$nm of specimen thickness along the c-axis), rather than domains lying perpendicular to the direction of the incident beam.

The structural modulation along the c-axis also was examined by using VG HB-5 Scanning Transmission Electron Microscope (STEM) from a much thinner crystal (10-20nm, smaller than the correlation length along the c-axis). The advantage of STEM is that the real-spacing resolution is limited only by the size of the probe so allowing a direct correlation to be made between real and reciprocal space. In our case, the probe diameter is about 1nm (0.7nm width at half-maximum of the intensity). Fig.6.33 shows a series of nano-diffraction patterns of the ($hh0$) systematic row; no splitting of the spots was seen. Instead, an oscillation of intensity of the (110) spot was observed when we traversed a 1nm beam across the tweed. The period of the oscillation of the (110) Bragg reflection was about twice of the size of the domain suggesting that the (110) lattice planes in the neighboring domains were slightly tilted. From a simple geometric construction, we estimate that the average tilting angle is about 7 mrad, as shown schematically in Fig.6.33(g), where the horizontal lines represent the possible lattice or structural discontinuity associated with the modulation along the c-axis, and t is the thickness of the sample. Fig.6.34 shows a strain-sensitive annular dark-field (ADF) image from the same sample, viewed along the (100) projection. The (001) lattice fringes show a quasi-periodic modulation of the contrast. There are bands of fringes which appear clearer and straighter than elsewhere at contours of 43nm (36 times the c-axis lattice constant). This suggests a modulation of the structure along the c-axis, as observed from the diffraction patterns. In addition, there are irregular patches of dark and light contrast of average dimension about 6nm (5 times the c-axis lattice constant) which can be due to the modulation in the a-b planes. The significant change in contrast over a small region can be due to the overlapped domains with a small change of crystal orientation, stacking sequence, or lattice parameter; this is consistent with our previous image simulation of tweed, based on a structure with domains displaced on the top of each other [Zhu et al. 1993f].

6.7.2 $0.10 < x < 0.33$

The structural modulation along the c-axis is much more apparent in heavily doped $YBa_2(Cu_{1-x}M_x)_3O_{7-\delta}$. When x reaches 0.10, instead of observing streaks of diffuse scattering, we observed intensity maxima or satellite spots in the $(001)^*$ orientation, indicating an enhanced periodicity in the modulation. With a further increase in x, the intensity maxima split along the c-axis. Fig.6.35(a) shows the diffraction pattern of the $(001)^*$ zone for $x = 0.33$; no distinct streaks or intensity maxima can be seen around the fundamental reflection in the $(001)^*$ zone. However, both are clearly visible when the crystal is tilted away from the

6.7. Three Dimensional Structural Modulation

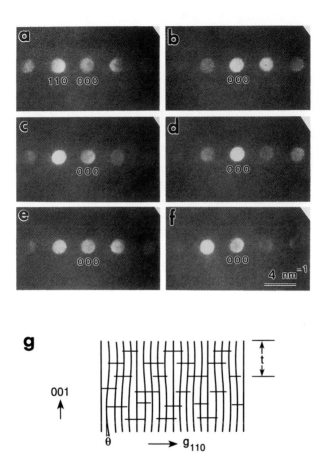

Figure 6.33 Nanodiffraction of the $(hh0)$ systematic row near the $(001)^*$ projection for $x = 0.03$ using STEM with a probe size of about 10Å: (a)-(f) the intensity oscillation when the electron beam is moved across the tweed domains; (g) the variation in the intensity of the $(110)^*$ disc suggests the presence of a small tilt of the (110) plane along the c-axis.

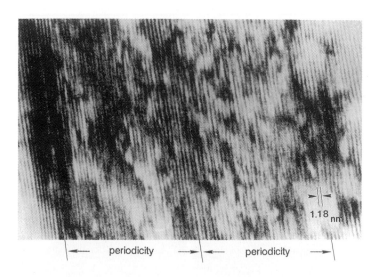

Figure 6.34 A STEM ADF high-resolution image near the a-axis using an ADF detector showing (001) lattice fringes, and the tweedy modulation in the a-b planes as well as the c-axis.

zone axis, especially when the $(101)^*$ or $(111)^*$ orientation is parallel to the electron beam, as shown in Fig.6.35(b) and (c). A diffraction pattern about 10° away from the $(100)^*$ zone is shown in Fig.6.35(d). It appears that the satellite spots are located above or below the Bragg spot, as illustrated in Fig.6.35(e). Sketches of our observations of the satellite spots in $(001)^*$, $(101)^*$, and $(100)^*$ are shown in Fig.6.35(f)-(h), respectively. It is clear that the intensity maxima, or satellites, are visible in the $[101]^*$ zone, but are invisible in the exact $(001)^*$ and $(100)^*$ zones. The satellites tend to point in the $[11l]^*$ directions (where $l \approx 1$), or are located at $hkl + q$, where q is the superstructure diffraction around the fundamental reflection, ($q \approx 1/6[11l]^*$ for $x = 0.15$, $q \approx 1/4[11l]^*$ for $x = 0.22$, and $q \approx 1/3[11l]^*$ for $x = 0.33$. Fig.6.36 shows the corresponding structure for $x = 0.33$ based on the stacking of $3d_{110} + d_{001}$ domains or orientation variants. The periodicity (or the size of the variants having the same oxygen ordering) along the [110] directions is about 0.8-0.9 nm ($\sim 3d_{110}$), while the periodicity along the [001] direction is about a c-lattice constant (by measuring the distance

6.7. Three Dimensional Structural Modulation

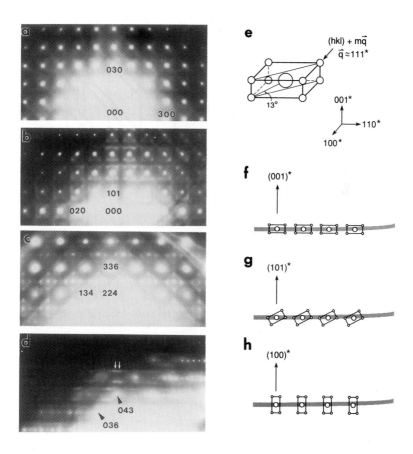

Figure 6.35 Selected area diffraction showing satellite spots surrounding the fundamental reflections for $x = 0.33$ in: (a) $(001)^*$ zone, (b) $(101)^*$ zone, (c) $(111)^*$ zone, and (d) 10° away from the $(100)^*$ zone. Sketch of an $(h00)$ reciprocal spot (e), where M is a fraction number $(1/6 - 1/3)$ varying with the dopant concentration. The intersection of the Ewald sphere (shaded curve) and the reciprocal lattice of $(001)^*$ zone (f), $(101)^*$ zone (g), and $(100)^*$ zone ((h), the position of the fundamental reflections and the satellites is not to scale). Note, the satellite spots are invisible in the exact $(001)^*$ and $(100)^*$ orientations, but visible when the crystal is tilted away from these orientations.

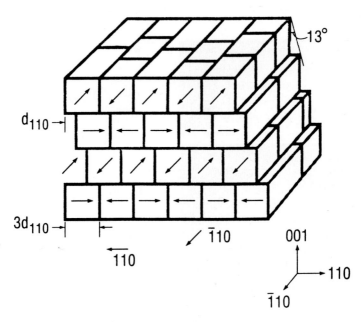

Figure 6.36 A structure model for $x = 0.33$. Domains with different orientations are stacked on top of each other along the c-axis. The correlation length along the c-axis is about a c-lattice constant. The domains in different layer are displaced by d_{110}. The average domain boundary is $13°$ away from the c-axis. This model agrees well with the superlattice spots being pointed at $hkl + m\mathbf{q}$ ($m\mathbf{q} \approx 1/3[111]^*$) for $x = 0.33$.

between the satellites along the c^*-axis, as indicated by arrow pairs in Fig.6.35(d)). This agrees with the fact, that for $x = 0.33$, almost all the Cu(1) atoms in the CuO planes were replaced by dopants resulting in periodic disorder on a scale of an unit-cell along the c- axis. The domain boundary may shift about d_{110} in the a-b plane. This results in an average boundary plane being $13°$ away from the c-axis, or equivalently gives rise to satellite positioning $\sim 1/3[111]^* + hkl$.

The detailed three-dimensional modulation for $x > 0.22$ is complicated, as demonstrated by observing the diffuse intensities between the satellites, and the "irrational" locations (i.e., non-exact superlattice positions meaning periodicity anomalies) of the satellites. However, we can focus on the main feature of the satellites to explore the modula-

6.7. Three Dimensional Structural Modulation

tion in that concentration range. In general, periodic or quasi-periodic structural modulation of a perfect crystal causes the Bragg reflections to be replaced by an array of satellites. If there is disorder in the periodicity of the modulation, the satellites broaden so that their intensity is smeared out, which may cause the higher-order satellites to disappear [Amelinckx and van Dyck 1993]. The position of the satellites, which corresponds to the reciprocal lattice of the modulation, can be determined by kinematic diffraction theory. If a crystal is modulated by displacement, the lattice potential at the position $\mathbf{r}$ can be treated as the same as that of position $\mathbf{r} - \mathbf{R}(\mathbf{r})$ in unmodulated crystal, where $\mathbf{R}(\mathbf{r})$ is displacement field, a continuous function of $\mathbf{r}$. The lattice potential can be written as:

$$\begin{aligned} V(\mathbf{r}) &= V[\mathbf{r} - \mathbf{R}(\mathbf{r})] \\ &= \sum_{\mathbf{g}} \exp(2\pi i \mathbf{g} \cdot \mathbf{r}) \exp[-2\pi i \mathbf{g} \cdot \mathbf{R}(\mathbf{r})] \\ &\approx \sum_{\mathbf{g}} V_{\mathbf{g}} \exp(2\pi i \mathbf{g} \cdot \mathbf{r}) \\ &\quad - 2\pi i \sum_{\mathbf{g}} \mathbf{g} \cdot \mathbf{R}(\mathbf{r}) V_{\mathbf{g}} \exp(2\pi i \mathbf{g} \cdot \mathbf{r}). \end{aligned} \quad (6.30)$$

where $\mathbf{g}$ are reciprocal lattice vectors of unmodulated crystal and the displacements $\mathbf{R}(\mathbf{r})$ are assumed to remain small, i.e. $|\mathbf{R}(\mathbf{r})| \ll |\mathbf{g}|^{-1}$.

Assuming that the displacements are periodic with a large periodicity, the product $\mathbf{g} \cdot \mathbf{R}(\mathbf{r})$ may be expanded as a Fourier series with coefficients $T_{\mathbf{Q},\mathbf{g}}$ so that the final term of Eqn.(6.30) becomes

$$\sum_{\mathbf{g}} \sum_{\mathbf{Q}} V_{\mathbf{g}} T_{\mathbf{Q},\mathbf{g}} \exp[2\pi i (\mathbf{g} + \mathbf{Q}) \cdot \mathbf{r}]. \quad (6.31)$$

Then the diffraction pattern of the displacively modulated crystal is given by Fourier transform of Eqn.(6.30) as:

$$\mathbf{A}(\mathbf{h}) = \sum_{\mathbf{g}} V_{\mathbf{g}} \delta(\mathbf{h} - \mathbf{g}) - 2\pi i \sum_{\mathbf{g}} \sum_{\mathbf{Q}} V_{\mathbf{g}} T_{\mathbf{Q},\mathbf{g}} \delta(\mathbf{h} - \mathbf{g} - \mathbf{Q}), \quad (6.32)$$

where $\mathbf{h}$ is the position vector in reciprocal space, and $\delta(x)$ is a Dirac delta function. The position of the Bragg peaks and the satellites are given by $\mathbf{h} = \mathbf{g}$ and $\mathbf{h} = \mathbf{g} + \mathbf{Q}$, respectively. Equation (6.32) is consistent with our TEM observations that the satellites are located at

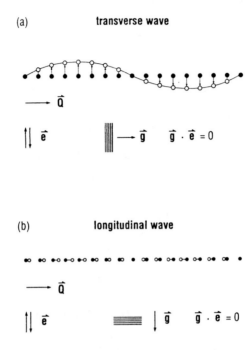

Figure 6.37 Relationship of the displacement wave vector **Q**, polarization vector **e**, and the corresponding lattice planes which cause the satellites (or diffuse scattering) to become extinct when **g** · **e** = 0. (a) transverse wave; (b) longitudinal wave.

g + **Q**, i.e., at the Bragg reflection, hkl, of the basic lattice by adding vector **Q**, where **Q** = m**q** ≈ $m[11l]$ ($l ≈ 1$, and M are irrational values, $m ≈ 1/6$ for $x = 0.15$, $m ≈ 1/4$ for $x = 0.22$, and $m ≈ 1/3$ for $x = 0.33$, see Fig.6.35), and suggests that displacive modulation can explain the observed tweed in $YBa_2(Cu_{1-x}M_x)_3O_{7-\delta}$ for both low and high x. The periodicity of the modulation is inversely proportional to the value of **Q**. The intensity of the satellites at **g** + **Q** are proportional to $V_g^2 T_{Q,g}^2$. The coefficients $T_{Q,g}$ have the form of $T_{Q,g} = A\mathbf{e} \cdot \mathbf{g}$, where A and **e** describe the amplitude and the polarization of the displacement wave with wave vector **Q**. The satellites become extinct for those reflections for which the reciprocal vector is perpendicular to the polarization vector of the wave, i.e., **g** · **e** = 0 (see Fig.6.37). In cases where

6.7. Three Dimensional Structural Modulation

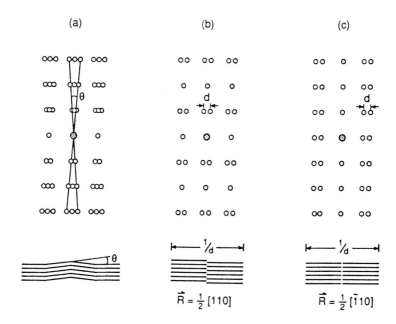

Figure 6.38 Illustrations of the splitting of the diffraction spots. (a) rotation of crystal lattice from an ideal lattice position; (b) an antiphase boundary with a displacement vector **R** parallel to the boundary; (c) an antiphase boundary with a displacement vector **R** perpendicular to the boundary.

the modulation is sinusoidal, two strong satellites per Bragg reflection will be visible, whereas if the modulation is quasi-periodic, or near non-periodic, the satellites broaden so that their intensity is smeared out over a larger area (e.g. to form reciprocal rods), or is redistributed in the entire Brillouin Zone which causes the satellites to disappear.

An equivalent way to understand crystallographic aspects of a structural modulation without detailed calculations is to analyze the splitting of the diffraction spots (fundamental or satellite spots). As we demonstrated before (Fig.6.38), a local lattice rotation from the ideal lattice position splits the Bragg spots: the magnitude of the split increases with increasing the distance from the origin (Fig.6.38(a)). On the other hand, splitting can also result from a structure with lattice discontinuity, especially when the discontinuity is planar [Zhu and Cowley 1982]. The nature of the discontinuity determines whether some or all of the spots are split. In the case of the lattice discontinuity at a domain boundary, splitting always occurs in the direction of the boundary normal. The index of the split spots obey the $\mathbf{g} \cdot \mathbf{R}$ criterion ($\mathbf{g}$: a diffraction vector, $\mathbf{R}$: a displacement vector). The spots for which $\mathbf{g} \cdot \mathbf{R} \neq n$ (n: an integer) are split; those for which $\mathbf{g} \cdot \mathbf{R} = n$ are not split. An example of $n = 1/2$ (an anti-phase boundary) is depicted in Fig.6.38(b) and (c) showing splitting of diffraction spots in every second row. In general, for $\mathbf{g} \cdot \mathbf{R} = n$, only every nth row remains unsplit. If $\mathbf{R}$ is small ($1/n$ is large) with the presence of double diffraction, or if the displacement is incommensurate, all the diffraction spots would be seen as split. This is the case for splitting of the satellite spot along the c^*-axis (lattice discontinuity along the [001] direction) observed for $x > 0.22$. The magnitude of the splitting is inversely proportional to the size of the domains (or equal to the inverse of the average spacing between the lattice discontinuities), i.e., the distance between the Bragg spots and the projection of the satellite spots along the [110]* corresponds to the domain size in the a-b plane, while the magnitude of the splitting along the c-axis corresponds to the domain size along the c-axis. A very large domain with lattice discontinuity at the boundary leads to the superposition of intensity with slightly different positions, i.e., gives rise to intensity broadening rather than splitting. This conclusion is consistent with the nano-diffraction observations (Fig.6.32). If the periodicity (or coherence length) along the c-axis was small for $x = 0.03$, the fundamental reflection would split into two discrete spots, lying above and below the zero-order Laue zone, and no split spots would be visible along the [110]* direction near the (001)* zone axis.

6.7.3 Simulation of Tweed Image Contrast

As we have demonstrated that tweedy modulation is three dimensional with a coherence length of 10-30 nm along the c-axis. Thus, they are expected to overlap for an average thickness of 20-100 nm of a TEM sample. Several domains stacking on top of each other contribute to the contrast when the incident electron beam is nearly parallel to the c-axis in two-beam imaging [Zhu et al. 1993f, Zhu et al. 1993d].

Tweed contrast can be calculated by considering the displacement

6.7. Three Dimensional Structural Modulation

of the lattice planes, **R**, relative to each other in the stacked domains. The so-called kinematical calculation gives the same result as two-beam dynamical calculations when the deviation from the Bragg position, s, is sufficiently large. The amplitude of the scattered wave Φ_g at the bottom surface of the crystal is:

$$\Phi_g = \frac{i\pi}{\xi_g} \int_0^t \exp(-2\pi i \mathbf{g} \cdot \mathbf{R}) \exp(-2\pi i s z) dz, \tag{6.33}$$

where ξ_g is the extinction distance, t is the total thickness of the foil, **g** is the diffraction vector used for imaging, and z is the distance from the center of the foil to the fault, measured along the direction of the incident beam. The intensity of the reflected beam, i.e. the dark-field image, can be expressed by:

$$\begin{aligned} I &= \frac{1}{(\xi_g s)^2} [\sin^2(\pi t s + \pi \mathbf{g} \cdot \mathbf{R}) + \sin^2(\pi \mathbf{g} \cdot \mathbf{R}) \\ &\quad - 2\sin(\pi \mathbf{g} \cdot \mathbf{R}) \sin(\pi t s + \pi \mathbf{g} \cdot \mathbf{R}) \cos(2\pi s z)]. \end{aligned} \tag{6.34}$$

Fig.6.39 shows examples of the contrast calculations. For a perfect crystal of constant thickness there is no contrast [Fig.6.39(a)], while for a stacking fault with constant displacement vector **R** running from top to bottom in a foil of constant thickness [Fig.6.39(b)], we see the oscillation of contrast with depth of the fault. When **R** varies as a function of position x, we obtain the modulated contrast shown in Figs.6.39(c)–(e), with various deviations from the Bragg position, $s = 3.25/t$ [Fig.6.39(c)], $s = 3.5/t$ [Fig.6.39(d)], and $s = 5.5/t$ [Fig.6.39(e)].

In these calculations, we used a simple sinusoidal wave as a [110] displacive modulation for **R**. The intensity profiles were calculated for a modulation with a period of 8 nm (the domain size is half of the period). The tweed contrast calculated for a different period (4 nm, $s = 3.25/t$) is shown in Fig.6.39(f). Here, the length of the tweed is limited by the effective extinction distance, ξ_g, which is of the same order as the domain size on the $a-b$ plane. By using more complicated wave functions for **R**, a quasi-periodic tweed contrast can be generated, as shown in Fig.6.39(g).

Our image calculation generated strain contrast similar to that observed from tweed structure, also confirming that: (1) the intensity of the tweed image decreases with period; (2) the width of the tweed is

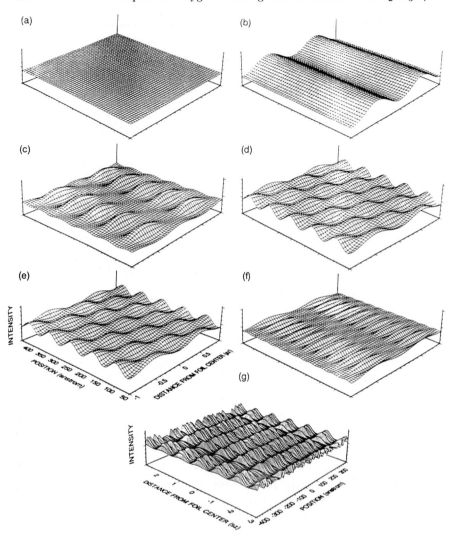

Figure 6.39 Two-beam dark-field images from domain interface with displacement, $\mathbf{R}$, when interface runs from top to bottom: (a) $|\mathbf{R}| = 0$, (b) $|\mathbf{R}| =$ constant; (c)-(f) $|\mathbf{R}| = f[\sin(x)]$, for different period p and different deviations from Bragg position s; (c) $p = 8$nm, $s = 3.25/t$; (d) $p = 8$nm, $s = 3.5/t$; (e) $p = 8$nm, $s = 5.5/t$; (f) $p = 4$nm., $s = 3.25/t$; (g) $|\mathbf{R}| = f_1[\sin(x)] + f_2[\sin(x)]$, $s = 1/t$.

independent of the deviation from the Bragg position, s, being consistent with the observations of Fig.6.14, although here the situation is more complicated for small s where the kinematical treatment does not apply. Thus, the contrast caused by more than one interface lying on top of another within the foil may not be separated when s becomes small. For most imaging conditions, the number of domain interfaces on top of each other may be unimportant. The observation of only minor changes in the tweed contrast with thickness suggests that the regions near the surfaces mainly contribute to the contrast, similar to the case for stacking faults, where absorption tends to reduce the contrast in the interior of the foil.

6.8 Summary

In this chapter we explored the microstructure of $YB_2Cu_3O_{7-\delta}$ that is related to oxygen ordering and doping of trivalent impurities.

Starting with experimental observations, we build a simple theoretical model that fits well the observed properties related to the orthorhombic to tetragonal phase transitions of $YBa_2Cu_3O_{7-\delta}$ under various pressure. This model is then used to study the structures of thermally quenched samples as well as samples with trivalent impurities such as Fe. Tweed-like modulations in thermally quenched sample as well as samples doped with Fe impurities were studied using transmission electron microscope as well as theoretical modeling techniques.

While we explored many aspects of the microstructure in YBCO system, we have not talked about the twin boundary structure in detail. The twin boundary is one of the most important structural features of YBCO systems. The formation of twin boundary is due to the interplay between the elastic strain due to the orthorhombic distortion and interface energy of the boundary. We will discuss the twin boundary structure in detail in the next chapter.

Chapter 7

The Twin Boundary Structure in $YBa_2Cu_3O_{7-\delta}$

7.1 Introduction

Since the discovery of high-T_c superconductivity in $YBa_2Cu_3O_{7-\delta}$ the relationship between the twin boundaries and the oxide's superconducting properties, such as critical temperature, T_c, and the critical current density, J_c, has been explored extensively. For example, Fang *et al.* discussed the possibility that the boundary region is an area of higher T_c than the matrix [Fang *et al.* 1988]. However, Deutscher and Müller proposed that the boundary layer is weakly superconducting, and that the magnetic flux lines flow freely along the boundary [Deutscher and Müller 1987]. On the other hand, Chaudhari and Kes argued that the twin boundaries act as strong pinning sites for the flux lines [Chaudhari 1987, Kes 1988]. Evidence for preferential pinning of the vortex lattice along the twin boundaries was observed in experiments on flux lattice decoration in $YBa_2Cu_3O_{7-\delta}$ crystals [Gammel *et al.* 1987, Vinnikov *et al.* 1988]. More recently, magnetic and transport measurements have shown that the twin boundaries can pin flux lines if the magnetic field is applied accurately parallel to the boundary planes [Gyorgy *et al.* 1989, Gyorgy *et al.* 1990, Kwok *et al.* 1990]. Otherwise, the boundaries do not act as effective pinning centers. Furthermore, the effectiveness of boundary pinning is observable only in cases where no other strong pinning centers are present. If other strong pinning centers coexist with the boundaries, as in thin films, the boundaries will add very little to the pinning strength of that materials [Lairson et al. 1990]. Thus, now it is believed that the twin boundaries do not play a primary role in determining the critical current densities in $YBa_2Cu_3O_7$. However, the boundaries can be used as a model system to study the nature of the interaction between the flux lines and a planar defect in $YBa_2Cu_3O_{7-\delta}$, and can provide useful insight in flux pinning mechanisms. In addition, Cu and O play an important part in the superconductivity of the $YBa_2Cu_3O_{7-\delta}$ system. Replacement of Cu by two valent (Zn, Ni) and three valent (Fe, Al) transition-metal elements, was shown to change the structures of the boundaries and may affect the flux pinning mechanisms. Thus, detailed investigations of the structure of the twin boundaries, including the effects of substitution on the boundary structure, are of interest and have been extensively studied.

In characterizing the microstructure of the $YBa_2Cu_3O_{7-\delta}$ system, conventional transmission-electron-microscopy (TEM) [Dou et al. 1987, Pande et al. 1987, Sarikaya et al. 1988, Shaw et al. 1989] and high-resolution-electron-microscopy (HREM) [van Tendeloo et al. 1987b, Barry 1988, Hiroi et al. 1989] have been used widely to study the twinning and twin boundary structure although systematic studies as a function of the stoichiometry of samples are not plentiful. Using various electron microscopy techniques, we have systematically studied the structure of twin boundary in carefully prepared $YBa_2Cu_3O_{7-\delta}$ ($\delta \approx 0.0 - 0.4$) and $YBa_2(Cu_{0.98}M_{0.02})_3O_{7-\delta}$ (M = Fe, Al, Ni, or Zn) samples. Here, we review our studies on the twin boundary structure, based on TEM and HREM as well as electron diffraction observations [Zhu et al. 1989a, Zhu et al. 1990c, Zhu et al. 1991a, Zhu and Suenaga 1992, Zhu et al. 1993g]. In Sec. 7.2, we discuss the gross features of the twin boundary of $YBa_2Cu_3O_{7-\delta}$ and cation substitution effects using two-beam, selected-area-diffraction (SAD) and HREM techniques. In Sec. 7.3, we focus our attention on the observed displacement at twin boundaries by HREM *in situ* experiments, on analyses of fringe contrast and diffuse scattering of the twin boundaries; and we present evidence of the variable nature of the twin boundaries. In Sec.7.4, we illustrate the geometry of the twin boundary and twin boundary dislocation using coincidence-site-lattice (CSL) and displacement-shift complete-lattice (DSCL) theories. In Sec.7.5, we

7.2 Gross Features of the Twin Boundary

7.2.1 Conventional TEM Observations

Twin boundaries are the most commonly observed defects in $YBa_2Cu_3O_{7-\delta}$. It is well known now that in $YBa_2Cu_3O_{7-\delta}$ crystals, twins are formed in two equivalent orientations ([110] and [-110]) to accommodate the strain energy of the tetragonal to orthorhombic phase transformation as increasing oxygen is absorbed by the oxide when it is cooled from high temperature in an oxygen atmosphere. Each twinning forms two twin variants. Visually, the morphology of the twinning consists of alternating lamellas contrast from the twin variants, as shown in Fig.7.1(a). The width of the lamellas of $YBa_2Cu_3O_{7-\delta}$ usually depends on the grain size, which is controlled by processing parameters. The difference in contrast between neighboring lamellas arise from their difference of deviation from the Bragg reflection. When the twin boundary is viewed inclined, it shows fringe contrast under most two-beam conditions. A close inspection of the boundary shows that the boundary has a perceptible thickness. Figs.7.2(a) and 7.2(b) are typical multibeam images of the twin boundaries seen edge-on along the (001) zone axis in $YBa_2Cu_3O_7$ and $YBa_2(Cu_{0.98}Al_{0.02})_3O_{7-\delta}$, respectively. Double thin lines of contrast at the twin boundary are clearly visible. Such a contrast suggests that the crystal structure at the boundary may be differ from the orthorhombic twin matrix. The thin layer between the lines represents a transition region between the two neighboring twin variants. The width of the twin boundary ranges from 7Å to 39Å, depending on stoichiometry (cation substitution) of the samples, level of oxygenation, and heat treatment. The observation of the twin boundary as a transitional region, rather than as a monoatomic plane, is consistent with the X-ray studies of Laue patterns, X-ray oscillating patterns, and angular scanning topograms [Ossipyan *et al.* 1988].

Figure 7.1(b) shows a well-aligned (001)* selected area diffraction

150 Chapter 7. The Twin Boundary Structure in $YBa_2Cu_3O_{7-\delta}$

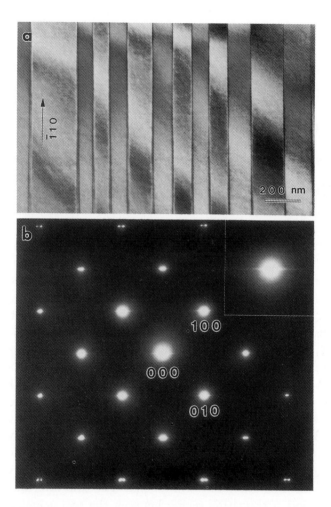

Figure 7.1 (a) A typical two-beam image of the twinning morphology in Y-Ba-Cu-O systems showing alternative lamellas. The contrast between neighboring lamellas is twin boundaries. (b) A well-aligned $(001)^*$ selected area diffraction pattern covering several twin boundaries from a $YBa_2Cu_3O_7$ sample. Note that sharp streaks superimpose on all the diffraction spots in a direction perpendicular to the twin boundary. The inset shows an enlarged reciprocal spot.

7.2. Gross Features of the Twin Boundary   151

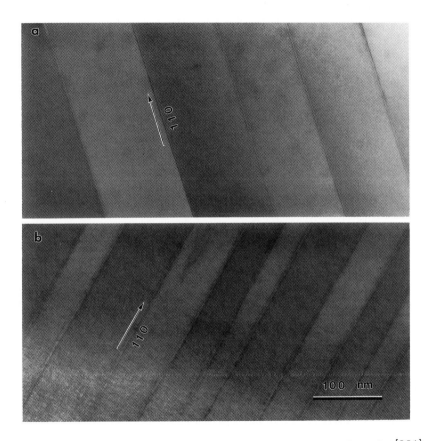

Figure 7.2 Multi-beam images of twin boundaries viewed edge-on along the [001] direction from a YBa$_2$Cu$_3$O$_7$ sample (a), and a YBa$_2$(Cu$_{0.98}$Al$_{0.02}$)$_3$O$_7$ sample (b). The twin boundary image from YBa$_2$(Cu$_{0.98}$Al$_{0.02}$)$_3$O$_7$ is wider and fuzzier than that of YBa$_2$Cu$_3$O$_7$.

152   Chapter 7. The Twin Boundary Structure in $YBa_2Cu_3O_{7-\delta}$

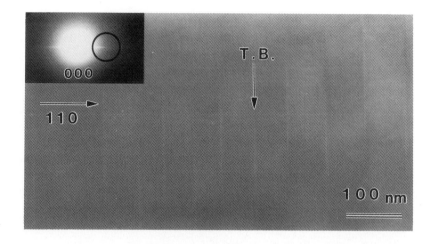

Figure 7.3 A dark-field image obtained by allowing half of a streak, marked by a circle on the inset, to pass through the objective aperture. The enhanced contrast of the twin boundaries unambiguously demonstrates that the streaks result from the boundaries.

(SAD) pattern obtained by using a nearly parallel electron beam, which extends over an area of 0.3 µm in diameter covering several (110) twin boundaries, such as that shown in Fig.7.1(a). Because of twinning, all diffraction spots are split, except for the (110) systematic row in which the reflection of the two twin variants is common. However, the main feature to be noted in Fig.7.1(b) is that sharp streaks are superimposed on all reciprocal spots including the origin in a direction perpendicular to the boundaries. The streaks are straight and narrow, the latter feature suggesting a coherence parallel to the twin boundaries in excess of 500 Å. Although the intensity of the streaks is not strong, they can easily be observed by having the second condenser lens of the electron microscope well over-focused.

To explore the origin of the streaks, a near parallel-beam dark-field image technique was used allowing half of a streak to pass through the objective aperture (marked by a circle on the inset of Fig.7.3) to form an image. The enhanced contrast of the twin boundary unambiguously suggests that these streaks result from the boundaries rather than from

7.2. Gross Features of the Twin Boundary

the shape of the twin lamellas. Such streaks can be due to a thin boundary layer with a different structure, and/or a phase shift at the boundary. These two possibilities can be distinguished by comparing the intensity distribution at different reflections in reciprocal space [Zhu et al. 1991c]. However, to analyze such a diffraction pattern in a highly symmetric orientation from a thick specimen, one must bear in mind that there are strong multiple scattering effects, as commonly observed in electron diffraction when many Bragg beams are excited simultaneously; this complicates the interpretation.

7.2.2 HREM Observations

Figure 7.4(a) is an example of the structural image of a twin boundary of a fully oxidized $YBa_2Cu_3O_{7-\delta}$ ($\delta \approx 0.0$) sample viewed along the [001] direction with a point-to-point resolution better than 1.9Å. The bright dots arranged in a near-square cell correspond to the projection of the fundamental perovskite unit cell. The approximate 3.82Å × 3.88Å periodic dot patterns correspond, respectively, to the a and b lattice parameters of the crystal. Computer-image simulations under our imaging conditions [Zhu et al. 1990b] suggest that the heavy atom of the Ba(Y) columns appears in strong white contrast, and the Cu(O) columns in weak white contrast. The plane-oxygen (chain-oxygen) columns are shown between the Ba(Y) columns. The sharp, uniform atomic image (except for the twin-boundary region) implies that the columns of atoms are accurately aligned in the direction of the incident beam on both sides of the twin boundary.

Several observations can be made about the structural image of the twin boundaries in $YBa_2Cu_3O_7$: 1) The twin boundary consists of several severely distorted atomic layers (10~30 Å), in which the exact positions of the atoms are difficult to define; 2) Some twin boundaries show steps perpendicular to the boundary plane; and 3) There is a lattice discontinuity at the boundary, or a lattice shift along the boundary. The interchange of the a and b planes across the twin boundary is accomplished not only by a rotation of the planes in one twin variant with respect to the other by an angle of 90° - Θ, where $\Theta = 2(b-a)/(a+b)$, but also by a translation of the lattice plane by approximately $(1/3 - 1/2) \cdot 2d_{110}$ along the twin boundary. The

7.2. Gross Features of the Twin Boundary

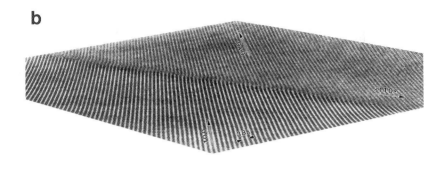

Figure 7.4 A structural image of a twin boundary of YBa$_2$Cu$_3$O$_7$ viewed along the [001] direction. The bright dots arranged in a near-square cell correspond to the projection of the fundamental perovskite unit cell. Computer-image simulations suggest that the atom of the Ba(Y) columns appears in strong white contrast, and the Cu(O) columns in weak white contrast. The oxygen columns are shown between the Ba(Y) columns. The sharp, uniform atomic image (except for the twin-boundary region) implies that the columns of atoms are accurately aligned in the direction of the incident beam on both sides of the twin boundary. (b) A photograph taken from near glancing angle of the twin boundary image shown in (a). It can be clearly seen that the a and b lattice planes interchanged at the twin boundary involve a shifting along the boundary as well as a rotation across the boundary.

lattice shift is most clearly seen by tilting the micrograph and observing the series of the dots along the [100] or [010] direction across the boundary, as demonstrated in Fig.7.4(b) which was taken by placing an optical camera at a near-glancing angle along the [100] direction from a micrograph of a twin boundary shown in Fig.7.4(a). Unless the twin boundary had been damaged (including the loss and/or disordering of oxygen) by electron beam irradiation, such a rigid-body translation at the twin boundary is readily observable in YBa$_2$Cu$_3$O$_{7-\delta}$ ($\delta \approx 0.0$), except for oxygen-reduced cases ($\delta \approx 0.4$). A preliminary computer simulated image, which focused on the observed lattice shift rather than the exact atomic position at the boundary, using a large supercell (81.72Å × 21.72Å × 11.70 Å, 1492 atoms) with a lattice translation along the boundary, matches well with the experimental images confirming the existence of a $(1/3 - 1/2) \cdot 2d_{110}$ lattice translation [Zhu et al. 1990b].

7.2.3 The Effects of Cation Substitution

To study the effect on the structure of the twin boundary of replacing Cu in YBa$_2$Cu$_3$O$_{7-\delta}$ by another element, we studied YBa$_2$(Cu$_{0.98}$M$_{0.02}$)$_3$O$_{7-\delta}$ for $\delta \approx 0.0$ and M=Ni, Zn, Fe, and Al. Examples of the HREM images of these twin boundaries viewed along the [001] orientation are shown in Figs.7.5 (a), (b), (c), and (d), respectively. The horizontal distorted areas are the twin boundary regions. It is generally agreed that, at least at low concentrations, Fe and Al substitute predominately for Cu on chain sites. In the case of M=Zn and Ni the sites are more controversial. We believe that Zn substitutes on plane sites and Ni occupies both sites [Xu et al. 1990]. Compared with pure YBa$_2$Cu$_3$O$_7$, the widths of the twin boundaries of YBa$_2$(Cu$_{0.98}$Ni$_{0.02}$)$_3$O$_7$ (Fig.7.5(a)) and YBa$_2$(Cu$_{0.98}$Zn$_{0.02}$)$_3$O$_7$ (Fig.7.5(b)) are essentially unchanged, but are widened considerably with Fe (Fig.7.5(c)) and Al (Fig.7.5(d)) substitutions. We also noted that the image of the boundary in Zn- and Ni- doped YBa$_2$Cu$_3$O$_{7-\delta}$ is very similar to that of the pure YBa$_2$Cu$_3$O$_7$, i.e. straight and narrow. On the other hand, the addition of Fe and Al causes the boundary not only to widen but also to become diffuse and wobbly. The twin boundary thickness varies from ~1 nm for the pure and the Zn- and

7.2. Gross Features of the Twin Boundary

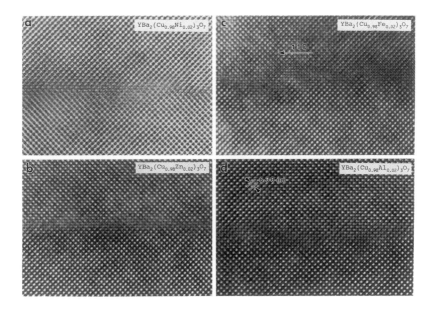

Figure 7.5 High resolution structure images of twin boundaries observed along the [001] direction for YBa$_2$(Cu$_{0.98}$Ni$_{0.02}$)$_3$O$_7$ (a), YBa$_2$(Cu$_{0.98}$Zn$_{0.02}$)$_3$O$_7$ (b), YBa$_2$(Cu$_{0.98}$Fe$_{0.02}$)$_3$O$_7$ (c), and YBa$_2$(Cu$_{0.98}$Al$_{0.02}$)$_3$O$_7$ (d). The horizontal distorted region in the middle of each micrograph is a twin boundary.

Ni- substituted YBa$_2$Cu$_3$O$_{7-\delta}$ ($\delta \approx 0.0$), to 2.5~3 nm for the Fe- and Al- substituted oxides and for oxygen-deficient YBa$_2$Cu$_3$O$_{7-\delta}$. The samples with wider twin boundaries usually have lower orthorhombicity than those with narrow ones as measured by X-ray and electron diffraction. Table 7.1 lists the values of twin boundary thickness (as measured by the width of the distorted boundary region in HREM images and the length of the streaks in diffraction pattern) together with the values of twin spacing, orthorhombicity, and critical temperature, T_c (as measured by an ac technique) [Zhu et al. 1990d].

It is notable that the twin boundary width changes with cation substitution. It has been speculated that some dopants (such as Fe, Al or Co, which form cross-links) tend to segregate at the twin boundary [Bordet et al. 1988c, Hiroi et al. 1988]. However, after carefully examining a large number of high-resolution twin-boundary images in the

Table 7.1 Examples of twin boundary thickness, orthorhombicity, and superconducting critical temperature in YBa$_2$(Cu$_{0.98}$M$_{0.02}$)$_3$O$_{7-\delta}$, where M=Cu ($\delta \approx 0.0$, 0.45), Zn, Ni, Fe, and Al ($\delta \approx 0.0$).

	Thickness of twin boundary	Orthorhombicity X-ray	SAD	T_c(K) midpoint
YBa$_2$Cu$_3$O$_7$	1.0 to 1.5 nm	0.0171	0.0170	90.5
YBa$_2$Cu$_3$O$_{6.55}$	2.0 to 3.0 nm	0.0060	0.0100	40.0
YBa$_2$(Cu$_{0.98}$Zn$_{0.02}$)$_3$O$_7$	0.9 to 0.4 nm	0.0173	0.0166	67.0
YBa$_2$(Cu$_{0.98}$Ni$_{0.02}$)$_3$O$_7$	0.7 to 1.0 nm	0.01765	0.0169	82.5
YBa$_2$(Cu$_{0.98}$Fe$_{0.02}$)$_3$O$_7$	2.0 to 3.0 nm	0.0117	0.0115	86.5
YBa$_2$(Cu$_{0.98}$Al$_{0.02}$)$_3$O$_7$	2.6 to 3.9 nm	0.0119	0.0114	88.0

Y-Ba-Cu-O system, we could reach no conclusions about the clustering of the dopants at the boundaries. Instead, we found that the general feature of these twin boundaries is the existence of a lattice translation (see Figs.7.5 (a)-(d)). The magnitude of the translation may vary with the amount and type of substitution. The equilibrium width of the twin boundary appears to be determined by a balance among the chemical potential of the O-Cu-O chain, the internal stress present at the twin boundary, and the Coulomb repulsion force between like-atoms on opposite sides of the twin boundary, during the twin formation. A clear indication that chain-oxygen dominates the nature of the distortion is seen from the effects on the width of the boundary of the substituting elements Ni, Zn, Fe, and Al. Zn (at a level of 2%) is thought to substitute for Cu in the CuO$_2$ plane and, thus, is not likely to influence the location of the chain oxygen. On the other hand, for Fe and Al, additional oxygen is incorporated in the chain region. This extra oxygen disorders the O-Cu-O chains at the boundary region and creates a diffuse and wider boundary. A similar broadening in the boundary was observed in oxygen-reduced specimens, YBa$_2$Cu$_3$O$_{7-\delta}$ ($\delta \approx 0.3 \sim 0.4$).

7.2.4 The Effects on Superconducting Properties

The observation of the finite width (10~30 Å) of the distorted, antiphase-like, twin boundary region is of significant interest because it may influence the superconducting properties of the bulk material. For instance, there might be a depression in the superconducting energy gap at the boundary. As pointed out by Deutscher and Müller, such a depression would form a weakly coupled junction at the boundary, and magnetic flux lines could move freely along the boundary [Deutscher and Müller 1987]. Thus, for a sharp boundary, such as fully oxidized $YBa_2Cu_3O_{7-\delta}$ or one formed with the addition of Ni or Zn, the gap should not be degraded drastically since the width of the boundary is ~ 10Å, compared to the coherence length in the $a - b$ plane of 16~30Å. Conversely, specimens with Fe or Al substitutions or with insufficient oxygen are expected to have a large degradation of the gap at the boundary because the boundary width is approximately equal to or greater than the value of the coherence length. Thus, the critical current density, J_c, in these specimens is likely to be lower than in those with a sharp boundary, if the above theory applies. On the other hand, Kes, Chaudhari, and Matsushita *et al.* have argued that pinning by twin boundaries is strong, and high-critical currents, which are observed in single crystals and in $YBa_2Cu_3O_{7-\delta}$ thin films, are caused by flux pinning by the boundaries [Kes 1988, Chaudhari 1987, Matsushita *et al.* 1987]. For his argument, Kes used a mechanism of pinning based on electron scattering by the boundary, while Chaudhari pointed out a possible pinning of the flux lines through core pinning at the boundary. In either case, both the width and sharpness of the boundary will determine the pinning strength. Thus, the variations in the twin boundary thickness should also influence the pinning strength. A preliminary study of intragrain critical currents, measured by the magnetization of oriented powders [Finnemore *et al.* unpublished, Shelton *et al.* unpublished] suggested that those alloying elements which widen the boundary layer decrease J_c by nearly an order of magnitude. However, substitutions which narrow the boundary tend not to decrease J_c in comparison to that for pure $YBa_2Cu_3O_{7-\delta}$ ($\delta \approx 0.0$). This result indicates that an increase in the boundary width reduces J_c. However, Wördenweber *et al.* pointed out that J_c is also determined by other factors, such as H_{c2}, in addition to the size of the defects

[Wördenweber et al. 1989]. Further study is needed to elucidate the actual relationship between the pinning strength of the boundary and its width in this high T_c material.

7.3 Studies of the Displacement at the Twin Boundary

7.3.1 Evolution of the Twin-Boundary Structure During Electron-Beam Irradiation

Observations by TEM *in situ* experiments provide an useful insight into understanding the structural changes in a crystal. Electron beam irradiation and/or beam heating in a microscope has been used to study the evolution of twinning in $YBa_2Cu_3O_{7-\delta}$ [Iijima et al. 1987, Mitchell et al. 1988, Roy et al. 1989, Sasaki et al. 1987]. During the orthorhombic-tetragonal phase transformation under beam irradiation, the crystal evolves from a twinned structure through an ortho-II phase (a partially ordered structure) to a tweed structure [Iijima et al. 1987]. Twin contrast, fading irreversibly or reappearing reversibly, was also observed under beam irradiation [Smith and Wohlleban 1988, Zou et al. 1988]. These structural changes are attributed to oxygen reduction and to oxygen disordering through a knock-on [Roy et al. 1989] or an ionization process [Eaglesham et al. 1988]. The aim of our HREM in-situ experiment was to elucidate the evolution of the twin-boundary structure on an atomic scale.

Figure 7.6 is a series of micrographs showing the structural changes in a twin boundary during irradiation by 200 kV electrons with a beam diameter ~ 600Å and current density $\sim 30 A/cm^2$. The HREM in-situ observations start with a $YBa_2Cu_3O_7$ sample, viewed along the c axis. As also seen in Figs.7.6(a) and (b), the twin boundary of $YBa_2Cu_3O_7$ shows a characteristic lattice shift at the boundary. Two minutes after the start of irradiation, the lattice shift has disappeared preferentially (comparing the reference lines of $(100)_{t1}/(010)_{t2}$ lattice in Fig.7.6(b) and Fig.7.6(a)) at a relatively thin area (left side of the micrograph) without destroying the twinning ($\sim 89.1°$ lattice rotation). After the area was irradiated for about 10 minutes, the lattice shift

7.3. Studies of the Displacement at the Twin Boundary

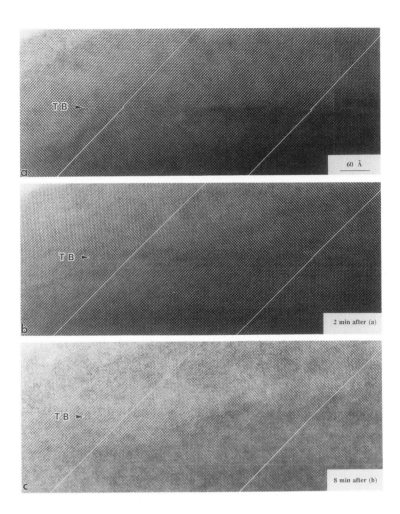

Figure 7.6 HREM *in-situ* observations of structural changes of a twin boundary during electron beam irradiation. (a) original structural image of a $YBa_2Cu_3O_7$ twin boundary. The two reference lines indicate the existence a lattice shift across the boundary. (b) Two minutes after the start of the electron irradiation ($30A/cm^2$, 200kV). (c) Ten minutes after the start of the irradiation; Note that there is no lattice shift across the boundary (see the reference lines), however, the lattice rotation due to the twinning is still visible.

at the twin boundary became completely invisible (Fig.7.6(c)) over the whole area available for structural image observations. Despite the fading of boundary contrast and the enhancement of lattice distortion, the ~0.9° rotation of the $(100)_{t1}$ and $(010)_{t2}$ lattices (or ~89.1° rotation of O-Cu-O chain) caused by twinning was still visible across the boundary, suggesting that, at an early stage in electron irradiation, only the $(1/3-1/2) \cdot 2d_{110}$ lattice shift at the boundary is eliminated. The evolution of twin structure during irradiation occurs preferentially from the twin boundary. This finding is in good agreement with electron diffraction observations using a special convergent-beam technique, which also shows the orthorhombic-tetragonal transformation occurring first at the twin boundary. The structural changes in the twin boundary during irradiation appear to result from a gradual loss of oxygen at the boundary.

7.3.2 The Variability of the Twin Boundaries

The above *in-situ* experiments suggest that the twin boundary in $YBa_2Cu_3O_{7-\delta}$ exhibits variability (with or without a lattice shift). Additional evidence of such variability was obtained by HREM observations of samples with different oxygen stoichiometries. Fig.7.7 compares the structural images of $YBa_2Cu_3O_7$ (Fig.7.7(a)) and $YBa_2Cu_3O_{6.6}$ (Fig.7.7(b)). Disregarding the details of the twin boundary region, we consistently observed an important difference between pure samples of $YBa_2Cu_3O_{7-\delta}$, with $\delta \approx 0.0$ and oxygen-deficient one with $\delta \approx 0.4$, as indicated schematically in Fig.7.8. For $\delta \approx 0.0$, the (100) and (010) planes are shifted $\sim d_{200}$ and $\sim d_{020}$ ($a/2$ or $b/2$) across the twin boundaries, and, by measuring the distance normal to the twin boundary between the corresponding rows of dots in the interior of the twins, we found it to be $(n + 1/2)2d_{110}$ (n is an integer) for $\delta \approx 0.0$, while, for $\delta \approx 0.4$, the shift is not present and this distance is $n2d_{110}$. As also discussed later in this section, these observations are consistent with the twin boundaries being centered at the (110) planes through the oxygen atoms in CuO planes for $\delta \approx 0.0$ (Fig.9(a)) and at the cations for $\delta \approx 0.4$ (Fig.7.8(b)). Both types of boundaries have same crystallographic symmetry; in other words, in both twin boundaries the neighboring twins are related by a mirror operation across the twin boundary. However,

7.3. Studies of the Displacement at the Twin Boundary

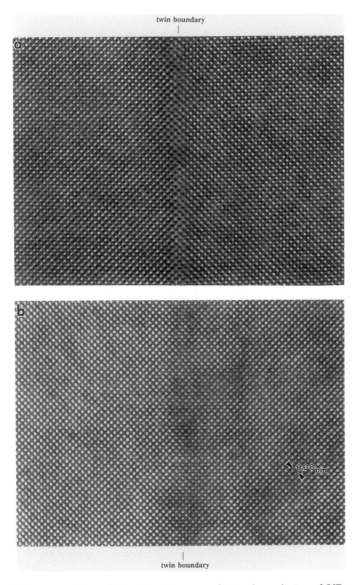

Figure 7.7 A comparison of structural images of twin boundaries of $YBa_2Cu_3O_7$ (a) and $YBa_2Cu_3O_{6.6}$ (b).

164 Chapter 7. The Twin Boundary Structure in $YBa_2Cu_3O_{7-\delta}$

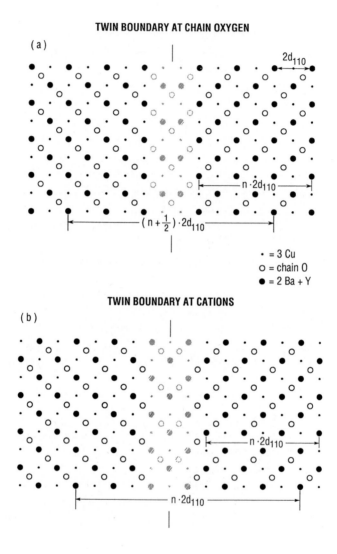

Figure 7.8 Idealized drawing of the twin boundary region consistent with the observations of Fig.7.7(a) and Fig.7.7(b), respectively. Most of the oxygen atoms are omitted, except for chain-oxygen atoms.

7.3. Studies of the Displacement at the Twin Boundary

because the O atoms in the CuO chains contribute negligibly to the electrostatic potential, other atoms-mainly the cations-define the correspondence between the structure and the image. Thus, the structure of Fig.7.8(a) appears to be an anti-phase boundary.

Previous HREM observations and discussions of the structure of the twin boundary were limited to the twinning symmetry, strain contrast, and possible lack of oxygen at the boundary region [Barry 1988, Shaw et al. 1989, Jou and Washburn 1989]. Although a model of an oxygen-centered twin boundary was proposed before, the earlier HREM observations of the twin boundary supported the Cu-centered boundary model [Hewat et al. 1987]. The rigid-body translation at the twin boundary was first observed by Zhu et al. [Zhu et al. 1989a], then confirmed by other groups [van Tendeloo et al. 1990, Kulik et al. 1991, Cheng et al. 1991]. The observations were puzzling, since a such lattice translation would likely be a high energy process. A recent HREM study [van Tendeloo et al. 1990] suggests that the displacement at the twin boundary may be a consequence of small differences in the orientation of crystal that is not ideally oriented. However, for a reflection twin-boundary, the regions away from a twin boundary should give a similar image on either side of the boundary (except for a rotation around the [001] axis) when the thickness and the orientation of the crystal relative to the electron beam are the same. This is what we saw in HREM when there was proper alignment; the lattice shift was repeatedly observed at the twin boundary in perfectly-aligned twinned crystals in $YBa_2(Cu_{0.98}M_{0.02})_3O_7$ (M=Cu, Ni and Zn). The inclination of the boundary plane normal, and misalignment of the microscope, would not generate a lattice shift across the boundary. Therefore, we believe that the displacement is an intrinsic feature of the twin boundary in $YBa_2Cu_3O_7$ system. The discrepancies, which exist in the literature on the detailed structure of the boundary, may be due to oxygen loss or surface relaxations at the boundary region, as we also observe the twin boundary without lattice shift in a very thin area. To avoid any misleading "imaging artifacts" from HREM observations, complementary methods to study the displacement from a thick area are desirable.

7.3.3 Analyses of Fringe Contrast of the Twin Boundary

Fringe contrast analysis has been very successful in studying the nature of boundaries or interfaces. The symmetry of the intensity variation, or "rocking curve", from a tilted boundary provides important information about the crystallographic features of respective crystals across the boundary. Figs.7.9(a) and (b) show the fringe contrasts of an inclined twin boundary of a $YBa_2Cu_3O_{6.6}$, obtained with the $(\bar{1}10)$ reflection being strongly excited. The outer fringes (the fringes from the top and bottom surfaces of the foil) of the boundaries in a bright-field (BF) are asymmetric [Fig.7.9(a)], i.e., dark-bright and bright-dark, while in a dark-field (DF) they are symmetric [Fig.7.9(b)], i.e., bright-bright and dark-dark. For a boundary with pure twinning characteristics, these are the results we expect to observe [Amelinckx and van Landuyt 1976].

In contrast, the symmetry of the twin-boundary fringes from $YBa_2(Cu_{0.98}M_{0.02})_3O_{7-\delta}$ (M=Cu, Zn, Ni, and $\delta \approx 0.0$) samples, which exhibit a lattice displacement seen in HREM, is different. In a BF image of the boundaries [Fig.7.10(a)], the outer fringes of the boundaries are symmetric. In this particular case, $S_1 = -S_2$ (where S_1 and S_2 are the deviation from the Bragg reflection of the adjacent twin domains), the intensity of the background and the fringes in each image of the boundaries are the same. On the other hand, in a DF image in Fig.7.10(b), the outer fringes of the boundaries are alternatively symmetric and asymmetric for the series of the boundaries. Such contrast is not caused by overlapping boundaries because the spacing between the boundaries (see Fig.7.10(c) where the twin boundary is edge-on) is substantially larger than the image widths of the tilted boundaries. Similar alternative fringes was observed in Nb_2O by van Landuyt et al. [van Landuyt et al. 1966].

Gevers et al. and Amelinckx and van Landuyt subdivided the types of planar interfaces into three categories, depending on the nature of the interface: (1) α-interface ($\alpha = 2\pi\mathbf{g}\cdot\mathbf{R}$, where $\mathbf{g}$ and $\mathbf{R}$ are a diffraction and a displacement vector); the contrast of the interface is due to the phase difference, i.e., a displacement, or to a small misfit at the boundary, such as a stacking fault or an anti-phase boundary; (2) δ-interface ($\delta = \xi_{g1}S_1 - \xi_{g2}S_2$); the contrast of the interface is due to

7.3. Studies of the Displacement at the Twin Boundary 167

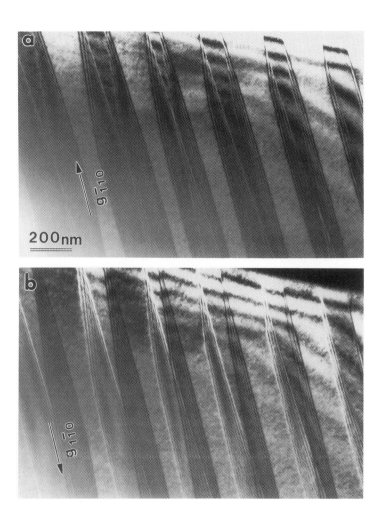

Figure 7.9 Fringe contrasts of an inclined twin boundary of YBa$_2$Cu$_3$O$_{6.6}$, obtained with the $(\bar{1}10)$ reflection being strongly excited. The outer fringes of the boundaries in a bright-field image are asymmetric (a), while, in a dark-field image, they are symmetric (b).

168 Chapter 7. The Twin Boundary Structure in $YBa_2Cu_3O_{7-\delta}$

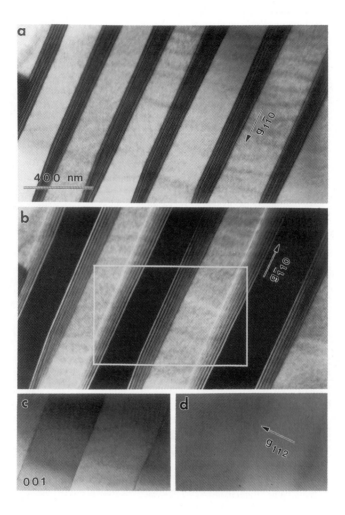

Figure 7.10 Twin boundaries observed in $YBa_2(Cu_{0.98}Zn_{0.02})_3O_7$. (a) and (b) are of the same region of the specimen, and (c) and (d) are from the region in (b) which is enclosed by a white border. (a), (b), and (d) were imaged in an orientation tilted about 15° away from the [001] zone (c). Note that the outer fringes of the inclined boundaries are symmetric in bright-field (a), and alternatively symmetric and asymmetric in dark-field (b). Also, when a **g** perpendicular to the twin boundary is excited, the boundary fringes are out of contrast (d).

7.3. Studies of the Displacement at the Twin Boundary

a slight misorientation, i.e., the different deviation S from the Bragg reflection of the adjacent crystals, together with a possible difference in extinction distance ξ_g, such as in ordered domains, or a twin boundary; and (3) $\alpha - \delta$ interface; a mixture of (1) and (2) [Gevers et al. 1965, Amelinckx and van Landuyt 1976]. These authors also calculated the fringe intensity profile from these three interfaces in a thick foil in terms of anomalous absorption and two-beam dynamics theory. They found that the outer fringe for an α-interface is symmetric in BF but asymmetric in DF, while the δ-interface is asymmetric in BF but symmetric in DF. For a $\alpha - \delta$ mixed interface, the outer fringe has an intermediate characteristic between that of pure α and pure δ fringes, and depends on the relative contributions from each of these types. According to the above criteria, the boundary fringes from $YBa_2Cu_3O_{6.6}$ are pure δ fringes, while the fringes observed in Figs.7.10(a) and (b) are not pure α or pure δ types, but have both α and δ characters. The δ character could arise from the twinning, whereas the α character arises from the displacement at the boundary.

The conventional $\mathbf{g} \cdot \mathbf{R} = 0$ invisibility criterion can be used to determine the nature of the observed displacement. Figs.7.10(c) and 7.10(d) are the same as the marked area of Fig.7.10(b). Fig.7.10(c) is imaged in an exact [001] zone axis, which was tilted about 15° from Figs.7.10(a) and 7.10(b). The fringes from the inclined twin boundary do not always show contrast. We tilted these boundaries ±35° from [001] through [$\bar{1}\bar{1}1$] to [$\bar{3}\bar{3}2$] zone axes. When $\mathbf{g}$ = 110, 112, and 113 in these zone axes (i.e., $\mathbf{g}$ perpendicular to the boundary), the fringes are out of contrast, as shown in Fig.7.10(d) for $\mathbf{g} = [112]$. This finding implies that the phase angle $\alpha = 2\pi \mathbf{g} \cdot \mathbf{R}$ is either 0 or $2n\pi$, where n is an integer. Then, unless the component of $\mathbf{R}$, which is perpendicular to the twin boundary, is $n/2 \cdot [110]$, this two-beam contrast study suggests that the displacement at the twin boundary is along the boundary. However, this does not completely rule out the possibility of a displacement perpendicular to the twin boundary. If such $\mathbf{R}$ exists, it has to be a multiple of d_{110}. To examine whether there is such volume expansion at the boundary, the average d_{110} spacings along the [110] direction were determined in two regions: an area containing a boundary, and the undistorted matrix, as shown in Fig.7.8(a). There was no difference between the measured d_{110} spacing in these two regions. All these observations suggest that the displacement is $[\bar{1}10]$ type, which is parallel to the twin boundary, and this is consistent with the HREM observation discussed above.

Recent works based on the use of the symmetry properties of the α and δ characters in diffraction contrast images of pure and Fe doped $YBa_2Cu_3O_{7-\delta}$ has confirmed that the twin boundary is an $\alpha - \delta$ interface (the α component is $\mathbf{R} = 1/2[110]$) [Cheng et al. 1991]. The observed displacement is also in good agreement with our calculations from diffuse scattering and electron diffraction observations (see the next section).

7.3.4 Analyses of the Streaks of Diffuse Scattering

The sharp streaks seen in the SAD pattern ((Fig.7.1(b)) reflect the important features of the twin boundary structure, as shown in a dark-field image produced by a streak (Fig.7.3). These streaks of low intensity are longest and most clearly visible for $YBa_2(Cu_{0.98}M_{0.02})_3O_{7-\delta}$ for $\delta \approx 0.0$ and M=Cu, Zn, Ni, where the streak lengths are typically 1/4 of $\mathbf{g}_{110}$. Much shorter streaks are observed for pure specimens with oxygen content of $\delta \approx 0.4$, and for fully oxidized specimens with M=Al, Fe. The length is then typically 1/10 of $\mathbf{g}_{110}$. Streaks observed in reciprocal space can be primarily attributed to a crystallographic shape effect (such as a thin layer) and/or a strain effect (such as lattice displacement) in real space. From a kinematical point of view, the main basis for the distinction between these two follows from the fact that the extension of the streaks due to the shape effect extend equally on all diffraction spots, including the origin. For a lattice with displacement vector $\mathbf{R}$, streaks in the diffraction pattern will be visible through reflection $\mathbf{g}$ for which $\mathbf{g} \cdot \mathbf{R} \neq n$ ($n = 0, 1,...$) and invisible for $\mathbf{g} \cdot \mathbf{R} = 0$. Originally, the streaks were interpreted as evidence for the existence of a thin non-orthorhombic layer. The shortening of the streaks were attributed to the broadening of the distorted boundary region, as confirmed by HREM.

Similar streaks were also observed in a diffractogram produced by Fourier transforming a digitized HREM micrograph of a single twin boundary (Fig.7.11). The diffractograms, obtained from an HREM im-

7.3. Studies of the Displacement at the Twin Boundary

age from a ~100 Å thin area (having less multiple scattering), show that the streaks at different reflections have a different intensity distribution. To minimize the multiple scattering effects in SAD observations it is desirable to reduce the number of reflections. By tilting the specimens slightly about the axis normal to the twin boundaries, we found that the streaks at the direct beam and the $(hh0)$ reflections disappear, suggesting that they are caused by multiple scattering, because the $[hh0]$ line of reciprocal space was observed also after tilting. From careful observations of these streaks for different incident-beam directions, it appears that the strongest streaks are through the (100) and (010) type reflections. The diffraction pattern is schematically shown in Fig.7.12(a).

As we discussed earlier, a twin boundary with a cation lattice shift acts like an anti-phase boundary for electrons. Anti-phase boundaries result in a splitting of reflections for which $\alpha = 2\pi \mathbf{g} \cdot \mathbf{R} \neq 2\pi n$ ($n = 0, 1,...$) when a small electron probe is used or the domain sizes are small [Zhu and Cowley 1982, van Dyck et al. 1984]. When the domain size increases, a streak normal to the anti-phase-like boundary is more pronounced than the splitting. Two reflections which are most strongly perturbed by the anti-phase-like boundary are those for which $\alpha = 2\pi \mathbf{g} \cdot \mathbf{R} = 2\pi(n + 1/2)$, e.g., the (100) and the (010) type reflections when $\mathbf{R} = 1/2[1\bar{1}0]$, which is the displacement for the cations lattice in Fig.7.8(a).

To support our arguments made from diffraction observations, we calculated the intensity of the diffuse scattering of twin boundaries using a periodic boundary condition. With a kinetic treatment, the intensity of diffuse scattering from the periodic twin boundaries of $YBa_2Cu_3O_7$ can be expressed by

$$I \approx \left\{ \frac{1}{2} + \sum_{n=1}^{D/2} \cos 2\pi(nh + n\phi' k) + \sum_{n=D/2+1}^{D-1} \cos 2\pi[(nh + (D\phi'/2 - n\phi' + \alpha')k] \right\}^2, \quad (7.1)$$

where ϕ' is the orthorhombicity, α' is the lattice displacement ($\alpha' = 0.5$ for $|\mathbf{R}| = d_{110}$), and D is the twin spacing. The calculated intensities of $(1\bar{1}0)$, (100) and (110) reflections are shown in Fig.7.12 (b) for $\mathbf{R} = 0$

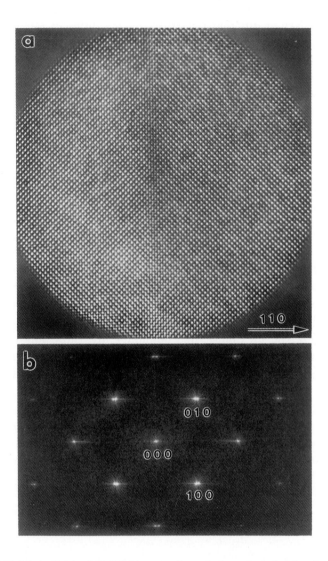

Figure 7.11 (a) A digitized HREM image of a twin boundary (with a lattice shift) from a thin area with less multiple scattering effects. (b) A diffractogram Fourier transformed from (a) showing sharp streaks. Note the difference in intensity distribution of the streaks at different reciprocal spots.

7.3. Studies of the Displacement at the Twin Boundary

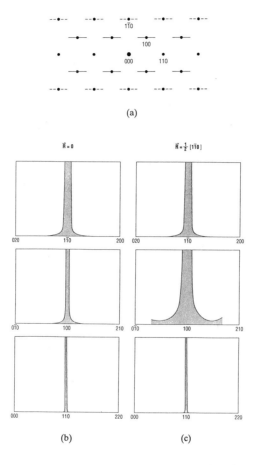

Figure 7.12 (a) Schematics of the observed streaks in the (001) diffraction patterns of $YBa_2(Cu_{0.98}M_{0.02})_3O_{7-\delta}$ for $\delta \approx 0.00$ and M=Ni, Zn, and Cu. The dashed streaks may be caused by multiple scattering. (b) and (c) are calculated intensity profiles along the streaks using a twin spacing of 270Å and periodic boundary conditions for two different displacement vectors $\mathbf{R}$. (b) $\mathbf{R}$=0; (c) $\mathbf{R} = 1/2(\bar{1}10)$. Note the difference in streak length (intensity profile) for the (100) reflection. The difference in the intensity profile width of $[110]$ and $[1\bar{1}0]$ reflection for $\mathbf{R} = 0$ is caused by twinning.

and Fig.7.12(c) for $\mathbf{R} = 1/2[1-10]$; note the difference in streak length for (100) reflection in Fig.7.12(b) and (c). The intensity profile for $\mathbf{R} = 1/2[1\bar{1}0]$ is consistent with the SAD observation of $YBa_2Cu_3O_7$ when the multiple scattering effect is eliminated. Similar observations were made by laser simulation using optical masks. For a pure twin-boundary configuration (only a $\sim 89.1°$ rotation about the [001] axis without lattice distortion at the boundary), no streaks are seen in the diffractogram, although streaks were observed for a twin with a lattice shift at the boundary. Thus, the diffuse scattering experiments support the interpretation of the HREM observations and the proposed models shown in Fig.7.8(a) and Fig.7.8(b).

The analyses of the diffuse scattering results above indicates that a major fraction of the intensity of the streaks is also consistent with a shift of the cation lattice parallel to the twin boundary. Then, considering the effect of multiple scattering on the intensities of the streaks, and weak streaks observed in oxygen-deficient samples (without a lattice shift), it appears that the streaks observed in SAD (Fig.7.1(b)) result from a combination of both effects, the lattice displacement at the twin boundary and the distorted boundary with a limited width. The increased length of the streak observed in M=Zn, Ni, and Cu appears to be caused by the abrupt lattice shift occurring within a ~ 10 Å thin boundary region.

7.3.5 Interfacial Energy of the Twin Boundaries

To interpret the lattice displacement at the twin boundary and the difference between fully oxygenated and oxygen-deficient $YBa_2Cu_3O_{7-\delta}$, we proposed two twin boundary models, illustrated in Fig.7.8(a) and Fig.7.8(b). The twin boundary centered at oxygen would appear to be energetically unfavorable because the cations and the O atoms in the CuO_2 and BaO_2 layers would only be separated by 2.72 Å for an ideal twin boundary rather than 3.82Å in the matrix region. Nevertheless, the alternative of a 90° bend in the O-Cu-O chains at the Cu atoms which accompanies a cation-centered twin boundary is also thought to be very energetically unfavorable, as was pointed out by de Fontaine et al. from considerations of effective pair interactions [de Fontaine et al. 1987]. In addition, strong experimental evidence against 90° bends

7.3. Studies of the Displacement at the Twin Boundary

is provided by the observed short-range order diffuse scattering from oxygen-deficient $YBa_2Cu_3O_{7-\delta}$, where narrow rods of diffuse scattering are observed, suggesting long unbroken O-Cu-O chains [Werder et al. 1988b, Zhu et al. 1990a]. This type of oxygen-centered boundary may require considerable local displacement of the atoms at the boundary to reduce the energy. Thus, the HREM images suggest that the shift of the corresponding features on one side of the twin boundary relative to the other side of the twin boundary may not be exactly as shown in Fig.7.8(a) (which gives a shift of d_{110} along the twin boundary); rather, the observed shift is $(0.80 \pm 0.15)d_{110}$ (while it is zero when the twin boundary is at the cations). The local structure disorder and compositional variation at the twin boundary regions were also confirmed by field-ion microscopy observations [van Bakel et al. 1990].

In contrast, for oxygen-deficient samples, the number of O-Cu-O chains is reduced, thus reducing the strain energy and also the net energy penalty for the 90° bends in the chains. Hence, forming an oxygen-centered twin boundary becomes unnecessary, consistent with our observation that the twin boundary occurs at the cations for lower oxygen content (Fig.7.8(b)). The energy associated with the formation of the twin boundary through the cations may be further reduced if the twin boundary region has a lower oxygen content than the interior of the twins. It was observed that, during electron irradiation, $YBa_2Cu_3O_{7-\delta}$ becomes tetragonal first at the twin boundaries, suggesting that the energy associated with the reduction of oxygen content is lowest at the twin boundaries. When Fe or Al is added, the presence of these elements causes the formation of oxygen "cross links" and disrupts the linear Cu-O chain structure [Cai and Mahanti 1989]. Hence, these elements are likely eliminate the chain-bending energy penalty and favor the formation of cation-centered twin boundaries. Thus, our observations suggest that these modifications reduce the energy associated with the formation of the twin at the cations below that required for forming the boundary at the O-atoms in the CuO plane. The twin boundary forms at cations below a certain oxygen content or above a certain Al or Fe content.

The transformation from tetragonal to orthorhombic is a disorder to order transition which occurs when oxygen is localized to (0 1/2 0) sites, leaving the (1/2 0 0) sites vacant. During the transformation, if no constraint is present, such as in single crystals, the lattice is free to relax and no twinning results [Thomsen et al. 1988]. With constraints, the shear strains associated with the transformation can be relieved by twinning at the cost of twin boundary energy, which mainly comes from the Coulomb repulsion. The spacing of the twin is determined by minimizing the total energy of the system associated with the elastic strain energy and twin boundary interfacial energy [Khachaturyan 1983, Roitburd 1969]. The relationship between twin spacing D, the twin boundary energy γ, the elastic modulus M, the grain size g, and the orthorhombicity, $\phi = 2(b-a)/(a+b)$, is given by:

$$D \approx (g\gamma/CM\phi^2)^{1/2} \qquad (7.2)$$

where C is a constant ~ 1. The validity of this relationship was shown for bulk [Chandrasekar unpublished, Chumbley et al. 1989, Roy and Mitchell 1991], and also for thin films, in which case the critical length is the thickness of the film rather than the grain size [Streiffer et al. 1991]. Using $M = 2 \times 10^{12}$ dynes/cm^2 as reported [Baetzold 1988, Baumgart et al. 1989]. and measuring the twin boundary spacing, we find $\gamma \approx 80$ ergs/cm^2 for YBa$_2$Cu$_3$O$_{7-\delta}$. It is interesting to note that substitution with Al or Fe at a high oxygen content or reducing the oxygen content in pure YBa$_2$Cu$_3$O$_7$ both substantially reduce the average twin-boundary spacing D [Zhu et al. 1989b, Xu et al. 1989, Chandrasekar unpublished, Wördenweber et al. 1989], and the twin boundary energy γ was found to decrease by an order of magnitude [Chandrasekar unpublished]. (Interestingly, these values of γ are in the range of the twin boundary energies which are found in metals and alloys.) This indicates that the Cu-centered boundaries are lower in energy than the O-centered ones, if the boundary regions can be broadened by the deficiency or the excess of oxygen, by low oxygen contents, or by substitution with Fe, Al, etc.

7.4 Crystallographic Analysis of the Twin Boundary

7.4.1 A Σ64 Coincidence Boundary

An elegant way to elucidate a boundary geometry is by using a coincidence site lattice (CSL) model. CSL can be produced by rotating two crystals relative to one another about a common axis, using a lattice site as the origin. The reciprocal density of CSL (or the ratio of the unit-cell volume of CSL to that of the crystal) is usually denoted by Σ. Considerable evidence shows that low-energy interfaces are associated with a high density, short-period coincidence lattice, i.e., a boundary with low Σ value [Sun and Balluffi 1982, d'Anterroches and Bourret 1984, Babcock and Balluffi 1987, Chen and King 1987, 1988]. Furthermore, based on CSL theories, boundary dislocations have been successfully used to explain the boundary structure [Balluffi 1979, Carter 1988, Balluffi et al. 1982]. To shed light on this geometry, crystallographic analysis of the fully oxygenated twin boundary ($YBa_2Cu_3O_7$) was carried out. Fig.7.13(a) shows the superposition of the neighboring twin crystals, produced by a rotation of 89.1° of the near-square lattice ($a = 3.82$Å and $b = 3.88$Å) – one to another – along the [001] axis. The perfect-match positions, or coincidence sites, form a systematic row, and can be considered as the twin boundary. The lattice mismatch increases with increasing the distance from the coincidence sites and changes periodically. According to the definition, we have

$$\Sigma = [\mathbf{C}_1 \cdot (\mathbf{C}_2 \times \mathbf{C}_3)]/[\mathbf{a} \cdot (\mathbf{b} \times \mathbf{c})] = 64, \quad (7.3)$$

where $\mathbf{a}$, $\mathbf{b}$, and $\mathbf{c}$ are lattice vectors of a $YBa_2Cu_3O_7$ crystal, and $\mathbf{C}_1$, $\mathbf{C}_2$, and $\mathbf{C}_3$ are three primitive vectors of the CSL unit cell, as also indicated in Fig.7.13(a). Thus, the twin boundary can be described by a CSL system of Σ64, 89.1°/[001].

The relative lattice shift due to the twinning shear, τ, is schematically shown in Fig.7.13(b), where Θ_{100} is the rotation angle between $\mathbf{a}_1$ and $\mathbf{b}_2$ or $\mathbf{a}_2$ and $\mathbf{b}_1$. By examining the geometry of the twinning, the minimum lattice-twinning-shear, τ, can be given by

$$\tau = \mathbf{a} - \mathbf{b} \quad (7.4)$$

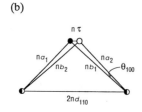

Figure 7.13 (a) A superposition of two neighboring twin of $YBa_2Cu_3O_{7-\delta}$ ($a = 3.82$Å, $b = 3.88$Å), or a coincidence site lattice of the twin boundary, produced by rotating $89.1°$ of the lattice one to another along the $[001]$ axis. The CSL unit cell ($32 \cdot 2d_{110} \times 2d_{110}$ in size and 64 lattice sites) is also depicted. b): Atom rearrangement under the twinning shear.

7.4. Crystallographic Analysis of the Twin Boundary

The magnitude of the twinning shear, $|\tau|$, is then

$$\begin{aligned} |\tau| &= |\mathbf{a} - \mathbf{b}| \\ &= (a^2 + b^2 - 2ab\cos\Theta_{100})^{1/2} \\ &= \left\{a^2 + b^2 - 2ab\cos[\pi/2 - 2\tan^{-1}(a/b)]\right\}^{1/2}. \end{aligned} \quad (7.5)$$

(Note $|\mathbf{a}-\mathbf{b}| \neq |a-b|$). For $a = 3.82$Å and $b = 3.88$Å, $|\mathbf{a}-\mathbf{b}| = 0.085$Å, which is the shear displacement at the distance of d_{110} from the twin boundary. At a distance of $64 \cdot d_{110}$ from the boundary, the lattices reach a 2π phase shift (0.085Å $\times 64 = 5.44$Å $= 2d_{110}$), which gives a period of 64 (110) lattice planes in the direction of the boundary plane normal.

7.4.2 Twin-Boundary Steps and Twinning Dislocations

Careful inspections of the twin boundaries reveal that some have steps (ledges), where the center of the twin boundary leaves one (110) lattice plane and moves to a neighboring (110) lattice plane (Fig.7.14). This observation of a twinning step is believed to be the first in $YBa_2Cu_3O_7$, although a twinning-step model was proposed to explain the broadening of the twin boundary by an oxygen diffusion mechanism [van Tendeloo et al. 1990]. The twinning step runs perpendicular to the micrograph with a d_{110} step parallel to the twin boundary plane normal. The step is easily visible by looking at the twin boundary under grazing incidence along the (110) lattice planes. As the step propagates, atoms near the step rearrange themselves in the manner of a dislocation motion. The dislocation associated with such a step is called a twinning dislocation, since it exists only in a twin boundary. A twinning dislocation can glide in the twin boundary plane, and, as it does so, the amount of one twin domain grows at the expense of the other. The Burgers vector of a twinning dislocation is not a lattice vector of the twinned lattices; its magnitude should be equal to the distance moved by each lattice plane due to the twinning shear and be proportional to the height of the twinning step.

A twinning dislocation can be defined rigorously by a line integral along the so-called Frank circuit, which is equivalent to the Burgers

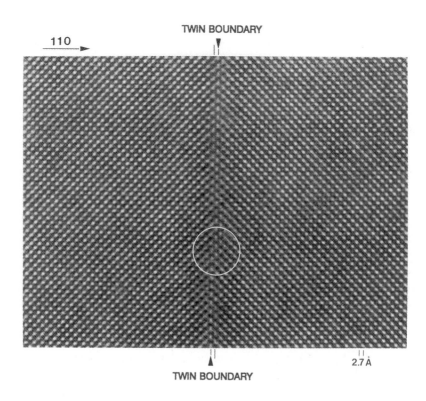

Figure 7.14 High-resolution image of a (110) twin boundary associated with a d_{110} twinning step, indicated by a circle, in YBa$_2$Cu$_3$O$_7$ view along the [001] axis. Note that the center of the twin boundary (marked by an arrow) shifts to a neighboring (110) plane at the step. The step is most visible by looking the twin boundary under grazing incidence along the (110) lattices.

7.4. Crystallographic Analysis of the Twin Boundary

circuit for a lattice line-defect, surrounding the interfacial dislocation. Fig.7.15(a) is a scheme of a (110) twin boundary associated with a twinning step (step height 2.7 Å) in $YBa_2Cu_3O_7$. The rectangular lattice represents the Cu-O sublattice in the CuO plane and the open circles represent chain-oxygen atoms; the copper atoms are located at the corners. The differences between a and b lattices and the rotation angle are exaggerated to emphasize the characteristic features of the boundary. The d_{110} lattice shift can be seen from the Cu-O sublattice across the boundary.

To characterize the Burgers vector of the twinning dislocation, the circuit must begin and end on the interface, as suggested by Frank and Read [Frank 1951, Read Jr. 1953]. First, consider a reference twin boundary without steps. The Frank circuit is closed whether or not the boundary has a d_{110} cation lattice shift. Then, we construct the same circuit in the real system containing a step (shown as a circuit of thick dash lines in Fig.7.15(a)). The local Burgers vector of the twinning dislocation, **B**, defined by an SF/RH convention, can be given by the closure failure of the circuit. Let routes "1" and "2" be positive and "3" and "4" be negative; then we have

$$\begin{aligned} \mathbf{B} &= \oint_l \frac{\partial \mathbf{u}}{\partial \mathbf{l}} d\mathbf{l} \\ &= [5\mathbf{b}(1) - 5\mathbf{a}(2)] + [6\mathbf{a}(1) - 6\mathbf{b}(2)] \\ &= \mathbf{a}(1) - \mathbf{b}(2), \end{aligned} \qquad (7.6)$$

where **u** is the elastic displacement around the dislocation, **l** is taken in a right-handed sense relative to the dislocation line direction, ξ, and (1) and (2) denote the twin 1 and 2, respectively. As shown in Fig.7.15(b), the directions of both $\mathbf{b}(1) - \mathbf{a}(2)$ and $\mathbf{a}(1) - \mathbf{b}(2)$ vectors are along the twin boundary with the same magnitude, but with an opposite sense. Therefore, the direction of resultant vector, **B**, also lies along the boundary. Reversing the sense ξ causes **B** to reverse its direction. Thus, the geometry of the observed twinning step at a (110) twin boundary can be associated with a perfect edge dislocation, with a line direction of [001], and a Burger vector of $|\mathbf{a} - \mathbf{b}|[\bar{1}10]$.

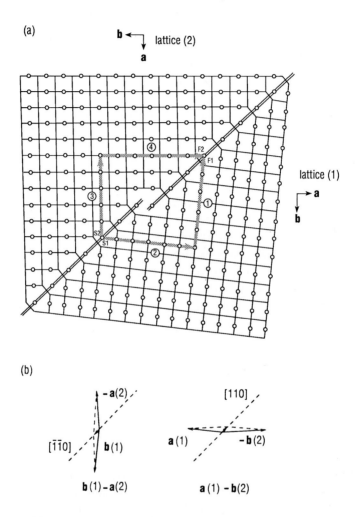

Figure 7.15 (a): Schematic drawing of a twinning step in $YBa_2Cu_3O_7$. The rectangular lattice represents Cu-O sublattice in the basal plane. The open circles represent chain-oxygen atoms, and the copper atoms are located at the corners. The thick shade-lines are the Frank circuit. (b): Determination of the resultant direction from the Frank circuit. Note the differences between the a and b lattices and the rotation angle are exaggerated to emphasize the characteristic feature of the boundary.

7.4.3 A Displacement-Shift-Complete-Lattice (DSCL) Treatment

The magnitude of the Burgers vector of the observed twinning dislocation is less than 1.6% of the primitive lattice vector of a $YBa_2Cu_3O_{7-\delta}$ unit cell. Such a small interfacial dislocation is usually called a secondary dislocation which can be characterized in a more graceful but equivalent circuit construction using a DSCL [Hirth and Balluffi 1973]. The DSCL defines all the vector displacements of the two crystals relative to one another under conditions where the overall pattern of lattice points produced by the two interpenetrating lattices in the coincidence orientation remains unchanged. Therefore, this DSCL defines all the possible Burger vectors of interfacial dislocations.

Figure 7.16(a) illustrates the DSCL of a CSL unit cell ($32 \cdot 2d_{110} \times 2d_{110}$, 64 lattice sites) of the twin boundary. The two primitive DSCL-vectors can be geometrically derived as $|\mathbf{a}-\mathbf{b}|[\bar{1}10]$ and $[110]$, as shown in Fig.7.16(a) with different scales. With the DSCL, the twinning dislocation can be demonstrated by extra DSCL planes on one side of the twin boundary, which is lacking in Fig.7.14(a). For a step at the twin boundary in $YBa_2Cu_3O_{7-\delta}$, the Burgers vector of a twinning dislocation can be expressed by

$$\mathbf{B} = h(\mathbf{a} - \mathbf{b})/d_{110} \qquad (7.7)$$

where h is the step height measured by the number of (110) lattice planes. The equation suggests that the magnitude of the Burgers vector, or the number of the extra half-plane of DSCL is proportional to the height of the step, regardless the lattice shift at the boundary. For a twinning step with a step height of $h = d_{110}$, as seen in Fig.7.14, the Burgers vector of the twinning dislocation is one primitive DSC vector, i.e., $|\mathbf{a} - \mathbf{b}|[\bar{1}10]$ [Fig.7.16(b)], which is consistent with the result from a Frank construction using a crystal lattice [see Eqn.(7.6)]. The step vector is then [010], as shown in Fig.7.16(a). A Frank circuit using DSCL for a twinning dislocation with a Burgers vector of $2|\mathbf{a}-\mathbf{b}|[\bar{1}10]$, associated with a $2d_{110}$ twinning step (step vector $\mathbf{s} = [110]$, step height $h = 2d_{110}$) is shown in Figs.7.16(c) and (a). However, one would not expect to see such an infinitesimal dislocation by transmission electron microscopy.

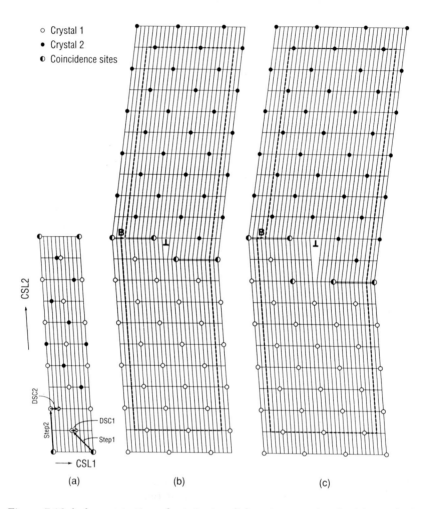

Figure 7.16 A demonstration of a twinning dislocation associated with a twinning step described by a DSC-lattice. (a): A CSL unit-cell of the twin boundary in $YBa_2Cu_3O_7$. Note that the difference of the scale between CSL1 and CSL2 is understated and the real lattice sites and CSL unit cell is shown in Fig.7.13. (b): A twinning dislocation ($\mathbf{B} = DSC1 = |\mathbf{a} - \mathbf{b}|[\bar{1}10]$) associated with a d_{110} step ($\mathbf{s} = [010]$, $h = d_{110}$). (c) A twinning dislocation ($\mathbf{B} = DSC2 = 2|\mathbf{a} - \mathbf{b}|[\bar{1}10]$) associated with a $2d_{110}$ step ($\mathbf{s} = [110]$, $h = 2d_{110}$).

7.4.4 Structure of Mixed Twin Boundaries

Our observations suggest that the twin boundaries form at the (110) planes through the oxygen atoms in the CuO planes for fully oxygenated samples, and at the cations for oxygen-deficient ones. The atomic configuration of the transition from one to the other, seen in *in situ* experiments (Fig.7.6), may be understood by mixing the two types of twin boundaries. Fig.7.17 depicts the mixture (combining Fig.7.8(a) and 7.8(b)) of the twin boundary centered at a chain oxygen plane (lower part) and at a cation plane (upper part). Here, we focus our attention only on the lattice shift and the position of the twin boundary. The change of the boundary center produces a $1/4 \cdot 2d_{110}$ twinning step ($h = 1.35$ Å). The step also can be related to a perfect edge dislocation at the (110) twin boundary, as shown in the middle of Fig.7.17. Constructing a Frank circuit results in a closure failure along the boundary. The Burgers vector of such a dislocation is determined to be

$$\begin{aligned} \mathbf{B} &= 1/2 < 110 > +6\mathbf{b}(1) + 5.5\mathbf{a}(1) - 6\mathbf{a}(2) - 5.5\mathbf{b}(2) \\ &= 1/2 < 110 > +\delta, \end{aligned} \quad (7.8)$$

where $\delta = 1/2|\mathbf{a} - \mathbf{b}| < 110 >$. In the real case, because of the lattice distortion at the twin boundary and the limit of the spatial resolution of the microscope, a clear 1.35 Å step may not be visible. However, by counting the number of the (110) planes on both sides of the twin boundary between the regions with and without a lattice shift (seen in Fig.7.18), an enlarged micrograph of Fig.7.6, we did find an extra (110) lattice plane in the lower part of the twin matrix, as indicated by the numbers shown in Fig.7.18. This finding strongly supports our dislocation model for the twin boundary structure and suggests that the dislocation is essential for the transition between the two types of twin boundaries in $YBa_2Cu_3O_{7-\delta}$. A dislocation, ($\mathbf{B} = 1/2 < 110 > +1/2|\mathbf{a} - \mathbf{b}| < 110 >$), passing through a coherent twin boundary, would produce a $\sim d_{110}$ lattice shift at the boundary.

The early stage of the evolution of the twin boundary in $YBa_2Cu_3O_7$ is found to be associated with switching the twin boundary centered from oxygen atoms to cation atoms, as suggested by HREM *in-situ* experiments. The kinetics of such a transition may be explained by the motion of twin boundary dislocations. The observed transformation-twinning-dislocations and the crystallographic analyses of the twin

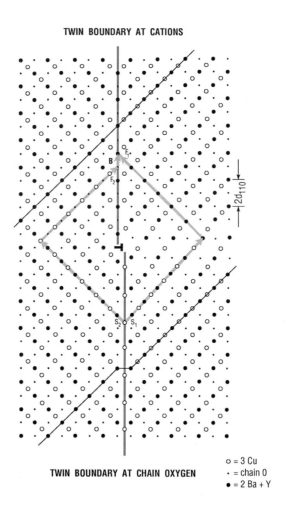

Figure 7.17 Schematic representation of a mixed twin boundary: a twin boundary through cation atoms with a cation sublattice shift (upper part) and through oxygen atoms without cation sublattice shift. Note the $1/4 \cdot 2d_{110}$ step and a perfect dislocation at the boundary.

7.4. Crystallographic Analysis of the Twin Boundary

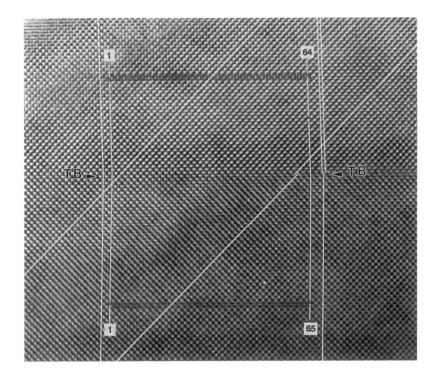

Figure 7.18 An enlarged micrograph of Fig.7.6 shows the mixed type of a twin boundary as proposed in Fig.7.8. The reference lines indicate that there is no lattice shift on left side of the boundary, but a lattice shift on the right. The difference in the number of the (110) planes across the boundary suggests the existence of an edge dislocation at the twin boundary.

boundary using CSL and DSCL concepts imply that dislocations play an important role in the growth of the twin. (The spiral growth of a $YBa_2Cu_3O_{7-\delta}$ thin film was shown to be related to the presence of screw dislocations by Scanning Tunneling Microscopy studies [Hawley et al. 1991, Jin et al. 1991]. Under irradiation, the strain energy can be reduced by the loss and/or disordering of the chain oxygen. The dislocations, associated with twinning steps moving through the (110) twin boundary plane, would eliminate the cation lattice-shift. This would create a cation-centered twin boundary, as shown in Fig.7.17, resulting in a lower boundary energy. For a transition from a cation-centered twin boundary to an oxygen-centered twin boundary during oxidation, a similar mechanism may apply. However, the origin of the formation energy of the dislocation will be different. Since such twin boundary dislocations have been determined to be edge dislocations with a Burgers vector of $\mathbf{B} = 1/2[\bar{1}10] + 1/2|\mathbf{a} - \mathbf{b}|[\bar{1}10]$, the (110) twin boundary plane is the dislocation glide plane, which contains both the dislocation line and its Burgers vector. Therefore, the dislocation is "glissile", and its motion is conservative. The motion of a glissile dislocation is a rapid process and can take place without any atomic diffusion. The formation of such dislocations may require a certain amount of energy. Once formed, however, the dislocation can move very fast even under the action of small internal shear stress.

7.5 Twin Tip

Twinning on the (110) and the equivalent ($\bar{1}10$) planes is frequently encountered within one crystal grain in $YBa_2Cu_3O_{7-\delta}$. In the region where they meet, there are four orientations of the twins (denoted as I, II, III, and IV in Fig.7.19), two associated with each of the orthorgonal twin boundaries. (Here, for simplicity, we consider that the twin boundaries are centered at the Cu-O plane, which was observed in oxygen deficient samples [Zhu et al. 1991a]. Presumably, the interface separating the orthorgonal twins takes a favorable orientation, either (110) or ($\bar{1}10$), yielding a common invariant twinning plane to both sets of the orthorgonal twins [Zhu et al. 1993g].

7.5. Twin Tip

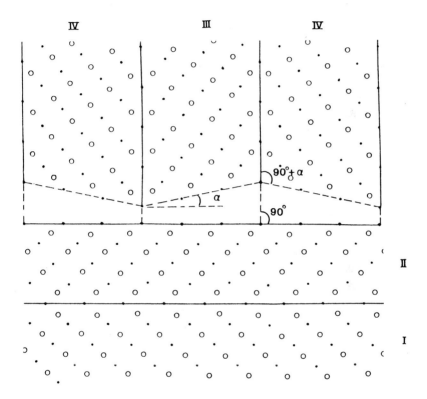

Figure 7.19 The four possible crystallographic orientation of twins in orthorhombic YBa$_2$Cu$_3$O$_{7-\delta}$ uninfluenced by constrain. The open circles are oxygen atoms and solid dots are copper atoms, a/b is not to scale. Here, we use $a/b = 0.8$ insteadof $3.82/3.88 = 0.985$, and thus, $\alpha = 12^o$ rather than 0.89^o.

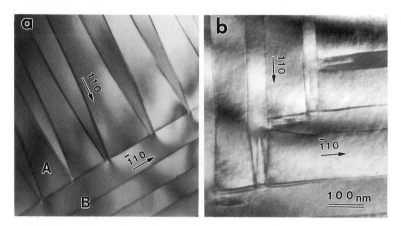

Figure 7.20 (a) Bright-field image of two orthorgonal sets of twins near the [001] projection. The upper twin is denoted as 110 twin and the lower twin as $\bar{1}10$ twin. (b) Splitting of the tapered 110 twin. Note the split twin tips have a similar wedge angle θ compared with unsplit twin tips.

7.5.1 Four Twinning Variants

Figure 7.20 shows typical bright-field images of orthorgonally orientated sets of twins. In fig.7.20(a), we see that two orthorgonal sets of twins meet in a planar region parallel to the (110) plane, and every second twin belonging to the set of twins with boundary normal perpendicular to this (110) plane have a wedge-shape. We denote this set of twins the 110 twins (marked as A), to distinguish them from the set of twins with a twin boundary parallel to this (110) plane. the $\bar{1}10$ twins (marked as B). The plane normal of the interface is [110] and yet the interface may be locally curved; a close look reveals that it is always faceted along the (110) plane. Fig.7.20(b) shows two orthogonal sets of twins that meet along the [010] direction; nevertheless, the termination of the impinging twins not only taper off but also split when they confront the boundaries of the other sets of twins. The spread of the lamella of any one twin is restrained by its impingement on another twin lamella.

Figure 7.21 shows the atomic image of the interface of the orthorgonal twins and the associated tapered, impinging twin. There are four twin variants and five interfaces. Four of the boundaries, the $1.87°$

7.5. Twin Tip

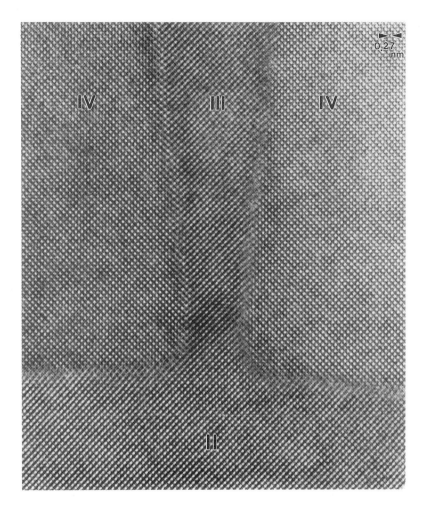

Figure 7.21 High-resolution image of a twin tip showing the interface where the orthogonal twins meet. There are five boundaries: four twin boundaries, and one structureless boundary. Note, the lattice in twin variant III does not align with that in II.

twinning rotation can be clearly seen. The core of the twin boundary is severely distorted. The curvature of the boundary is accommodated by twinning steps. The fifth boundary, which has no apparent structure, is a boundary between twin III and twin II, because the wedge-shaped twin III does not taper down to zero thickness, but is approximately 5nm at its narrowest point. According to Fig.7.1, if there is no constrain, the orientation of the twin III and twin II should be equivalent. Nevertheless, when viewed along a prominent row of dots, the lattice of twin II does not align with that of twin II; there is a small rotation.

A small lattice rotation can be enhanced by means of an optical analogue of moiré pattern (Moiré patterns are formed when two crystals, which have differences in lattice parameter or orientation, overlap) by placing the high-resolution image (A) of Fig.7.21 over a perfect lattice-grating (B), as shown in Fig.7.22. The rotation angle measured from the moiré fringes (the thick, dark line in Fig.7.22) between twin II and III is about 9°. By calculating the "moiré magnification", which is a function of the difference in spacing and the relative rotation between lattice A and B [Hirsch et al. 1965], we found that the real rotation angle between twin II and twin III is about 2°.

Electron diffraction is very useful in determining a crystal structure, especially combined with HREM. We analyzed the diffraction pattern from the region where two sets of twins meet. Fig.7.23(a) shows a diffraction pattern of the (001) projection from one set of twins, and Figs.7.23(b)-7.23(d) show enlarged portions containing the (100) reciprocal lattice row (also (010) reciprocal lattice row because these are nearly parallel). Figs.7.23(a)-7.23(d) were obtained by illuminating a $0.3\mu m$ region of the specimen in different areas of Fig.7.20(a). Fig.7.23(b) is from the $\bar{1}10$ twins (lower set of twins in Fig.7.20(a)), Fig.7.23(c) is from the region where the $\bar{1}10$ twins and the 100 twins meet, and Fig.7.23(d) from the 110 twins. In Fig.7.23(e) we have illuminated a much larger area of about $3\mu m$, thus including the whole image of Fig.7.20(a). We note the three Bragg spots forming a triangle in Fig.7.23(c), and the streaks through two of the spots in Fig.7.23(e).

One set of orthogonal twins (for example, the lower set of twins in Fig.7.20(a)) is a well-defined obstacle for the other set of twins, and every second twin becomes tapered (upper set of twins in Fig.7.20(a)). The four different twin orientations within one crystal grain, shown

7.5. Twin Tip

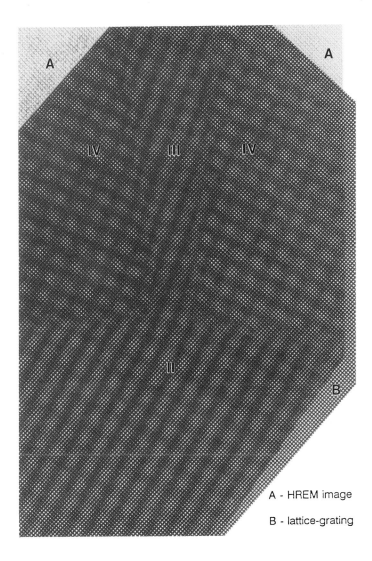

Figure 7.22 A small lattice rotation between III and II can be revealed clearly in terms of moir'e fringe by overlapping the HREM image (denoted as A) on a lattice-grating (denoted as B). The rotation of the moiré fringes between III and II is about 9°, which reflects a real lattice rotation of $\sim 2^o$.

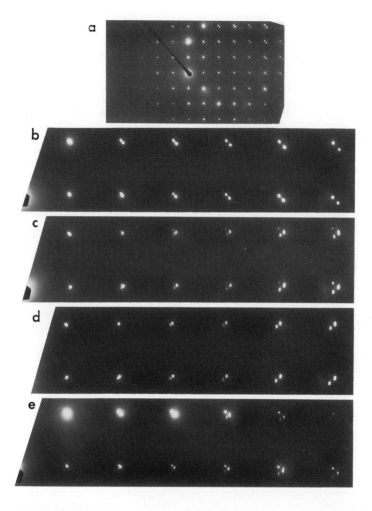

Figure 7.23 Diffraction patterns near the (001) projection of Fig.7.20(a). (a) overview of the low-set of twins, (b)-(e): changes in the splitting of the reflections when a 0.3μm diameter electron beam is moved from bottom to top. (b) from the lower set of twins, (c) from the region where the two sets meet, (d) from the upper set of twins where the twin boundaries start to become parallel, and (e) diffraction contribution from an area larger than the whole image of Fig.7.20(a).

7.5. Twin Tip

in Fig.7.19, would result in a diffraction pattern, as in Fig.7.24(a), if they were not disturbed by obstacles and the 110 and $\bar{1}10$ twins were illuminated simultaneously. In particular, this results in the [h00] and [0k0] type of reflections being split into four forming a square, Fig.7.24(b). However, where the 110 and $\bar{1}10$ twins meet we observed a triangle, Fig.7.23(c) and 7.24(c), rather than a square. By watching the diffraction pattern while moving the 0.3μm-diameter electron probe over the region where the orthogonal twins meet, we observed a gradual movement of the diffraction spots, consistent with the streaks seen in Fig.7.23(e), using the much larger electron probe. These streaks are also illustrated in Fig.7.24(d).

The observed (110) and ($\bar{1}$10) invariant planes for the four twin variants is shown in Fig.7.25(a) and (b). where ' denotes the local orientation of the impinging twins near the region where the orthogonal sets of twins meet. These observations suggest that the twins that were not tapered (twin IV' in Fig.7.25(a) and (b)) belonging to the 110 set have the same crystallographic orientation as every second twin (twinI in Fig.7.25(a) and (b)) belonging to the $\bar{1}10$ set. This configuration results in an angular mismatch of zero for a twin boundary between twin IV' and twin II in Fig.7.25(a) and (b) marked as TB, rather than α, $\alpha = 90^o - 2\arctan(a/b)$ as shown in Fig.7.19 and Fig7.25(c). On the other hand, the angular mismatch increased to 2α for the other twins (III'-II) to retain the twinnning rotation of $180^o - 2\alpha$ between twin IV' and twin III'. This scheme is consistent with the HREM observations. The change of the rotation angle appears to be a gradual process, because the streaks in Fig.7.23(e) and Fig.7.24(d) show that the tapered twins slowly rotate from 2α to α as we move away from the interface of the 110 and $\bar{1}10$ set of twins to the region where the twin boundaries between the (110) set become parallel.

7.5.2 The Shape of a Tapered Twin

In the region where two orthogonal sets of twins meet, every second twin lamella of the impinging twins always tapers off. We first discuss the geometry of these tapered, or wedge-shaped twins. As shown in Fig.7.21, HREM reveals that the curved twin boundary of a tapered twin is associated with twinning steps; where the existence of twinning

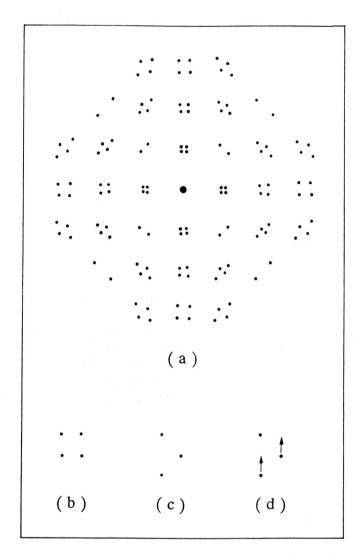

Figure 7.24 (a) Resulting diffraction pattern expected from the four orientations shown in Fig.7.19. Note, a/b is not to scale. (b) The splitting of the $(h00)$ and $(0k0)$ Bragg spots in (a). (c) Splitting where the twins meet, see also Fig.7.23(c). (d) Movement of the spots when moving upwards in Fig.7.20(a), see also Fig.7.23(e).

7.5. Twin Tip

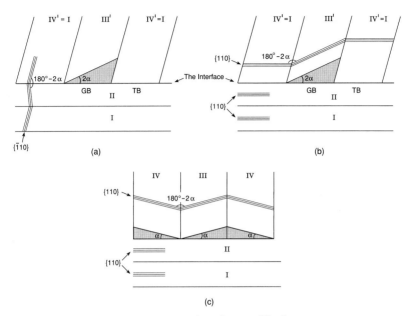

Figure 7.25 Relative orientation of the $\{110\}$ and $\{\bar{1}10\}$ twinning planes for four twin variants, if IV' and II become twinned by increasing the angular mismatch from α to 2α between III' and II. Note IV' and I are equivalent. The triple lines represent $\{\bar{1}10\}$ planes (a) and $\{110\}$ planes (b), respectively. (c) Angular mismatch if the two sets of twins meet according to the orientation in Fig.7.19; niether IV-II not III-II can have a twinning relationship.

steps implies the presence of twinning dislocations with discontinuities in shear stress. Although the shape of lenticular twins induced by mechanical deformation in simple metals has been studied for decades [Cahn 1953, Kosevich and Boiko 1971], only recently have interfacial dislocations and their motion been observed in twin structures caused by reduction of the symmetry of the crystal lattice in a complex material such as $YBa_2Cu_3O_{7-\delta}$ [Zhu and Suenaga 1992], The observed twinning dislocations and twinning steps was illustrated in Fig.7.15, based on the *in situ* HREM analysis. Using a Frank circuit, we obtained a closure failure, which yields a Burgers vector, $\mathbf{B}$ ($\mathbf{B} = |b - a|[110]$).

The associated step height, h, was found to be $h = \mathbf{n} \cdot \mathbf{s}$, where $\mathbf{n}$ is the twin boundary plane normal and $\mathbf{s}$ is the step vector (see section

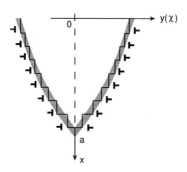

Figure 7.26 Sketch of lenticular twin consisting twinning steps and twinning dislocations. The shape of the twin is subjected to similar obstacles at two ends. a is the length of the wedge and D is the twin spacing.

7.4.3). Twinning dislocations share many of the properties of lattice dislocations; in particular, they can glide in the twinning plane under the action of a shear stress. They may be spontaneously generated in pairs under the influence of stress concentrations. Once the dislocations are there, their motion will cause the shape of a twin to change (enlarge or shrink by one lattice layer) under a much smaller stress than that would be needed in their absence. When the twin lamella has spread across the crystal, or has stabilized, the dislocations then become immobilized.

The shape of a lenticular twin can be quantitatively described as a simple geometrical corollary by using a dislocation model based on the mechanical equilibrium distribution of twinning dislocations in a double pile-up [Frank and van der Merwe 1949, Friedel 1964, Hirth and Lothe 1968]. The equilibrium edge-dislocation density distribution $\delta(x)$, along the twin length of $2a$ (Fig.7.26) is:

$$\delta(x) = \frac{2\sigma(1-\nu)}{\mu b} \frac{x}{\sqrt{a^2 - x^2}} \qquad (7.9)$$

where μ and ν are the shear modulus and Poisson's ratio, respectively. b is the magnitude of the dislocation Burgers vector, and σ is the total stress applied on the dislocations. Since one dislocation is usually associated with one twinning step (Fig.7.26), which is equal to the in-

7.5. Twin Tip

terplanar distance of the twinning plane d_{110}, the shape of a twin can be expressed through the dislocation density $\delta(x)$ by:

$$\frac{dy}{dx} = d_{110}\delta(x). \quad (7.10)$$

Integrating Eqn.(7.10) with $\delta(x)$, we have:

$$y(x) = d_{110}\int_{-a}^{x} \frac{2\sigma(1-\nu)}{\mu b}\frac{x}{\sqrt{a^2-x^2}}dx = \frac{D}{2a}\sqrt{a^2-x^2} \quad (7.11)$$

where D is the twin spacing, and $y(x)$ is the twin thickness at the coordinate x describing the shape of the lenticular twin (Fig.7.26). In interpreting the shape described by Eqn.(7.11), we note that the results for the very tip of the twin do not necessarily give the real geometry. The shape at the tip strongly depends on the obstacle which hinders its expansion.

7.5.3 Interfaces of the Orthogonally Oriented Twins

As we have demonstrated, in regions where the 110 twins and $\bar{1}10$ twins meet, the 110 twins takes a characteristic form because of the additional constraint due to the impediment and the local orientational difference. The crystal lattice of the 110 impinging twins rotates around the c-axis where they are blocked by the $\bar{1}10$ twins, and every second 110 twins tapers down nearly to zero thickness, while the other forms a coherent twin boundary with a twin variant belonging to the $\bar{1}10$ set (Fig.7.27(a)). The (110) planes of twin III' have a 2α angular mismatch with these planes in twin II, while the mismatch of the (110) plane is zero between twin IV' and twin II (Fig.7.25(b)). Twin IV' has the same crystallographic orientation as twin I; therefore, the boundary between IV' and II is a twin boundary, the same as that between II and I (Fig.7.25(b)). In other words, the interface of the 110 and $\bar{1}10$ twins comprises twin boundaries and grain boundaries. The former has $90°\pm0.98°$ twinning rotation, while the latter has about $2°$ tilting, both along the [001] axis. Becausethe grain boundary has a higher interfacial energy than the twin boundary, the only way of avoiding a high energy surface is for the grain boundaries to shrink and the twin boundaries

200   Chapter 7. The Twin Boundary Structure in $YBa_2Cu_3O_{7-\delta}$

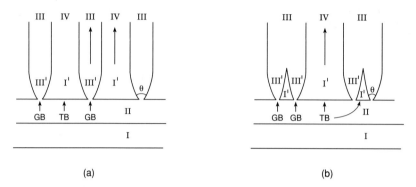

Figure 7.27 (a) The observed interface and its vicinity where two orthogonal sets of twins meet. IV' has the orientation defined by Figs.7.25(a) and 7.25(b). The gradual change of the orientation of teh impinging twins is indicated by arrows. The interface consists of twin boundaries and grain boundaries, (b) Splitting of the tapered twins. Note, the additional twin variants (I') due to the splitting have a twinning relationship with III' and II.

to expand; this results in the tapering off of one set of the impinging twins, terminating at the grain boundary.

The misorientation at a grain boundary is known to be accommodated by arrays of grain boundary dislocations and can be described using the coincidence site lattice model [Bollmann 1970]. However, in the present case, the length of the small-angle tilt boundary is so short (< 5nm) that adding even a single dislocation overcompensates for the $\sim 2^o$ misorientation by a factor of three (the calculated dislocation Burgers vector is 0.16 nm, about one third of the shortest available lattice vector perpendicular to the boundary). Such a small misorientation of a grain boundary with a finite length can be well described as wedge disclination dipole of small rotation angle, which is geometrically equivalent to a single edge dislocation [King and Zhu 1993].

Finally, it is interesting to note that the splitting of the twin tip (Fig.7.20(b) and Fig.7.27(b)), which can be attributed to the accommodation of strain at the tip. Similar observation was made during electron beam heating/irradiation [Iijima and Ichihashi 1990]. After carefully observing dozens of twin tips, we found that the split and unsplit twin tips have almost the same wedge slope, or wedge angle θ (see Fig.7.20(a) and (b)), suggesting that a larger wedge angle is energetically unfavorable, and that there is a threshold for a possible smooth change of slope of the tapered twin boundary. Based on a calculation of stress and strain near a lenticular twin, Lifshitz [Lifshitz 1948, Cahn 1954] showed that there is a limiting wedge angle θ; if θ is exceeded, the stress at the interface becomes locally infinite. This conclusion seems to hold true for the present case, and we estimated such limiting angle to be 15° for $YBa_2Cu_3O_{7-\delta}$. The splitting of the twin tip might begin at the interface of the orthogonal sets of twins when the resolved shear stress reaches a critical value, simultaneously giving rise an additional twin variant I' having twinning orientation with twin II and twin III' (Fig.7.27(b)). When the new twin variant grows away from the interface, it reduces the wedge angle of the tapered twin at the expense of increasing the length of the twin boundary. The equilibrium of the constrain at a large slope twin tip, the interfacial energy of twin boundary as well as the $\sim 2^o$ grain boundary, determines the splitting and the final shape of the twin tip, and hence, the structure of the interface where the two orthogonal twins meet.

7.6 Summary

In this chapter we found the following about the twin boundary structure in $YBa_2Cu_3O_{7-\delta}$.

1. The twin boundary in $YBa_2(Cu_{1-x}M_x)_3O_{7-\delta}$ ($\delta \approx 0.0$) consists of a distorted region with a width of several atomic layers. The width depends on substitution and on oxygen content; it varies from ~ 1.0 nm for M = Cu, Zn, Ni, $x = 0.02$, to ~ 3.0 nm for M = Fe, Al, $x = 0.02$.

2. The twin boundary in $YBa_2Cu_3O_7$ involves a twinning rotation of the respective lattice across the boundary, and also a translation along the boundary. The $(1/3 \sim 1/2) \cdot 2d_{110}$ lattice shift can be easily detected in samples for M = Cu, Zn, and Ni by HREM. Computer simulations confirmed such observations.

3. The narrow streaks of the diffuse scattering observed in the SAD pattern arise from the twin boundaries. Calculation of the intensity of the diffuse scattering from periodic twin boundaries

with a lattice translation $\mathbf{R} = 1/2[1\bar{1}0]$ shows consistent intensity profiles with the observed streaks.

4. Fringe contrast analysis of the twin boundary suggests that the boundary is an $\alpha - \delta$ interface. The δ characteristic arises from the twinning, whereas the α character arises from the lattice displacement.

5. HREM observations of twin boundaries with different oxygen stoichiometries (including *in situ* experiments) suggest that there are two types of twin boundaries - one with, and the other without, a lattice shift. For the former, the twin boundary is probably centered through the O atoms in the Cu-O layer, while, for the latter, it is through cation atoms.

6. Crystallographic analyses reveal that the twin boundary in the $YBa_2Cu_3O_{7-\delta}$ system is a $\Sigma 64$ interface with a twinning shear of $\mathbf{a} - \mathbf{b}$. The observations of twinning steps and twinning dislocations suggest that dislocations may play an important role in forming such a twin boundary with a cation lattice shift.

7. Twinning on the (110) and the equivalent ($\bar{1}10$) planes is frequently encounted within one crystal grain in $YBa_2Cu_3O_{7-\delta}$. The crystal lattice of the impinging set of twins rotates around the *c*-axis when they confront the orthogonally oriented twins. Every second one of the impinging twins is pointed, but does not taper off to zero width, yielding a small-angle grain boundary; the other twin forms a coherent boundary with a twin belonging to the orthogonal set. Thus, the interface of orthogonally oriented twins comprise of an alternation of twin boundaries and grain boundaries.

8. There is a limiting wedge-angle for the tapered twins, and if this angle is exceeded, the tip of the twin splits. The structure of the interface and its vicinity is determined by the equilibrium of elastic strain energy due to the tetragonal-orthorhombic phase transformation and the interfacial energy of the twin boundary and the grain boundary.

Chapter 8

Structural Modulation and Low-Temperature Phase Transformation in $La_{2-x}Ba_xCuO_4$

8.1 Introduction

Although $YBa_2Cu_3O_7$ and Bi-2223 compounds have been the center of attention of the high-T_c research community as the superconductors with T_c above the liquid nitrogen temperature, there is still much interest in the properties of the doped La_2CuO_4 superconductors discovered by Bednorz and Müller [Bednorz and Müller 1986]. Part of the reason is the relative structural simplicity of these materials compared to YBCO and Bi-related compounds, and the belief that a common mechanism of superconductivity is shared by all the high-T_c materials with copper-oxide structure.

As we discussed in Chapter 4, the lattice mismatch between the perovskite and rock-salt structures in the high-T_c superconductors causes structural modulations in these materials. In La-214 compounds, such modulation is commensurate, thus it is easier for us to analyze than the incommensurate modulations in Bi-2212 compounds, which we will discuss in detail in Chapter 10. The enormous internal stress caused by the lattice mismatch, on the other hand, results in structural instabilities associated with soft phonons in doped La-214 materials. It is therefore not surprising that inelastic neutron-scattering studies revealed a soft phonon mechanism for the high-temperature body-centered tetragonal (HTT) to low-temperature-orthorhombic (LTO) transformation in $La_{2-x}M_xCuO_{4-y}$ (M=Sr, Ba)[Axe et al. 1989a]. The transformation temperature, T_0, decreases strongly with alkaline-earth doping. It is widely believed that superconducting phases of this material $(0.05 < x < 0.03)$ have the LTO structure.

The phase diagram of $La_{2-x}Ba_xCuO_4$ and its relationship to the superconducting transition temperature, T_c, is intriguing. In the closely related system $La_{2-x}Sr_xCuO_4$, T_c for Sr content $x \geq 0.05$ rises nearly monotonically until it reaches its broad maximum of about 30 K at $x = 0.15$, then monotonically drops to zero at $x \approx 0.32$ [Torrance et al. 1988]. The general trend is similar for $La_{2-x}Ba_xCuO_4$ except for the important difference that a drastic reduction in T_c is observed for $0.10 < x < 0.15$ with a local minimum below 4 K at $x \approx 0.125$ [Axe et al. 1989a].

While $La_{2-x}Sr_xCuO_{4-y}$ and $La_{2-x}Ba_xCuO_{4-y}$ have similar structural features, the bond length of Ba-O is much larger than that of Sr-O (see section 4.3). This causes structural anormalies in $La_{2-x}Ba_xCuO_{4-y}$ materials for certain doping level x. At $x = 1/8$ a new low-temperature tetragonal (LTT) structure was discovered [Axe et al. 1989b]. The superconducting transition temperature T_c drops sharply when there is a majority phase of LTT structure present, indicating the close correlation of superconducting properties with the structural modulations of this material.

In this chapter we propose a lattice dynamical model with a two-component order parameter to study the consecutive phase transitions (HTT-LTO-LTT) for $La_{2-x}Ba_xCuO_{4-y}$ materials [Cai and Welch 1994a]. Such model is useful not only for the high-T_c materials, but for a wide range of materials undergoing entropy-driven first-order phase transitions.

The presence of LTT phase in $La_{2-x}Ba_xCuO_{4-y}$ also alters the structure of twin boundaries in LTO phase. Unlike the twin boundaries of YBCO we discussed in Chapter 7, the width and density of the twin boundaries in $La_{2-x}Ba_xCuO_{4-y}$ materials are greatly affected by the presence of LTT phase. We will discuss the major features of the twin boundary in this chapter using transmission electron microscope experiments [Zhu et al. 1994a] and theoretical modeling techniques [Cai and Welch 1994b, Cai and Welch 1995a].

8.2 Consecutive Structural Transformation and Lattice Dynamical Model for $La_{2-x}Ba_xCuO_{4-y}$

The HTT-LTO-LTT transition in $La_{2-x}Ba_xCuO_{4-y}$ materials was first investigated by Axe et al. using high-resolution X-ray diffraction techniques [Axe et al. 1989a]. Fig.8.1 shows the representative scans for a sample with $x = 0.1$ of the [110] La_2CuO_4 Bragg reflection. The splitting characteristic of the LTO phase, absent at 300K, is evident on cooling to 180K and 53K. The 52K scan reveals a recovered unsplit tetragonal component. On further cooling, the LTO peaks decrease, but they are still weakly visible at 10K, even after several days. When the sample were quenched (30 minutes) from room temperature to 10K, the LTO to LTT transformation occurs slowly over a period of several hours. Similar but less pronounced hysteretic effects are evident upon heating above 53K [Axe et al. 1989b]. Axe et al. interpreted this as evidences that LTO to LTT transition is of first order.

The theoretical study of these consecutive transitions up to now has been in the framework of Landau free energy models with a two- component order parameter defined as the amplitudes of the normal modes along (110) and ($\bar{1}10$) directions [Axe et al. 1989b, Ting et al. 1990, Ishibashi 1990]. In particular, Ishibashi [Ishibashi 1990] was able to reproduce the transition sequence upon cooling by expanding the Landau free energy to sixth order, and by assuming that the parameters depend on the doping level x, he was able to reproduce the phase diagram of $La_{2-x}Ba_xCuO_4$ correctly. A microscopic structural model of $La_{2-x}Ba_xCuO_4$ was later proposed by Whangbo et al. [Whangbo and Torardi 1991, Whangbo 1992]. in which the LTT structure was shown to have lower energy than the LTO structure in $La_{2-x}Ba_xCuO_4$, and the x−dependence of the parameters in Ishibashi model was justified.

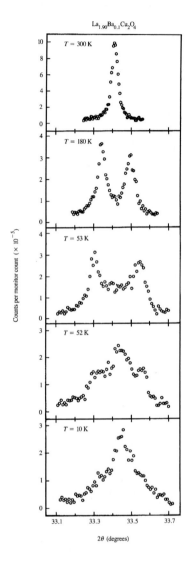

Figure 8.1 The temperature-dependent splitting of the tetragonal [110] reflection at several temperatures. The unsplit component, which re-emerges below 53K, is noticeably broader than 300K [Axe *et al.* 1989a].

8.2. Consecutive Structural Transformation and Lattice Dynamical Model

However, the temperature-dependence of the parameters in these models is not specified by a Landau-type analysis. A microscopic model is also needed to study the critical fluctuations and the dynamics of these phase transitions.

In this section we propose a lattice dynamical model with a two-component order parameter to represent the essential features of the successive structural phase transitions in $La_{2-x}Ba_xCuO_4$. Specifically, we will discuss a mechanism to stabilize the LTO phase at intermediate temperatures [Cai and Welch 1994a]. This work is inspired by the lattice dynamical model with a one-component order parameter recently proposed by Gooding et al. to study entropy-driven first-order phase transitions in alloys [Morris and Gooding 1990, Kerr et al. 1992, Kerr and Rave 1993, Gooding 1991].

Similar to the one-component model, we consider the following (temperature independent) Hamiltonian on a two-dimensional square lattice with sites labeled by index i:

$$H = \sum_i V_1(x_i, y_i) + \sum_{<ij>} V_2(x_i, y_i, x_j, y_j) \tag{8.1}$$

where the simplest functional form of on-site potential to describe the LTT, LTO and HTT phases is

$$V_1(x,y) = -\frac{A}{2}(x^2+y^2) + \frac{B}{4}(x^4+y^4) + \frac{C}{6}x^2y^2(x^2+y^2) \tag{8.2}$$

in which $A > 0$, $B > 0$ and $C > 0$. The two-body potential, which we restrict to nearest neighbor pairs, is given by

$$\begin{aligned} V_2(x_i, y_i, x_j, y_j) &= \frac{1}{2}k_1[(x_i-x_j)^2 + (y_i-y_j)^2] \\ &+ \frac{1}{4}k_2[(x_i^2-y_i^2)^2 + (x_j^2-y_j^2)^2] \\ &\quad [(x_i-x_j)^2 + (y_i-y_j)^2], \end{aligned} \tag{8.3}$$

where $k_1 > 0$ and $k_2 > 0$. (x_i, y_i) is the amplitude of the normal modes in the $La_{2-x}Ba_xCu_4$ which are relevant to the structural phase transitions we are interested in. It should be noted that if we change (x_j, y_j) to $(-x_j, -y_j)$ in the above equation (x_i, y_i) can then be viewed as the magnitude of the tilting of CuO_6 octahedral along (100) and (010) directions (in terms of the HTT unit-cell).

The parameters A, B and C in Eqn.(8.2) determine the equilibrium structure and the energy of LTT, LTO and HTT phases. In this letter we choose $A = 1$, $B = 1$ and $C = 6$ in the subsequent discussions.

Differentiating the above Hamiltonian with respect to x_i and y_i gives three distinctive structures which have extrema of energy:

HTT phase: in which $x_i = y_i = 0$, $E = 0$.

LTO phase: in which $x_i^2 = y_i^2 = 1/3$, $E = -11/54$.

LTT phase: in which $y_i^2 = 0$, $x_i^2 = 1$, $E = -1/4$.

The energy has a local maximum for the HTT phase and minima for the LTO and LTT phases. It is clear that the LTT phase is the ground state in this model and LTO phase is a energetically metastable state for this choice of parameters. The set of parameters we choose gives the relative ratio of the energy minima of LTO and LTT phase to be 1.2, while a first-principles calculation for $La_{1.9}Ba_{0.1}CuO_4$ [Pickett et al. 1991] gives the ratio to be 1.3. Therefore by properly scaling the proper energy unit, the chosen set of parameters can reproduce the microscopic energy surface of $La_{1.9}Ba_{0.1}CuO_4$ quite well. Of course, the simple form of this model potential precludes the treatment of disorder induced via Ba substitution and the detailed atomic structure of this system.

The two-body interaction energy shows different behavior in the LTT and LTO phases. The second term in Eqn.(8.3) represents an anharmonic intersite coupling which renormalizes the phonon dispersion curve in the LTT phase. The LTO phase, in which $x_i^2 = y_i^2$, has a "softer" vibration mode compared to the LTT phase. Therefore the difference of vibrational entropy between LTT and LTO phase is determined by the parameter k_2 in Eqn.(8.3). It was shown by Gooding et al. [Morris and Gooding 1990, Gooding 1991] that a phase with relatively high energy can be stabilized above a certain temperature if its vibrational entropy is large enough so that its free energy is lower than that of the low energy phase. The phase transition was shown by Kerr et al. to be of first order when the anharmonic coupling constant k_2 is large [Kerr et al. 1992, Kerr and Rave 1993]. In our model, the LTO phase can be stabilized in a certain temperature range, due to the enhanced vibrational entropy, when k_2 is large enough. We choose $k_2 = 2k_1$ for simplicity's sake. We know from the results of Monte Carlo simulation that the LTT-to-LTO phase transition will be of first order for this choice of parameters. At still higher temperature, the LTO phase will transform to HTT, similar to the ϕ^4 model [Schneider and Stoll 1976]. Since the intersite coupling in the LTO and HTT phases is essentially the same, the transition is likely to be continuous. Unlike the ϕ^4 model, where the order parameter has only one component and the transition belongs to the universality class of the Ising model, the order parameter in the model we propose has two components, and the LTO-to-HTT transition thus belongs to the universality class of the XY model with a four-fold symmetry-breaking field.

A change of phonon dispersion in $La_{2-x}Ba_xCuO_4$ of the type described by our model in which one X-point soft mode hardens below the transition temperature of LTO to Pccn phase (which is similar to the LTO to LTT transition in $La_{2-x}Ba_xCuO_4$) was found in the neutron scattering experiments of $La_{2-x}Nd_xCuO_4$ materials [Keimer et al. 1993], although the exact values of k_1 and k_2 are difficult to determine without detailed measurement of phonon dispersion curves for $La_{2-x}Ba_xCuO_4$ system. In this work we performed simulations for various values of k_1 to fit the value which can best fit the observed LTT to LTO to HTT transition temperatures.

Because of the existence of the metastable states for the model discussed here, it is very difficult to ensure ergodicity in conventional molecular dynamics simulations. Therefore to calculate the static properties related to the phase transitions in this model, we utilized Monte Carlo simulation, which is more efficient in sampling the phase space (see section 3.3).

Monte Carlo simulations were performed on a square lattice with periodic boundary conditions. 500,000 Monte Carlo Steps (MCS)/site were used to calculate the thermal average of various quantities after 50,000 MCS/site were run and discarded to attain thermal equilibrium. Near the phase transition temperature 5,000,000MCS/site were used to calculate the thermal averages after 500,000MCS/site were discarded to attain thermal equilibrium. Both heating and cooling runs are performed and, unlike the molecular dynamics simulation of Kerr et al. [Kerr et al. 1992], we found the hysteresis near the LTT to LTO transi-

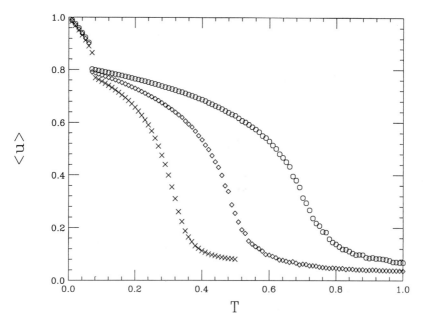

Figure 8.2 The temperature dependence of the total amplitude of the normal mode, for $k_1 = 1$ (×), $k_1 = 2$ (◇), $k_1 = 4$ (○), and $k_2 = 2k_1$. The Monte Carlo simulation was performed on a 16×16 lattice.

tion to be small and we can obtain the LTT ground state in the cooling runs. This difference may be due to the more efficient sampling of phase space using Monte Carlo method in our simulation.

Figure 8.2 shows the resulting temperature dependence of the average normal mode amplitudes as defined by

$$<u> = \frac{1}{N}\sqrt{<\sum_i x_i>^2 + <\sum_i y_i>^2}, \quad (8.4)$$

for three different values of k_1, where N is the total number of sites in the systems simulated. The temperature is defined in units of A/k_B. It is clear that there exist two transitions as the temperature is raised. The low-temperature transition is insensitive to the value of k_1, and u drops discontinuously at $T = T_1 = 0.08 \pm 0.01$. The high-temperature transition temperature increases with increasing values of k_1. For $k_1 = 1$, the high-temperature transition happens at $T = T_0 = 0.33 \pm 0.01$,

8.2. Consecutive Structural Transformation and Lattice Dynamical Model

and it appears to be continuous.

To determine the nature of the two phase transitions observed in the simulation, we have to first define the order parameter for the LTT as well as the LTO phase. The ordered structure of the model (LTT or LTO) has a fourfold degeneracy. The two-component order parameter is defined as

$$M_x = <\frac{1}{N}\sum_i x_i>$$
$$M_y = <\frac{1}{N}\sum_i y_i> \quad (8.5)$$

The order parameter for the LTT phase is defined then as

$$M_{LTT} = max(M_x, M_y) - min(M_x, M_y) \quad (8.6)$$

Similarly the order parameter for the LTO phase is defined as

$$M_{LTO} = max(M_1, M_2) - min(M_1, M_2) \quad (8.7)$$

where

$$M_1 = \frac{\sqrt{3}}{2}(M_x + M_y)$$
$$M_2 = \frac{\sqrt{3}}{2}(M_x - M_y). \quad (8.8)$$

Fig.8.3 shows the temperature dependence of the LTO and the LTT order parameters obtained by Monte Carlo simulation for 16×16 and 32×32 lattices. It is clear that the LTO phase is stable in the temperature range $T_1 < T < T_0$, and there is no LTT phase present in this temperature range. The sharpening of the slope of the LTO order parameters near T_0 with the increasing system size indicates that LTO to HTT transition is a continuous phase transition. It is interesting to note the small increase of the LTT order parameter near $T = T_0$. This small "peak" of the LTT order parameter near T_0 sharpens as the lattice size increases, indicating that this can not be simply dismissed as a finite size effect. Close inspection of lattice pictures indicates that this small increase of the LTT order parameter is due to the formation

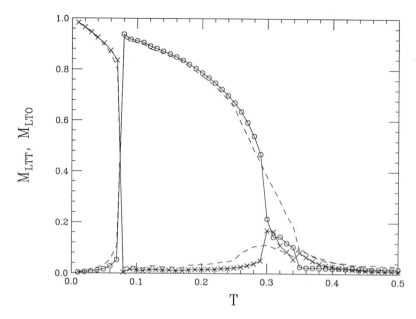

Figure 8.3 The temperature dependence of the order parameter for LTT phase M_{LTT} (×) and LTO phase M_{LTO} (○) for $k_1 = 1$, obtained by Monte Carlo simulation on a 32×32 lattice. The solid lines are guides for the eye. The dashed lines show the simulation results for M_{LTT} and M_{LTO} for a 16×16 lattice (symbols are omitted for clarity).

of small (110) and (1$\bar{1}$0) LTO domains near T_0. Because of the high interface energy between LTO domains, the domain wall consists of the intermediate $Pccn$ structure [Crawford et al. 1991] in which $x_i \neq y_i \neq 0$ with a single layer of the LTT structure in the center of the domain wall. The rapid increase of domain wall density near T_0 then caused the increase of LTT order parameter. Transmission electron microscope measurement (see section 8.3) shows that the twin boundaries in LTO phase consist of the $Pccn$ structure with the LTT structure at the center of the domain wall, which nucleates the formation of the LTT phase as the temperature is lowered below the LTT to LTO transition [Zhu et al. 1994a]. The appearance of LTO-LTT- LTO domain wall near the LTO to HTT transition temperature is due to the existence of low energy LTT phase. This is qualitatively different from the herterophase fluc-

8.2. Consecutive Structural Transformation and Lattice Dynamical Model 213

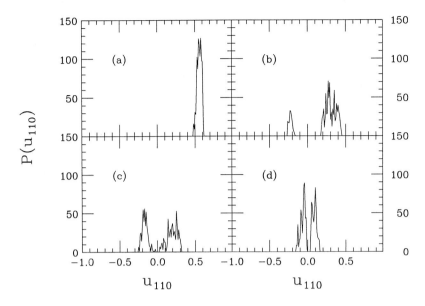

Figure 8.4 The distribution function of average local LTO order parameter near the LTO to HTT transition temperature $T_0 = 0.33$: (a) $T = 0.25$; (b) $T = 0.30$; (c) $T = 0.33$; and (d) $T = 0.40$ for the 32×32 lattice.

tuations near LTO to HTT transition temperature in $La_{2-x}Sr_xCuO_4$, in which the LTO phase has lower energy than LTT phase. In that case only LTO-HTT-LTO domain walls are possible.

Figure 8.4 shows the local order parameter distribution function near the LTO to HTT transition temperature T_0. The local LTO order parameter at each lattice site is defined as the amplitude of the (110) mode. We find that the local LTO order parameter distribution function broadens significantly near T_0. Unlike the LTT to LTO transition, which is driven by the vibrational entropy difference between the LTT and LTO phase, the LTO to HTT transition is driven by the formation of domains of the LTO phase with different orientations. Fig.8.4(d) shows that even though the average value of the LTO order parameter is close to zero at $T > T_0$, the local LTO order parameter remains nonzero even at temperatures well above LTO to HTT transition temperature. This further indicates that the LTO to HTT transition is of the order-disorder type, driven by the formation of domains, which is quite different from the entropy-driven first-order LTT to LTO transition. It has been shown by the XAFS experiments of Rechav et.al [Rechav et al. 1994] that the oxygen octahedra local tilting angles remain nonzero hundreds of degrees above the transition temperature for antiferrodistortive type perovskite such as $Na_{0.82}K_{0.18}TaO_3$. It will be interesting to perform the similar measurements on $La_{2-x}Ba_xCuO_4$ to see if the domain structures are also present in this type of material.

To compare the results of this model to experimental results for $La_{2-x}Ba_xCuO_4$, we also calculated the temperature dependence of the orthorhombic strain in the LTO phase. Since the strain energy of the orthorhombic strain η in the LTO phase is much smaller than the interaction energy due to the tilting of the CuO_6 octahedra, we can calculate the temperature dependence of η using perturbation theory. Taking into account only the lowest order coupling between the primary order parameter and η, the strain energy can be described by [Cowley 1980]

$$E_{ela} = \frac{c}{2}\eta^2 + \frac{d}{2}\eta M_{LTO}^2, \tag{8.9}$$

where c is the shear modulus c_{66} and η is the shear strain. Minimizing the energy with respect to the strain η yields the stress-free strain

$$\eta = -\frac{d}{2c}M_{LTO}^2. \tag{8.10}$$

The value of M_{LTO} can be obtained to a first approximation from the Monte Carlo simulation which ignores the contribution of the strain energy. Fig.8.5 shows the normalized orthorhombic strain calculated from the results of the simulation of a 32×32 lattice and $k_1 = 1$, compared with the experimental measurements [Cox et al. 1989]. The scaled temperature t is defined as

$$t = \frac{T - T_0}{T_0} \tag{8.11}$$

where T_0 is the LTO to HTT transition temperature. The orthorhombic strain is scaled by its value η_1 at the LTT to LTO transition temperature T_1. We also obtain the experimental values of η from the experimentally-measured temperature-dependence of lattice parameters a and b using the relationship

$$\eta = \frac{b - a}{b + a} \tag{8.12}$$

8.3. Low-Temperature Microstructure

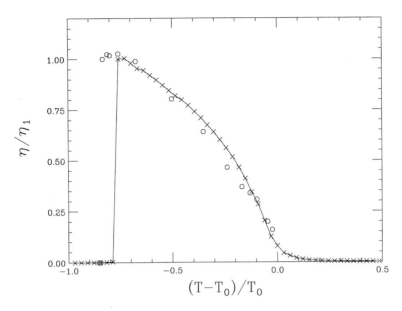

Figure 8.5 The temperature dependence of the normalized orthorhombic strain calculated from simulation results on a 32×32 lattice for $k_1 = 1$ ($\times$), compared with the neutron powder diffraction data for $La_{1.9}Ba_{0.1}CuO_4$ ($\bigcirc$) [Cox et al. 1989]. The solid line is a guide for the eye.

which is also shown in Fig.8.5. It is clear that with the proper choice of k_1 the model calculation can fit the experimental data very well.

The discontinuous drop of the orthorhombic strain η at LTO to LTT transition temperature as shown in Fig.8.5 is a strong evidence for that this transition is of first order. The first order nature of the LTO to LTT transition will have consequences on this systems low temperature microstructure if it is quenched, as we will discuss in the next section.

8.3 Low-Temperature Microstructure

As we discussed before, In the concentration range of the local T_c minimum in $La_{2-x}Ba_xCuO_4$ there is evidence of a low temperature tetragonal (LTT) structure with proposed space group $P4_2/ncm$ below

Table 8.1 A summary of the crystallography of $La_{1.88}Ba_{0.12}CuO_4$ and the expected superstructure reflections in the $(101)^*$, $(011)^*$ and $(001)^*$ projections. HTT and LTT are high- and low-temperature tetragonal, respectively; LTO and LTLO are low-temperature orthorhombic and less orthorhmbic, respectively.

Temperature	Phase	Space group	Allowed superstructure reflections for projection		
			$(101)^*$	$(011)^*$	$(001)^*$
$T > 200K$	HTT	$I4/mmm$[a]	none	none	none
$60 < T < 200K$	LTO	$Bmab$[b]	$h+l=2n$	none	none
$T < 60K$	LTLO	$Pccn$[c]	$h+l=2n$	none	$h+k=2n$
	LTT	$P4_2/ncm$[c]			$h,k = $ odd

[a]We use indexing for the expanded $\sqrt{2}$ cell for HTT.
[b]$Cmca$ is the standard setting for LTO, we use $Bmab$ for indexing consistency.
[c]Extinction rules are identical for LTLO and LTT.

about 60 K [Axe et al. 1989a, Axe et al. 1989b]. Above this temperature the structure is orthorhombic (LTO) with space group $Bmab$ and, for higher temperatures, again tetragonal (HTT) with the space group $I4/mmm$. For consistency we will characterize the structure of HTT phase using the LTO unit-cell ($a \leq b \approx 5.37$Å, $c \approx 11.2$Å). Table 8.1 serves as a guide to those reader not familiar with the extinction rules of the phases involved in this study. There the allowed reflections of the HTT phase are referred to as the fundamental reflections.

The transition from HTT to LTO during cooling results in twins [Chen et al. 1991, Chen et al. 1993, Zhu et al. 1994a]. The normal to the twin boundaries is the [110] direction or the equivalent [$\bar{1}$10] direction (Bmab setting). For a Ba-content of $x = 0.12$ the structure is HTT at room temperature, and transforms to orthorhombic at about 200 K. Below 60K, neutron diffraction studies show that there is a phase transition from LTO to LTT [Axe et al. 1989b]. It has been reported that there was little change in twin morphology, even when the sample was cooled down to 10K in a transmission electron microscope (TEM) [Chen et al. 1991]. This is inconsistent with pure LTT phase, since the twinning is associated with an orthorhombic symmetry. To address the issue of the LTO/LTT phase transition, we performed in-

8.3. Low-Temperature Microstructure

situ experiments with a 200 keV TEM equipped with a low temperature (liquid He) specimen holder. Samples are prepared by a procedure previously described [Moodenbaugh et al. 1988]. The TEM specimens were thinned by ion milling with a low-energy gun.

Figure 8.6 shows the temperature dependence of the microstructure and the corresponding diffraction pattern of a $La_{1.88}Ba_{0.12}CuO_4$ specimen, observed in-situ while being heated from 20K to 250K. At about 20K, fine twins are the predominant microstructural feature of the crystal (Fig.8.6(a)). With increasing temperature, first some needle-like twins disappear (Fig.8.6(d), $\sim$ 140K) with the specimen being twin-free above 200K (Fig.8.6(g), $\sim$ 250K), where the sample becomes HTT. It should be noted that the twin morphology is not always reversible after a thermal cycle. We often observed an increased density of fine twins at low temperature after cycling.

The structural transitions can be elucidated by analyzing the electron diffraction. Shown also in Fig.8.6 are the corresponding diffraction patterns of the $(001)^*$ [(b), (e), and (h)] and $(101)^*$ [(c), (f), and (i)] zone axes at 20K, 140K and 250K, respectively. At 250K, Fig.8.6(h) and (i) clearly indicate the I4/mmm space group (HTT phase), because only the fundamental reflections are present.

In contrast, at about 140K, superstructure reflections in accordance with the LTO phase (Table 8.1) appear in $(101)^*$ projection. Below about 60K, in $(001)^*$ projection, additional superstructure spots are visible for $h+k = 2n$ with h and k odd, e.g. the (110)-, (130)-reflection etc. The presence of (110) type reflections, which are forbidden in LTO phase, is allowed for the space-group $P4_2/ncm$ predicted for the LTT phase, and also for the space group Pccn (low temperature less orthorhombic (LTLO) phase), that was proposed to account for the presence of a considerable orthorhombic strain [Cox et al. 1989]. However, the $(101)^*$ diffraction pattern remains the same, because $h+l = 2n$ reflections are allowed in all of the $Bmab$, $P4_2/ncm$, and $Pccn$ space groups.

Compared to neutron and x-ray diffractions, one advantage of electron diffraction is the sensitivity to weak intensities. However, a disadvantage is that it also can produce double diffraction, when a reflected electron beam acts as a new incident beam. In $(001)^*$ projection [Fig.8.6(b)] at low temperature we also observe weak superstruc-

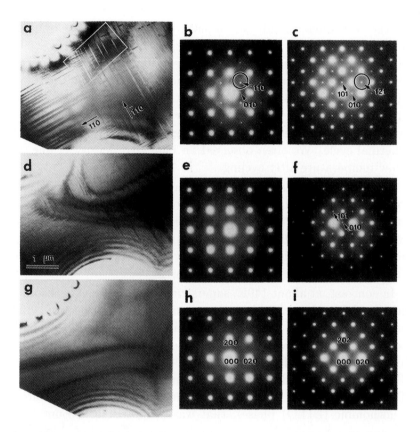

Figure 8.6 Temperature dependence of microstructure and diffraction patterns of $La_{1.88}Ba_{0.12}CuO_4$ at 20K ((a)-(c)), 140K ((d)-(f)), and 250K ((g)-(i)). (b), (e) and (h) are diffraction patterns of the $(001)^*$ zone, while (c), (f), and (i) are of the $(101)^*$ zone. Note the diffraction of the $(001)^*$ projection of HTT phase (h) and LTO phase (e) are identical, and the $(101)^*$ projection of the LTO (f) and LTT phase (c) are identical.

8.3. Low-Temperature Microstructure

ture spots at (100), (010), (210) etc. Some may be too weak to be reproduced here. According to the extinction rule of both space group $P4_2/ncm$ and $Pccn$, the reflections at $h+k = 2n+1$ should be absent. We attribute these spots to the double scattering involving large **g** vectors in the first or higher order Laue zones where $l = 1, 3, ..., 2n+1$. The intensity of the (100) type spots increases with a decrease of temperature, as is to be expected because of the increased coupling via large **g** vectors when the Debye-Waller factor is reduced.

Double scattering to the (110)-type reciprocal-lattice coordinates is, in principle, possible if twins stacked up along the incident electron beam are rotated 90°. However, the observation that the (110)- type reflections are much stronger than (100)-type reflection appears to rule out the possibility that double scattering is the major contributor to the [110]-type reflections in accordance with a previous electron diffraction [Chen et al. 1993]. Further evidence that the (110)-type reflections are real is that they appear in the electron diffraction patterns at about the temperature where the transition from LTO to LTT takes place according to neutron diffraction studies [Axe et al. 1989b, Axe et al. 1989a, Chen et al. 1991].

It is possible to use the distinctive superstructure reflections associated with LTO and LTT phases to image the location of the corresponding phases. A previous attempt to accomplish this was unsuccessful, possibly due to insufficient intensity and contrast [Chen et al. 1991]. With a specially designed objective aperture (5μm in diameter), we were able to record dark-field images using these weak superstructure reflections.

Figure 8.7 shows three enlarged micrographs from the rectangular area shown in Fig.8.6(a), for a nominal temperature of 20K. Fig.8.7(a) is imaged 22° away from the c-axis of the same area in alternatively the [101] and the twin related [011] orientation. Since the (121) superstructure spot in the (101) projection (marked by a circle in Fig.8.6(c)) was used to form the image, only every second twin ([101] oriented) is bright as is evident from the extinction rules, see Table 8.1. Fig.8.7(b) is a bright-field image showing the area morphology with twin boundaries edge-on in the (001) orientation. An intriguing observation in this orientation is the narrow bright lines in the dark field image of Fig.8.7(c), using the LTT-phase-related (110) reflection in the (001)*

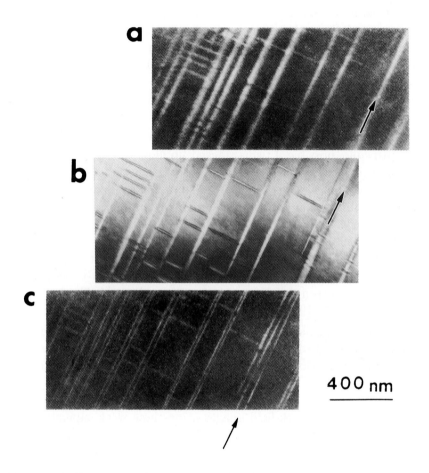

Figure 8.7 Diffraction contrast observed at 20K from the area bounded by a rectangular box shown in Fig.8.6(a). (a) The dark-field image using the LTO superlattice reflection (121) of the $(101)^*$ projection (see Fig.8.6(c)). Note that the twin boundaries are inclined and only one set of the twin domains shows contrast. (b) The bright-field image in the [001] orientation. The twin boundaries are edge-on. (c) The dark-field image viewed in the same orientation as (b) but using the LTT superlattice reflection (110) (see Fig.8.6(b)). Note that only the twin boundaries show the bright contrast suggesting that they are the corresponding LTT phase.

8.3. Low-Temperature Microstructure

projection (marked by a circle in Fig.8.6(b)). By comparison with Fig.8.7(b) these narrow lines occur at the twin boundary, suggesting that the LTT phase is located at the regions of twin boundaries (the arrow marked in Fig.8.7(a), (b), and (c) indicates the same twin domain). Similar images were also obtained at a nominal temperature of about 50 K (the overall specimen temperature, although we do not know the exact local temperature due to the electron beam heating) for a specimen with Ba content of $0.09 < x < 0.12$, but are not seen for pure La_2CuO_4 at room temperature. The intensity profile and thickness of the layer at the twin boundary are hard to estimate due to the intrinsic low intensity of the superstructure reflections. The thickness of the LTT phase is probably of the order of 100Å, which is very close to the value predicted (a few hundred Å) from an observation of the broadening of the superlattice peaks of the LTT phase by neutron diffraction.

The presence of the LTT phase at the location of the LTO twin boundary is consistent with the crystallography of the tipping of the CuO_6 octahedra [Zhu et al. 1994a]. In the LTO phase the CuO_6 octahedra are tilted a few degrees around the [100] axis. This results in a conflict at the twin boundary such that the two twin domains have displacements in orthogonal directions, as indicated in Fig.8.8(a).

Thus, from simple geometrical consideratifons, an ideally sharp twin boundary, which is likely to have a very high interfacial energy (especially near the LTO/LTT transition temperature, due to the increased orthorhombicity of the LTO phase), is incompatible with the required tilting of the octahedron. Possible configurations are either a localized twin boundary with severely distorted-octahedra at the boundary or extended twin-boundary regions where the structure gradually changes from orthorhombic to tetragonal. We propose a delocalized boundary with LTT structure at the center of the twin boundary, gradually changing via the orthorhombic Pccn (LTLO) to an orthorhombic LTO in the center of the twin matrix as shown in Fig.8.8(b). The LTT phase appears to nucleate at and grow out from the LTO twin boundaries, where the conflict of the tipping axis between the adjacent twins is resolved. Further cooling of the specimens of $La_{2-x}Ba_xCuO_4$ ($0.09 < x < 0.12$) through the LTO phase results in an increasing twin density, and thus a larger volume of the LTT phase. The observed hys-

222   Chapter 8. Structural Modulation in $La_{2-x}Ba_xCuO_4$

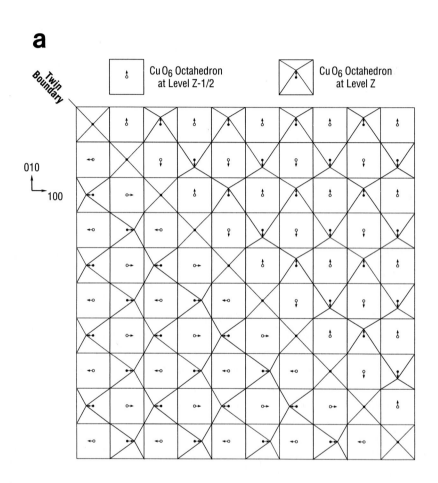

8.3. Low-Temperature Microstructure

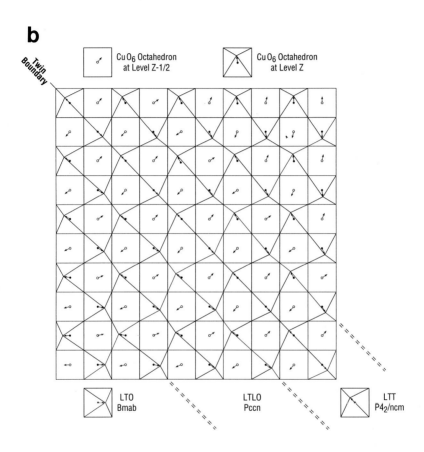

Figure 8.8 (a) An ideal twin boundary in the LTO structure results in a conflict between tilting of the CuO$_6$ octahedra around the [100] and [010] axes. The resultant is tilting around [110] axis. (b) The proposed model in which the LTT structure (tipping axis [110]) exists at the center of the twin boundaries with a gradual transition, via the LTLO structure, to LTO (tipping axis [100]) in the center of the twin domain.

teresis in developing the low-temperature twins during our in-situ thermal cycling implies that kinetics plays an important role in formation of the LTT phase. Fig.8.8(b) also suggests that the orthorhombic Pccn is located at the LTT/LTO interphase region and acts as a buffer structure. This is consistent with previous neutron studies [Cox et al. 1989, Axe 1992] which predict the existence of the LTLO phase and a continuous phase transition of LTO-LTLO-LTT in a finite temperature region. The present results appear to contradict the recent findings of Billinge, et al., of a predominantly tetragonal phase [Billinge et al.1993]; however, we note that the analysis on which that conclusion was based assumes a discrete orthorhombic phase, not the phase with varying orthorhombicity that has been observed in the TEM studies here and previously [Chen et al. 1993]. A twin boundary with a structure different from twin matrix has been observed at room temperature in $PbVO_4$ using high resolution electron microscopy [Manolikas et al. 1986].

To further study the structure, a theoretical model based on Ginzburg-Landau theory is used to analize the structure of twin boundary in LTO phase in the next section.

8.4 Theoretical Model of Twin Boundary Structure

As we discussed in Sec.8.2, the $La_{2-x}Ba_xCuO_4$ compound has a body-centered tetragonal structure at high temperatures (HTT structure). Upon cooling it undergoes a structural phase transition to low- temperature orthorhombic structure (LTO structure) which is caused by the tilting of the CuO_6 octahedra about the (110) and ($\bar{1}$10) directions. Depending on the doping level x, the LTO structure may go through another phase transition at even lower temperature to a low-temperature tetragonal phase (LTT phase) [Axe et al. 1989b].

The existence of the low-energy LTT phase also qualitatively changes the structure of twin boundaries in the LTO phase, as we see in the previous section. In the case of ferroelastic materials in which the structure of twin boundaries associated with a martensitic transformation has been extensively studied, the boundary often consists of structures of the high-energy phase [Wayman 1964, Christian 1965,

8.4. Theoretical Model of Twin Boundary Structure

Warlimont and Delaey 1974, Nishiyama 1978]. Therefore with decreasing temperature, the free-energy difference between the bulk phase and boundary phase increases. This causes a reduction of the twin boundary density, as well as boundary width. However in $La_{2-x}Ba_xCuO_4$ materials, it is possible that the boundary consists of the low energy LTT phase instead of the high energy HTT phase. In that case the free energy difference between the bulk phase and the boundary phase decreases with the decreasing temperature The twin boundary width, as well as its density, will then increase in order to minimize the gradient energy.

In this section we construct a Landau free energy model to study the structure of the twin boundaries in the LTO phase of $La_{2-x}Ba_xCuO_4$. We will show that the competition between the bulk and gradient free energy makes the temperature-dependence of the twin boundary structure very complicated.

8.4.1 Landau Model

In order to construct a theoretical framework under which the phase transitions and the related structural properties in these materials can be discussed, considerable effort has been made to establish a Landau free energy for this material [Axe et al. 1989b, Ting et al. 1990, Ishibashi 1990]. Axe et.al [Axe et al. 1989b] and Ting et al. [Ting et al. 1990] expand the Landau free energy in terms of order parameter up to fourth order term with the parameter v of the anisotropic fourth order term changes as a function of temperature. However the temperature dependence of v is not clear. Ishibashi [Ishibashi 1990] expanded the free energy in terms of order parameter up to sixth-order terms and all the parameter except that of the second order term are temperature independent. He was able to reproduce the transition sequence upon cooling and by assuming that the parameters depend on the doping level x, he was able to reproduce the phase diagram of $La_{2-x}Ba_xCuO_4$ correctly.

The Landau Free energy is expanded in terms of order parameter. The primary order parameter is defined as a two-component order parameter (Q_1, Q_2), where Q_1, Q_2 is the amplitudes of the normal modes along [100] and [010] direction (in terms of LTO unit cell). To consider the effect of strain in LTO phase, we also consider the secondary order parameter u, which is the shear strain orthorhombic phase.

The Free energy density can then be written as

$$F(Q_i) = F_G + F_L \tag{8.13}$$

where F_G is the gradient energy

$$F_G = D(\nabla Q)^2 \tag{8.14}$$

where D is a positive constant and $Q = \sqrt{Q_1^2 + Q_2^2}$. F_L is the Landau-type free energy which was given by Ishibashi [Ishibashi 1990]

$$\begin{aligned} F_L(Q_1, Q_2, \eta) &= \frac{A}{2}(Q_1^2 + Q_2^2) + \frac{\beta}{4}(Q_1^4 + Q_2^4) \\ &+ \frac{\gamma}{2}Q_1^2 Q_2^2 + \frac{\eta}{6}(Q_1^2 - Q_2^2)^2(Q_1^2 + Q_2^2) \\ &+ \frac{c_0}{2}u^2 + \zeta u(Q_1^2 - Q_2^2), \end{aligned} \tag{8.15}$$

where β, ϵ, and η are temperature-independent positive constants. The elastic constant c_0 represents C_{66} in the HTT phase and the u represents the shear strain u_6. Minimize free energy as respect to the strain u, we have

$$u = -\frac{\zeta}{c_0}(Q_1^2 - Q_2^2). \tag{8.16}$$

The only temperature-dependent coefficient A is defined as

$$A = \alpha - \left(1 - \frac{x}{0.21}\right) \tag{8.17}$$

where $\alpha = T/T_0(0)$ is the reduced temperature and x is the concentration of Ba. $T_0(0) = 500K$ is the LTO to HTT transition temperature for undoped La_2CuO_4 system. The x-T phase diagram is reproduced by assuming the x dependence of the coefficient γ as

$$\gamma = \beta - \frac{4\zeta^2}{c_0} + k\left(1 - \frac{x}{0.18}\right)\frac{2\zeta^2}{c_0}, \tag{8.18}$$

where k=3, $\beta = 0.8$, $\zeta = 0.5$, $\eta = 1$ and $c_0 = 1$ [Ishibashi 1990]. Fig.8.9 shows the $x - T$ phase diagram obtained by minimize Landau free

8.4. Theoretical Model of Twin Boundary Structure

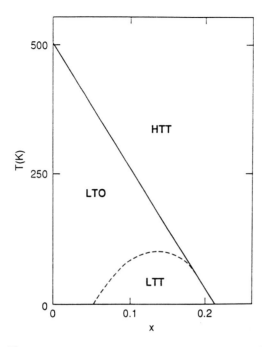

Figure 8.9 $x - T$ phase diagram of La$_{2-x}$Ba$_x$CuO$_4$ calculated from the Landau free-energy model given by Eqs.(8.15-8.18).

energy in Eqn.(8.15) using the value of the coefficients shown above. Please note that although for La$_2$CuO$_4$ there is no LTT phase even at $T = 0$, the Landau free energy still has a minimum at LTT configuration. It is just that the free energy of LTT is higher than that of LTO at all temperatures. Fig.8.10 shows the temperature dependence of Landau free energy difference between LTT and LTO phase for $x = 0.1$ as well as for $x = 0$. Therefore LTT is a *metastable* structure, not an *unstable* structure for undoped La$_2$CuO$_4$.

8.4.2 The Structure of Twin Boundary

The twin boundary is a domain wall between two energetically degenerate variants in the low-temperature-orthorhombic (LTO) phase. The

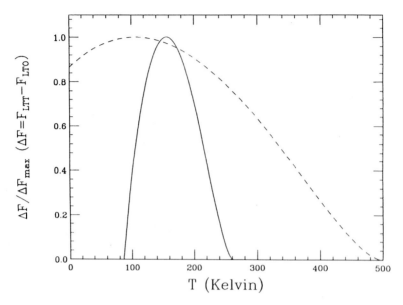

Figure 8.10 Temperature dependence of the bulk Landau free energy density difference between bulk LTT and LTO phase. Solid line: $x = 0.1$; Dash line: $x = 0$.

domain wall represents a transition between two orthorhombic variant, where the lattice structure is distorted, so that the formation of domain walls introduces inhomogeniety to the system. The twin structure in the LTO phase consists of two domains whose orthorhombic distortion are (nearly) perpendicular to each other.

Here we consider a twin structure of the following two variant: $Q_1 = (0, Q_O, 0)$ and $Q_2 = (Q_O, 0, 0)$, with the twin boundary oriented in [110] direction. It is convenient to work in a new coordinate system (s, t, r) which is a $45°$ rotation of the a-b plane about the c-axis. By the coordinate transformation

$$\begin{aligned} P_1 &= \frac{1}{\sqrt{2}}(Q_1 + Q_2) \\ P_2 &= \frac{1}{\sqrt{2}}(Q_1 - Q_2) \end{aligned} \quad (8.19)$$

8.4. Theoretical Model of Twin Boundary Structure

we find that the twin-boundary is in [100] direction in the new coordinate system and the two variants of the twinning are $P_1 = (P_O, -P_O, 0)$ and $P_2 = (P_O, P_O, 0)$, where $P_O = (1/\sqrt{2})Q_O$. We assume that the space profile of the order parameters is quasi-one-dimensional, i.e. it depends on s only.

Converting the Landau free energy into the new coordinate system, we obtain from Eqn.(8.15)

$$F_L(P_1, P_2) = A_1(P_1^2 + P_2^2) + A_2(P_1^4 + P_2^4) \\ + A_3 P_1^2 P_2^2 + A_4 P_1^2 P_2^2(P_1^2 + P_2^2) \quad (8.20)$$

where

$$A_1 = \frac{1}{2}[\alpha - (1 - \frac{x}{0.21})]$$
$$A_2 = \frac{\beta + \gamma}{8}$$
$$A_3 = \frac{3\beta - \gamma}{4} - \frac{2\zeta^2}{c_0}$$
$$A_4 = \frac{2}{3}.$$

The gradient energy becomes

$$F_G = \frac{D}{P_1^2 + P_2^2}\left(P_1 \frac{dP_1}{ds} + P_2 \frac{dP_2}{ds}\right)^2. \quad (8.21)$$

The total boundary free energy is defined as

$$E_{t.b} = \int_{-w/2}^{w/2} (\Delta F_L + F_G) ds \quad (8.22)$$

where w is the thickness of the domain wall and ΔF_L is the difference between the Landau free energy density in the domain wall and the Landau free energy density in the bulk. It is then possible to obtain the width of the domain wall w by minimizing the boundary free energy $E_{t.b}$.

It is clear that the twin boundary has excess free energy due to suppression of the LTO order parameter. For systems with one-component order parameter P, only $180°$ boundary is possible, and the order parameter varies continuously from $-P$ to P. For systems with two-component order parameter, the $90°$ twin boundary usually has lower boundary energy due to its smaller gradient energy. The $90°$ twin boundary has an interesting property, i.e., one of the components of the order parameter is the same on both sides of the twin boundary. For example, for the boundary we studied the order parameter changes from $(P_O, -P_O, 0)$ to $(P_O, P_O, 0)$, thus P_s is unchanged across the boundary. Therefore, at the center of the domain wall the gradient energy density is zero. However, the minimum of the Landau energy F_L is at $(P_T, 0, 0)$ (where P_T is the order amplitude of the normal mode of LTT phase), or $(0, 0, 0)$. Therefore, the center of the domain wall is unstable against the formation of LTT or HTT phase. Whether the LTT or HTT phase will form at the boundary will of course depend on which configuration has the lower boundary free energy $E_{t.b}$.

For simplicity's sake we approximate the kink solution of the order parameter profile at the boundary [Jacobs 1985] by a linear ansatz such that

$$P_1 = \begin{cases} P_O - \frac{P_O - P_T}{w/2}(s + w/2) & \text{if } -w/2 < s < 0 \\ P_T + \frac{P_O - P_T}{w/2} s & \text{if } 0 < s < w/2 \end{cases}$$

$$P_2 = \frac{P_O}{w/2} s \qquad \text{if } -w/2 < s < w/2 \quad (8.23)$$

where the twin boundary will be a LTO-LTT-LTO boundary if $P_T \neq 0$, and a LTO-HTT-LTO boundary if $P_T = 0$.

Figure 8.11 shows the equilibrium profile of P_s near a twin boundary for $T = 200K$ and $x = 0.1$. The solid line is obtained from the numerical minimization of Eqn.(8.22) using simulated annealing method. The dashed line is obtained from the minimization of Eqn.(8.22) using the linear ansatz defined in Eqn.(8.23). We can see that the the boundary width w defined in the linear ansatz agrees very well with the numerical results.

8.4. Theoretical Model of Twin Boundary Structure

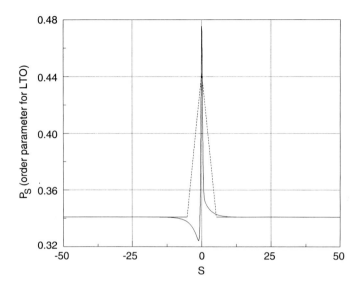

Figure 8.11 The profiles of the order parameter P_s near the twin boundary for $T = 200$K and $x = 0.1$. Solid line: obtained from simulated annealing of Eqn.(8.22); dashed line: obtained from using the linear ansatz defined in Eqn.(8.23).

Substitute Eqn.(8.23) into Eqn.(8.22) we have

$$\begin{aligned}
E = \ & (w/210) \ [70(-4P_O^2 + P_O P_T + P_T^2)A_1 \\
& + 42(-8P_O^4 + P_O^3 P_T + P_O^2 P_T^2 + P_O P_T^3 + P_T^4)A_2 \\
& + 7(-24P_O^4 + 2P_O^3 P_T + P_O^2 P_T^2)A_3 \\
& + (-360 P_O^6 + 30 P_O^5 P_T + 14 P_O^4 P_T^2 \\
& + 6 P_O^3 P_T^3 + 2 P_O^2 P_T^4) A_4] \\
& + (D/w) \ [8P_O^2 + (8+\pi)P_O P_T + 4P_T^2]. \quad (8.24)
\end{aligned}$$

where $P_O(T)$ and $P_T(T)$ can be calculated easily by minimizing the bulk free energy in Eqn.(8.20).

Minimizing E in Eqn.(8.24) with respect to boundary thickness w is trivial. By comparing the minimum boundary free energy E for LTO-LTT-LTO boundary to the boundary free energy of LTO-HTT-LTO boundary at a given reduced temperature α, we were able to calculate the temperature dependence of the equilibrium width of the twin

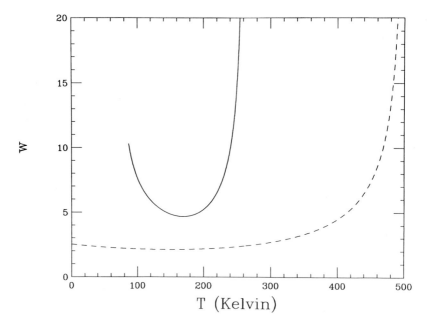

Figure 8.12 Temperature dependence of the twin boundary width in the LTO phase for $La_{2-x}Ba_xCuO_4$. Solid line: $x = 0.1$; dashed line: $x = 0$.

boundary $w(\alpha)$ as shown in Fig.8.12. The twin boundary width w diverges at the LTO-to-HTT transition temperature. This indicates that the LTO-to-HTT transition is a second-order transition in which the LTO order parameter decreases continuously to zero at the transition temperature. As expected, twin boundary with the lowest free energy is an LTO-LTT-LTO boundary for the entire temperature range where bulk LTO phase is stable, for both $x = 0.1$ and $x = 0$ (undoped) cases. The twin-boundary width w decreases as the temperature is reduced. As the temperature decreases further, the width of the boundary for $x = 0.1$ reaches a minimum at certain temperature and then begins to increase as the temperature is reduced further. This is quite different from the ferroelastic materials in which the boundary width decreases monotonically with decreasing temperature. In ferroelastic materials with single Martensitic transition, the decreasing boundary width with decreasing temperature is caused by the increasing free energy difference between a high-energy boundary phase and a low energy bulk

8.4. Theoretical Model of Twin Boundary Structure

phase [Khachaturyan 1983]. The decreasing boundary width decreases the volume fraction of the boundary phase, thus lowering the Landau free energy with the expense of increasing gradient energy. This is also the reason that we see the monotonic decrease of boundary width for La_2CuO_4 materials ($x = 0$ case in Fig.8.12). However for $x = 0.1$, below certain temperature the free energy difference between the LTT and LTO phase decreases with decreasing temperature, and eventually goes to zero at the LTT-to-LTO transition temperature T_1 (see Fig.8.10). Therefore the boundary width of an LTO-LTT-LTO boundary increases with decreasing temperature near the LTT-to-LTO transition temperature. There is a discontinuous jump of the boundary width as the temperature increases above the LTT-to-LTO transition temperature. This is because the LTT-to-LTO transition is a first-order transition, with a discontinuous change of the magnitude of the order parameters. Therefore the change of the gradient free energy F_G is discontinuous even though the bulk free energy F_L changes continuously.

It should be noted that the LTT phase at the center of a domain wall consists of only a single atomic layer since the extended LTT phase increases its free energy in the temperature range where the LTO phase is stable. However this is not true in the temperature range where the LTT phase is stable. The mechanism of growth of the LTT phase is that the size of the LTT structure in domain walls will widen and eventually becomes an LTT single crystal. If a systems is quenched from a temperature for which the LTO phase is stable, to a temperature below the LTT-to-LTO transition temperature T_1, the morphology of the system will consist of quasi-periodic domains of the LTO-LTT-LTO structure with wide domain walls (see the previous section).

8.4.3 The Size of the Twin Domains

Because the grain boundaries in polycrystalline $La_{2-x}Ba_xCuO_4$ act to restrain changes in the shape of the grain caused by the deformation inherent in the tetragonal to orthorhombic transformation, there will be a set of spatially oscillating forces exerted on the grain by the boundary after the transformation. On the average, these forces will produce no net force and no net torque. Therefore by Saint-Venant's principle [Love 1944], the strains resulting from the boundary will become negligible at distances large compared to the wave-length of the oscillatory forces (i.e. large compared to the $2L$, where L is the twin boundary spacing). The maximum value of the strain is of the order of the orthorhombic strain. Thus we may expect a strain of the form

$$\epsilon(x, y, z) = f\left(\frac{y}{2L}\right) \sin\left(2\pi \frac{x}{2L}\right) \tag{8.25}$$

where $f \to 0$ as $y/2L \to \infty$.

The strain energy density is

$$E_{strain}(x, y, z) = \frac{1}{2} c_{66} \epsilon^2(x, y, z). \tag{8.26}$$

The total strain energy is then

$$W_s = 2 \int_V E_{strain}(x, y, z) dx dy dz. \tag{8.27}$$

where $V = D^3$ is the volume of the whole grain. With Eqn.(8.25) we obtain

$$W_s = c_{66} \varphi^2 L D^2 \vartheta, \tag{8.28}$$

where φ is the orthorhombicity of the LTO structure at temperature T and

$$\vartheta = \int_0^\infty f^2(\xi) d\xi$$

is a constant which depends on (i) how many wave-lengths ($2L$) the strain-field extends into the grain and
(ii) what fraction of the maximum strain is removed by the constraining effect of the boundary(i.e. how flat the boundary is). Factor (i) is of the order of unity, while factor (ii) is less than or equal to unity.

The twin boundary energy $E_{t.b}$ in Eqn.(8.22) makes a contribution:

$$W_b = \frac{D^3}{L} E_{t.b}. \tag{8.29}$$

Thus the total energy is:

$$W = W_b + W_s = c_{66} \varphi^2 L D^2 + \frac{D^3 E_{t.b}}{L}. \tag{8.30}$$

8.4. Theoretical Model of Twin Boundary Structure

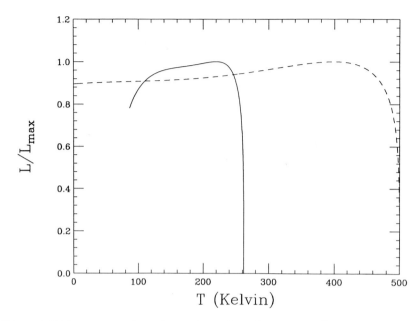

Figure 8.13 Temperature dependence of twin boundary spacing L in the LTO phase, normalized by its largest value $L_{\max}$, for $\text{La}_{2-x}\text{Ba}_x\text{CuO}_4$. Solid line: $x = 0.1$; dashed line: $x = 0$.

Minimizing the energy with respect to the twin boundary spacing L yields

$$L = \sqrt{\frac{DE_{t.b}}{\vartheta c_{66}\varphi^2}}. \tag{8.31}$$

Thus the relationship between the twin spacing L and grain size D obeys the classical square root law [Khachaturyan 1983]. Fig.8.13 shows the temperature dependence of L, normalized by its largest value, for $\text{La}_{1.9}\text{Ba}_{0.1}\text{CuO}_4$. The size of the twin domains increases continuously as the temperature is reduced below the HTT-to-LTO phase transition temperature. The small spacing of twin boundaries near the HTT-to-LTO transition temperature, coupled with the large width of the boundaries (see Fig.8.12) indicates that the morphology of the system will be tweed-like, as has been recently observed in the previous section. There are no drastic changes of L predicted to occur at the temperature where the HTT boundary changes into a LTT boundary, though the width of the boundary increases discontinuously (see Fig.8.12). Unlike the behavior of the twin domains in conventional systems [Khachaturyan 1983], in which L increases monotonically with decreasing temperature due to the increase of interface energy, the presence of the LTT phase in the twin boundary causes L to decrease near the LTO-to-LTT transition temperature. This unexpected result, however, can be clearly explained by the decrease of the energy difference between domain wall phase (LTT) and the bulk phase (LTO), as shown in Fig.8.10.

8.5 Summary

Electron microscopy provides evidence that, at low temperature, $La_{1.88}Ba_{0.12}CuO_4$ consists of a mixture of orthorhombic and tetragonal phases, the latter located in the twin boundary regions. The model proposed here features a gradual change from LTT at the twin boundary sites to LTO in the twin matrix.

The width and density of twin boundaries in LTO phase using a theoretical approach by considering the possibility that the twin boundary could be LTT or HTT. We compared the free energy of twin boundaries with LTT to those with HTT structure using a Landau-type free energy model. The numerical calculations showed that LTT boundary as shown in Fig.8.8(b) is favored at low temperature and LTT becomes the predominant structure if annealing time is long enough. However because of the low LTO-LTT transition temperature ($\approx$ 60K), the growth of the LTT domain may be very slow. The morphology of the system will then consist of quasi-periodic domains of the LTO-LTT-LTO modulated structure as shown in Fig.8.7. An anharmonic lattice dynamical model for $La_{2-x}Ba_xCuO_4$ with parameters obtained from the first principle calculations [Pickett et al. 1991] predicts that the LTO to LTT transition is a vibrational-entropy-driven first-order phase transition. A quenched sample can thus retain metastable LTO domains with the LTT twin boundaries serving as nuclei for the growth of LTT phase. The structure modulation is also consistent with the small energy differences ($\approx$ 15 meV per CuO_6 octahedron[Pickett et al. 1991]) between LTO and LTT phase.

Chapter 9

Kinetics of the Alignment and the Formation of Bi-2223 in Bi-2223/Ag Tapes

9.1 Introduction

The Bi cuprate 2223-phase is one of the most promising high-T_c superconductors for large-scale and high-current applications because of its high superconducting transition temperature ($\approx$ 110K), high critical current density due to the lack of weak links in the bulk material, and the relative ease with which it can be made into wires and tapes [IEEE Trans. 1997]. On the other hand, the superconducting properties of Bi-2223 depend sensitively on its microstructure, which in turn depends on the fabrication processes.

The most widely-used method of making Bi-2223 materials is to mix Bi-2212, $CaPbO_3$ and CuO powder and to anneal the mixture in a sealed tube at about 830^oC for 10–100 hours. The factors which affect the rate of conversion from Bi-2212 to Bi-2223 have been studied extensively. An understanding of the formation mechanism and reaction kinetics is obviously important for the efficient preparation of Bi-2223 superconductors.

Over the last several years, tremendous advances have been made in fabrication of $Br_2Sr_2Ca_2C_3O_{10}$/Ag (Bi-2223/Ag) tapes by the powder-in-tube process. Presently, self field critical currents of 15-20 kA/cm^2 (for the superconducting area) at 77K are often reported for long lengths of rolled tapes [IEEE Trans. 1997]. To further improve the critical current properties of Bi-2223 tapes, numerous microstructure studies have been performed to characterize the formation of the Bi-2223 phase from various precursor powders in a Ag sheath and to correlate the observations with tapes' critical currents [Morgan et al. 1991 1992, Luo et al. 1993a, Grivel and Flukiger 1996, Grasso et al. 1995, Feng et al. 1993, Wu et al. 1997a]. From these studies, it is generally known that the alignment of the c-axis of Bi-2223 platelets as well as a high fraction (nearly 100%) of the Bi-2223 phase in the core are critical to achieving high critical current densities. However there are different mechanisms proposed for the formation and alignment processes for Bi-2223 in tape. For example, based primarily on the observation of the droplets in the interior cavities of the sintered powder mixture and of the importance of a liquid in Bi-2223 formation, it was suggested that Bi-2223 forms by deposition of dissolved cations from the liquid onto the ledges of Bi-2212 or existing Bi-2223 platelets [Morgan et al. 1991 1992, Morgan et al. 1994]. On the other hand, some of have used Avrami relationship [Hulbert 1969] between the fraction of Bi-2223 and the heat treatment time to conclude that the conversion process is by two-dimensional diffusion-controlled growth, or a standard nucleation-and-growth mechanism [Wang et al. 1993a, Hu et al. 1995, Grivel and Flukiger 1996]. More recently, by careful and consecutive observations of the grains on the surface of a sintered compact during the formation of Bi-2223, it was also concluded that the conversion process is by a nucleation-and-growth mechanism. In contrast the insertion of Ca and Cu ions in Bi-2212 was suggested as a possible process for the phase conversion by others [Luo et al. 1993a, Wang et al. 1993a]. In particular, Luo et al. [Luo et al. 1993a] arrived at this conclusion based on the analysis of the temporal phase conversion rate of Bi-2223 by the Avrami relationship and on the observations of the Bi-2223 faults in Bi-2212 grains in transmission electron microscopy. However, in a later article [Merchant et al. 1994], this same group discounted the significance of the intercalation mechanism, but instead favored the so-called reaction induced alignment

9.1. Introduction

process for the alignment and formation. We believe that the intercalation of the Ca/CuO$_2$ layers into the existing Bi-2212 platelets was indeed the primary phase conversion process. We will lay out our argument in this chapter based on extensive studies using transmission electron microscopy and *ex-situ* transmission X-ray diffraction techniques in conjunction with a detailed theoretical analysis [Bian et al. 1995, Cai et al. 1995, Cai and Zhu 1997, Wang et al. 1996b, Wu et al. 1997a]. We would not pretend that this is the last word on this topic, as more extensive studies are currently being conducted and there are still many aspects of this phenomena that remain a mystery. It is, however, a good example of how we attack a complicated problem using both the theoretical as well as experimental tools discussed in Chapter 3.

The detailed mechanism for the alignment process of the Bi-2223 platelets (*c*-axis alignment oriented along the normal of the tape surface) is less thoroughly studied than the conversion mechanism, even though the alignment is crucial in achieving high critical currents in Bi-2223 tapes. It is generally believed that the alignment is facilitated by the surrounding Ag matrix which confines the growth of the platelets along the tape plane. Also, since the pressing or rolling operations between the heat treatment sequences generally improve the values of I_c, it was commonly believed that the alignment was developed by the repeated deformation and heating cycles [Grasso et al. 1995]. However, this was later shown not likely to be the correct mechanism. In fact, mechanical deformation actually tends to worsen alignment; the benefit of pressing or rolling processes is primarily the compact of the cores [Thurston et al. 1996, Thurston et al. 1997]. In addition, it was earlier suggested that the Ag matrix assists the texturing based on the preferential formation of Bi-2223 at the Ag-superconductor interface [Feng et al. 1993, Luo et al. 1993b]. In another case, it was suggested that the texture was developed during the reaction to form Bi-2223 in the planar constrain imposed by the sheath [Merchant et al. 1994]. This suggestion was based on a detailed study of the inter-platelet angles between adjacent Bi-2223 grains across the cross-section of a tape during the formation of Bi-2223. This was also called "reaction induced texturing". In contrast, results of rocking curve measurements of the Bi-2223 tapes during heat treatment by a transmission X-ray diffraction technique have shown that most of the alignment of the platelets takes place at a very early stage. In some cases alignment occurs in Bi-2212 platelets well before Bi-2223 begins to form [Thurston et al. 1996, Thurston et al. 1997]. These sometimes conflicting suggestions demonstrate the need for a close examination of the processes leading to aligned Bi-2223 during heat treatment.

More recently, in order to study the kinetics of the Bi-2223 formation in the composite tapes, a number of investigations focused on the early stages of the phase evolution in the Ag-coated precursor powder pellets during the temperature ramp to the reaction temperatures. These were carried out employing conventional hot stage X-ray diffraction techniques [Xu et al. 1994, Xu et al. 1995, Polonka et al. 1991]. Bi-2212 in the pellet became unstable well below the reported Bi-2223 formation temperature as evidenced by the decreasing intensities of Bi-2212 lines and by the emergence of the lines associated with secondary phases. One very important and consistent observation from these studies is that these instabilities in the Bi-2212 phase only occur if the pressed compacts are coated or are in contact with Ag. Without Ag, the Bi-2212 phase remains stable until the melting temperature is reached. Similarly, the reduction of the melting temperature of the compacted Bi-2212 by an addition of Ag has been extensively documented [Morgan et al. 1991 1992, Luo et al. 1993b, McCallum et al. 1993, Guo et al. 1993, Grivel and Flukiger 1994a, Sun and Hellstrom 1995, Grivel et al. 1993b, Grivel and Flukiger 1994b]. These results show that the reaction kinetics of the composite tape is quite different than those of the pressed (Ag free) compacts of precursor powders.

In addition, X-ray diffraction analysis of quenched tapes and pressed powders has shown that the crystal structure of Bi-2212 changes above $400^{\circ}C$, but redevelops its original structure at higher temperatures ($> 835^{\circ}C$) [Grivel et al. 1993b, Grivel and Flukiger 1994b]. These changes in the structure were attributed to the instability of Pb in the Bi-2212 in the precursor at intermediate temperatures. Pb segregates out from Bi-2212 above $400^{\circ}C$ into a Pb rich phase, then is taken back up by Bi-2212 above $800^{\circ}C$ [Oh and Osamura 1991, Grivel et al. 1993a, Majewski et al. 1994, Zhang and Hellstrom 1996, Luo et al. 1996]. Since Pb plays an important role in the formation of Bi-2212 [Grivel et al. 1993a, Luo et al. 1996], these studies further point out the importance of

9.2. The Fabrication Procedure

Table 9.1 The precursor processes and compositions of the tapes. Tape A was heat-treated to 845^oC. Other tapes were heat-treated to 840^oC.

Tape	Core	Precursor	Composition	Second Phases
A	Mono	Spray Pyrolysis	$Bi_{1.8}Pb_{0.4}Sr_2Ca_{2.2}Cu_3O_x$	$CaCu_2O_3$
B	Mono	Spray Pyrolysis	$Bi_{1.8}Pb_{0.4}Sr_{1.9}Ca_2Cu_3O_x$	CaO, CuO
C	Multi	Mechanical Milling	$Bi_{1.8}Pb_{0.4}Sr_2Ca_{2.2}Cu_3O_x$	Ca_2CuO_3
D	Multi	Mechanical Milling	$Bi_{1.8}Pb_{0.39}Sr_{1.9}Ca_2Cu_{3.05}O_x$	Ca_2CuO_3

the microstructural changes taking place during the initial period of the heat treatment process, as well as those taking place during the reaction treatment.

In order to develop a comprehensive understanding of the microstructure evolution in powder-in-tube processed Bi-2223 tapes and to reconcile some of the conflicting observations, we carried out a thorough study of the microstructure and texture in the tapes as a function of various processing and heat treatment conditions. This study particular addresses kinetic processes of both the formation and the alignment of the Bi-2223 platelets for mono-cored and multicored Ag sheathed tapes. We rely extensively on transmission electron microscopy as well as *ex-situ* and *in-situ* transmission X-ray diffraction techniques for characterization of temporal variations in the texture, microstructure, and secondary phase assemblage in the tape cores.

9.2 The Fabrication Procedure

Four precursor powders obtained from three sources, three of them having slightly different average compositions, but with major phase (Bi,Pb)-2212, were used to fabricate the four series of powder-in-tube tapes (designated as Tape A-D) whose properties we will discuss in the following sections. Table 9.1 includes an overview of their characteristics and the techniques used to characterize them. Note that in addition to the methods of the powder preparation, the main differences among these powders are the primary secondary phase(s) after the major phase, (Bi,Pb)-2212. Tapes A and B were fabricated from two in-house powders prepared by an aerosol spray pyrolysis process (see [Wang et al. 1995] for details of the precursor fabrication by a spray pyrolysis method), while Tape D utilizes a commercially purchased precursor which was fabricated by a solid state method followed by a mechanical milling process. For each of these series, Ag-sheathed, mono-core tapes were manufactured using a standard process. The cross-sectional dimensions of the tapes after mechanical reduction were $\sim$ 3mm wide by $\sim$ 0.2mm, with $\sim$ 0.1mm core thickness. Segments $\sim$ 40mm in length were used for heat treatments. Tapes C was a 37-core (multicore) tape, with the precursor processed similarly as that of Tape D and with the same average composition as for Tape A. This tape was rolled to a cross- section $\sim$ 3mm by $\sim$ 0.28mm, with a typical core diameter approximately 25μm.

Tape A was originally used for a study of superconducting properties [Wang et al. 1995] and thus was pressed using $\sim$ 2Gpa at intermediate stages of heat treatment, after 50, 100, and 150 hours at 845°C in air. For this study, the same specimen of Tape A used for the above study was used primarily to examine the phase conversion process, i.e. Bi-2212 to Bi-2223. Thus, the specimen were heat treated in a horizontal furnace and either air quenched or furnace cooled (Those having very long heating duration were furnace cooled). After the superconducting properties were measured, each specimen was characterized for its alignment and percentage of phase conversion by *ex-situ* transmission X-ray diffraction techniques [Thurston et al. 1996] (see below). Then, the same specimens were examined by transmission electron microscopy for microstructure evolution in the superconducting core as a function of heat treatment duration.

Tapes B, C, and D were primarily used to study the kinetics of the alignment and the formation of Bi-2223 at early stages of the heat treatment. These tapes received no further mechanical processing. Initially, *in situ* transmission X-ray diffraction was performed for the measurement of the degree of the alignment and the percent of Bi-2223 while temperature was raised using the following schedule [Thurston et al. 1997]; a temperature ramp to 600°C at 300°C/h and a final soaking period of 6 hours, then a ramp to 840°C at 840°C/h and a final soaking period of up to $\sim$ 80 hour.

Then, the segments of Tapes B and C for TEM examination were heat treated following the same schedule for the X-ray measurement (Tape D was examined only by *in situ* X-ray studies). During the heat

9.2. The Fabrication Procedure

treatment, a series of tape segments were oil-quenched at preselected temperatures. They were 600°C after 6h soak, 780, 800, 820, 840, and after soaking times up to 40h at 840°C. The temperature in all of the furnaces was monitored by using calibrated thermocouples. However, some differences in reaction rates among samples of the same series of the tapes in different furnaces were observed. These discrepancies we ascribe to sample temperature variations of less than 3°C, which are attributed to the difficulty in attaining a precise temperature match in the vertical furnaces which were required for oil-quenching and for *in situ* X-ray diffraction measurements.

Since the description of transmission X-ray diffraction experiments is given in Refs. [Thurston et al. 1996, Thurston et al. 1997], we will only briefly describe them here. The experiments were performed using a 25 kV X- ray beam at the Brookhaven National Laboratory's National Synchrotron Light Source to determine the degree of alignment and fraction of phase conversion during heat treatment. The relative abundance F_{2223} of Bi-2223 relative to that of the total Bi superconductor (sum of Bi-2223 and Bi-2212) is estimated from the ratio of intensities, designated I_{2223} and I_{2212} respectively, of the $[115]_{2223}$ and the $[105]_{2212}$ reflection peaks after a correction for absolute intensity,

$$F_{2223} = \frac{1.65 I_{2223}}{1.65 I_{2223} + I_{2212}}.$$

The measurement, both *in-situ* and *ex-situ*, of transmission X-ray rocking curves reveal the development of *c*-axis texturing during heat treatment. The FWHM of the [200] peak obtained from the rocking curve is used as a measure of texturing (Note that the [200] lines for Bi-2212 and Bi-2223 are very closely spaced and for this experiment, these were not resolved). An advantage of this X-ray technique over standard laboratory conditions is that detected X-ray penetrates the entire thickness of the tape, including Ag sheath. Thus the entire superconductor thickness is sampled, and overlying Ag is not disturbed.

Since the primary purpose of this study is to clarify the kinetics of microstructure development taking place in the early stages of the powder -in-tube processed Bi-2223/Ag tapes prior to intermediate p1ressing of rolling, detailed investigations of critical currents for these tapes were not performed. However, some segments of these tapes (Tapes A and

B) were pressed and heat treated for critical currents measurements and a brief summary of the maximum values of J_c measured for these tapes are given as follows. The maximum critical current density, which was attained for Tape A, under self magnetic field at 77K is $\sim 20,000 \text{A}/\text{cm}^2$ when heated at 845 and 840°C in air [Wang et al. 1995]. On the other hand, when a tape similar to Tape B was heated at 825°C in 8% O_2 atmosphere, its J_c was $\sim 21,000 \text{A}/\text{cm}^2$ with one pressing while it was $\sim 18,000 \text{A}/\text{cm}^2$ if it was heat treated at 840°C in air and with three intermediate pressings. Although Tape C was not tested for critical currents, similar rolled tapes have achieved J_c of $\sim 15,000 \text{A}/\text{cm}^2$ when they were further heat treated with intermediate rolling processes. No measurement for J_c was carried out of Tape D.

9.3 Experimental Results on Microstructural and Compositional Evolution at Early Stages of Heat Treatments

As made clear in earlier publications [IEEE Trans. 1997, Luo et al. 1993a, Thurston et al. 1997, Xu et al. 1995, Polonka et al. 1991], extensive and important microstructure changes take place very early in heat treatment, including the ramp-up portion. In particular, the *in situ* transmission X-ray diffraction study [Thurston et al. 1997] clearly demonstrated (as confirmed here) that the major fraction of the alignment of the Bi-cuprate platelets takes place before the formation of Bi-2223. As shown below, the details of the alignment process in the tape depend on the precursor powders used. Thus, for the microstructure examination of the cores at this early stage of the heat treatment, we have chosen three tapes, Tape B (mono-cored) containing the spray pyrolysis precursor and Tape C (multicored) and D (mono-cored) with the mechanically milled powder for the *in situ* X-ray study. Tapes B and C were used for TEM examinations. First we describe the *in situ* X-ray measurements from these tapes. Then we present the detailed microstructural analysis. Since a more exhaustive study was carried out for Tape C than for Tape B, we will present the results from Tape

9.3. Evolution at Early Stages of Heat Treatment

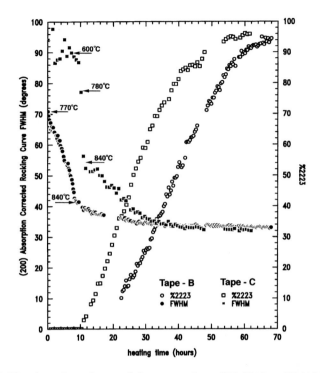

Figure 9.1 The time dependence of the conversion of Bi-2212 to Bi-2223, and the alignment (i.e. FWHM) of the Bi-2212 and Bi-2223 platelets are shown for Tapes B and C. These were determined by *in situ* transmission X-ray diffraction measurements.

C first, and then compare that to Tape B's results. Finally, the earlier findings relative to the alignment process for Tape A [Wang et al. 1996a] is briefly described.

9.3.1 *In situ* X-ray Diffraction Measurements

In Fig.9.1, the FWHM and the percentage of Bi-2223 in Tapes B and C are shown as a function of heat treatment duration [1]. Both tapes

---
[1] The heat treatment schedule for Tape C was what is stated in the previous section. However, Tape B was ramped to 770°C within two hours and then ramped to 840°C at 10C/h. Thus, unfortunately, the heat treatment schedules for these tapes were not the same. However, the differences in the general characteristics of

were heated in air. In both cases, the majority of the alignment occurs in the ramp-up portion of the heat treatment. However, a significant difference in the temporal dependence of the alignment process are seen in these two tapes. In Tape B, which contains a spray pyrolysis precursor powder, the alignment is essentially completed before the Bi-2223 formation temperature is reached. On the other hand, in Tape C the alignment is approximately one half completed when the formation temperature is attained. Then, the alignment continues to improve with time. Very interestingly, in contrast to Tape B, Bi-2223 in Tape C begins to grow immediately after the reaction temperature is reached, and the alignment improves as the Bi-2223 phase grows.

We have found in all of these *in situ* X-ray rocking curve measurements for the tapes that the degree of the alignment for a given tape is primarily dependent on heat treatment temperature rather than soaking duration at the temperature [Thurston *et al.* 1997, Wang *et al.* 1995]. Also, it was found that the characteristic temporal dependence of the alignment for tape D (as shown in Fig.9.2) is essentially the same as for Tape C and that for Tape A (heated at 840°C) was the same for Tape B. Thus, within the extent of materials and processing in this study, it appears that neither the small differences in the "average" compositions of the precursor nor the size of the filaments influences the time dependence of the alignment. In the as-rolled condition, the powder in Tapes C and D (mechanically milled) were randomly orientated while some alignment is seen in Tape B (spray pyrolysis). (Compare the values of FWHM before heat treatments in Figs.9.1 and 9.2) Thus, the factor which is controlling these differences in the temporal alignment characteristics shown in Fig.9.1 is likely to be the processing details of the precursor powders, i.e. the aerosol spray pyrolysis vs. the mechanical milling. In addition, as shown in Table 9.1 and as described below, the phase assemblages in these powders are quite different, (the powders in Tape B contain CaO and CuO as the major additional phases after Bi-2212 while those in Tapes C and D are primarily Ca_2CuO_3). This factor may affect formation of liquid and

---

the alignment and the Bi-2223 growth between these tapes were intrinsic, i.e. all of the tapes with the spray pyrolysis completed alignment before the formation of Bi-2223 initiated under various temperature ramp schedules while the alignment for Tapes C and D completed with the growth of Bi-2223.

9.3. Evolution at Early Stages of Heat Treatment

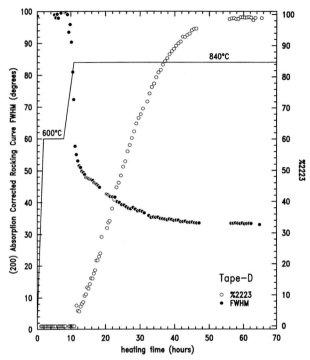

Figure 9.2 The time dependence of the conversion of Bi-2212 to Bi-2223, and the degree of the alignment (i.e. FWHM) of the Bi-2212 and Bi-2223 platelets during the heat treatment are shown for Tape D. Note that in this particular case, the temperature ramp rate between 600 and 840°C was 100°C/h.

therefore may also contribute to the different alignment behavior. The portion of the alignment for Tapes C and D occurring after the initiation of the Bi-2223 formation is likely to be the process which was earlier called the "reaction induced texturing" [Merchant et al. 1994]. It is also interesting to note that the difference in the characteristics of the alignment among these tapes can be easily observed in the results of the standard reflection X-ray diffraction of the tapes after short periods (1 ∼ 10h) of formation temperature soaking. The tape with a milled powder exhibits a relatively strong [115] line while the same line is absent in the tape with a spray pyrolysis powder.

9.3.2 Electron Microscopy

In order to further confirm that the difference in the alignment characteristics seem in Fig.9.1 is due to the difference in the preparation methods for the precursor powders, we have selected schedules for oil quenching of TEM specimens based on Fig.9.1. They are: for Tape C, as rolled, 600°C after 6h, 760, 780, 800, 820, and 840°C after 4, 12, 20, 30h; and for Tape B, as rolled, 780, 800, 820, 840, and 840°C after 20 and 40h. The results of the TEM studies from these tapes are presented comparing behaviors of these tapes relative to the alignment and the secondary phase assemblage in the early stages as well as the formation of Bi-2223. Finally, the earlier findings in Tape A relative to the alignment process is briefly summarized.

Tape C

i. As rolled

In the as-rolled condition, the powders were very heavily deformed, and we could only identify Bi-2212 in the filament areas even though it is known that the major secondary phase in the core is Ca_2CuO_3 [see Fig.9.3(a)]. The electron diffraction spots were quite diffuse, but those for Bi-2212 were very easily identified. As shown in Fig.9.3(b), many of the Bi-2212 platelets were very thin and extensively curved at this stage of tape fabrication. It is very interesting that the lattice images of Bi-2212 platelets were very easily obtained without tilting the specimen from the horizontal position in the microscope. This implies that these platelets are aligned along the axis of the tape in the rolling direction, but not necessarily along the flat face of the tape. Also, the X-ray energy dispersive spectroscopy revealed Pb in Bi-2212, but not in other areas. This is shown in Fig.9.4 for an energy range where L_α and L_β lines of Bi and Pb are clearly seen.

ii. After 6h at 600°C

After the specimen was heated for 6h at 600°C, the morphology of the filaments is altered significantly. The Bi-2212 platelets have polygonized into smaller grains [see Figs.9.3(c) and 9.3(d)]. This is believed to be due to the relief of high strain energies retained in the platelets by the mechanical deformation process during the precursor fabrication. An indication of the relieved strains was seen in the

9.3. Evolution at Early Stages of Heat Treatment 249

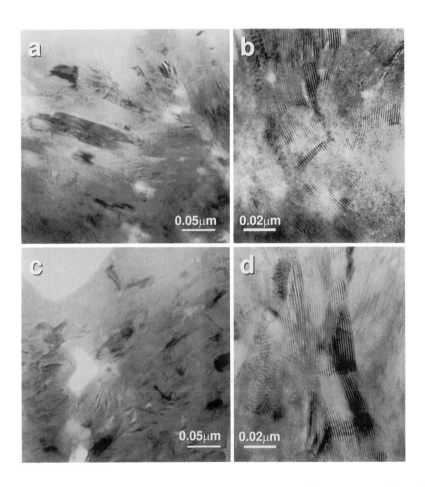

Figure 9.3 Cross-sectional TEM images of Tape C (a and b) as rolled and (c and d) after 6h at 600^oC are both shown at two magnifications.

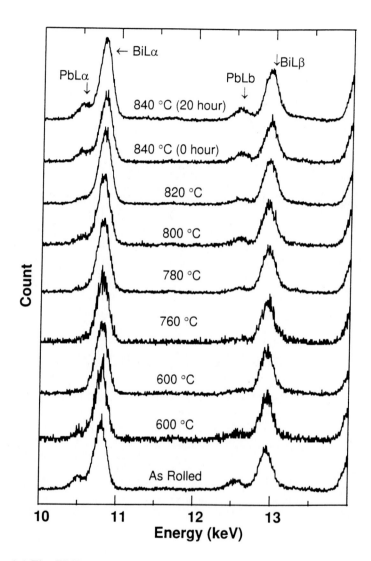

Figure 9.4 The EDX intensities corresponding to L_α and L_β lines of Bi, and Pb, taken from Bi-2212 grains, are shown as function of temperature for Tape C. Two data from 600^oC are shown since there were some variations in the relative intensities from one area to another.

9.3. Evolution at Early Stages of Heat Treatment

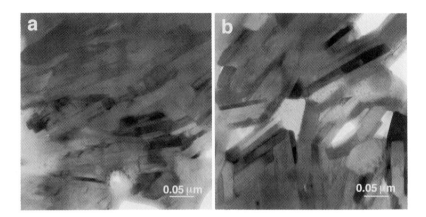

Figure 9.5 Cross-sectional TEM images (a and b) of Tape C after the tape specimens were quenched from 780 and 800°C, respectively.

sharper diffraction spots at this stage than those in the as-rolled condition. During this heat treatment, as reported previously [Grivel et al. 1993b, Grivel and Flukiger 1994b]and as shown in Fig.9.4, Pb from the Bi-2212 platelets has precipitated out to form very small secondary phase particles [see Fig.9.2(c)] which is often identified as "3221", or $(Bi,Pb)_3Sr_2Ca_2CuO_x$ [Dou et al. 1991]. (This is also sometimes called "451" or $(Bi,Pb)_4(Sr,Ca)_5CuO_x$. For a detailed discussion of the crystal structure and the composition of the phase, see [Wu et al. 1997a].) At this point, the precipitates are too small to be easily observed, and we could make no accurate determination of structure and composition. The EDX measurements of some of these precipitates indicate high Pb content. As in the case of the as-rolled specimen, the platelets are still axially aligned at this point [see Fig.9.3(d)].

iii. at 760°C

By the time the temperature had reached 760C, as shown in Fig.9.5(a), various precipitates had grown to typical dimension of 30nm and were quite uniformly distributed. Also, the ratio Pb/Bi of the new Pb containing precipitates is essentially the same as that for the large precipitates observed after higher temperature heat treatment [see Fig.9.6(a)]. The Bi-2212 platelets have also grown to the thickness of $\sim$ 40nm, with length several times the thickness. One also observes some appearance of local alignment among the platelets, but not necessarily along the tape face, as there is little change in the values of HWHM in Fig.9.1. Furthermore, Pb is essentially absent from the Bi-2212 phase as shown in Fig.9.4.

  iv. at 780°C

At this temperature, both Bi-2212 and the secondary phases have continued to increase in size as shown in Fig.9.5(b). Also, in the EDX measurements, there is some indication that Pb is being reabsorbed back into the Bi-2212 phase (see Fig.9.4). This temperature is significantly lower than the 835°C reported earlier for the incorporation of Pb back into Bi-2212, which was estimated based on an x-ray diffraction observation of structural changes that are associated with the incorporation of Pb [Grivel and Flukiger 1994b]. This difference may possibly be due to the different techniques used to detect the changes or to the fact that the latter work was done in a sintered pellet rather than in an Ag composite. Also, at 780°C, other precipitates in addition to 3221 are clearly identified due to a higher degree of segregation. These are $(Ca,Sr)_2CuO_3$ along with small amounts of SrO, CuO (often attached to 3221), $(Sr_{0.8},Ca_{0.2})_{14}Cu_{24}O_{41}$, and Ca_2PbO_4. While relative amounts are difficult to determine in TEM, we estimate that $(Ca,Sr)_2CuO_3$, CuO and 3221, in decreasing order, were the most prominent secondary phases. The main new development in specimens quenched from 780°C is the appearance of small amorphous regions, not necessarily limited to the Ag/powder interface. These regions are thought to be a liquid at high temperature. Since the amorphous region becomes much more substantial as the temperature is raised, we will defer the discussion on the partial melting to the specimens quenched from higher temperatures.

  v. at 800, 820, and 840°C (0h, without soak)

In this temperature range, as expected, the Bi-2212 grains grow significantly. However, the most notable change is a drastic increase of amorphous areas initially observed at 780°C. There are three clearly identifiable types of amorphous phases. Two of them are associated with the two precipitates, the Pb containing immediate phase 3221 and $(Ca,Sr)_{14}Cu_{24}O_{41}$, often called (14,24). (Note that only a small amount of this latter phase was found at this temperature.) Since the thin amorphous layers surround each precipitate as shown in Figs.9.6(a)

9.3. Evolution at Early Stages of Heat Treatment

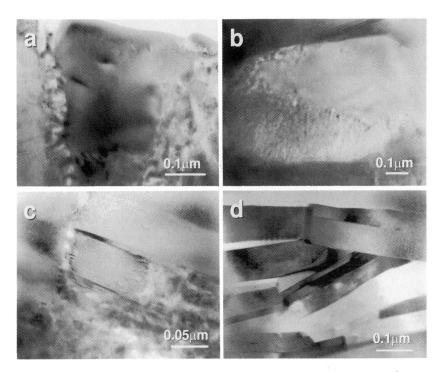

Figure 9.6 Cross-sectional TEM images from Tape C quenched from 820°C are shown for three types of amorphous regions associated with (a) the intermediate ("3221") phase, (b) $(Ca,Sr)_{14}Cu_{24}O_{41}$, and (c) Bi-2212. Also shown in (d) is locally aligned Bi-2212 platelets.

and 9.6(b), and the compositions of the layers are similar to the associated precipitates, as also shown in Fig.9.7, we speculate that these amorphous regions are formed by partial melting of the subject phases in contact with Bi-2212. (The precipitates and the amorphous regions shown in Figs.9.6 and 9.7 are from the tapes quenched from 820°C.) This is a reasonable proposal for the case of the intermediate 3221 phase since this phase disappears at higher temperatures, as discussed below. Furthermore, at this temperature, the Bi-2212 phase begins to reabsorb Pb (see Fig.9.4) and the melting of this 3221 precipitate certainly is expected to facilitate this process. On the other hand, the so-called (14,24) phase, which has a melting temperature of 955°C is known to be a major remaining secondary phase after the tape fabrication process [Hwang et al. 1990, Ströbel et al. 1992]. Thus, it is possible that (14,24) grows when in contact with a liquid of a similar composition rather than dissolving and shrinking in size. However, it is not obvious how such a liquid is formed. The presence of Ag and its contact with Bi- 2212 may promote the formation of the liquid.

The third type of amorphous region [see Fig.9.6(c)] is most common. This may be due to partial melting of Bi-2212 since the composition is close to the Pb-containing Bi-2212 phase (See. Fig.9.7), although some variation in composition is seen. Even though Ag was not detected in any of these amorphous areas, the presence of Ag is believed to be crucial in the formation of the liquids [Morgan et al. 1991 1992, Luo et al. 1993a, Merchant et al. 1994, Grivel and Flukiger 1994a, Sun and Hellstrom 1995, Grivel et al. 1993b, Grivel and Flukiger 1994b]. A number of authors have previously considered the possible lowering of the melting temperature of Bi-2212 in contact with Ag. For example, Morgan et al. [Morgan et al. 1991 1992]reported an eutectic temperature of $> 822^oC$ in the Bi-2223 mixture in porous pellet using a thermal gradient furnace. On the other hand, McCallum et al. [McCallum et al. 1993] reported an eutectic temperature at $\sim 830^oC$ for Bi-2212 in contact with Ag. Furthermore, they noted an endothermic event at a temperature as low as 810^oC, where a deviation from the base line was observed in a differential thermal analysis of the Bi-2212/Ag in air. This temperature is very close to that at which a significant amorphous region was first observed in this study (800^oC). Perhaps, the apparently low temperatures at which the liquid is observed in this study are due to the fact that these TEM examinations are much more sensitive than bulk techniques such as differential thermal analysis when it comes to detecting liquid in these specimens. On the other hand, Luo et al. reported an amorphous region having a composition of $Bi_{2.2}Pb_{1.8}Sr_{1.3}CaCu_{2.0}O_x$ which begins to melt at $\sim 650^oC$ [Luo et al. 1993a, Luo et al. 1993b, Luo et al. 1997]. We have not detected any sign of partial melting below 780^oC. We found that composition of the amorphous phase was lower in Pb and Bi in our study than that observed in their work. Also, there are other reports, including a number of high temperature x-ray measurements on the pellets of the Bi-2223 mixtures, that

9.3. Evolution at Early Stages of Heat Treatment

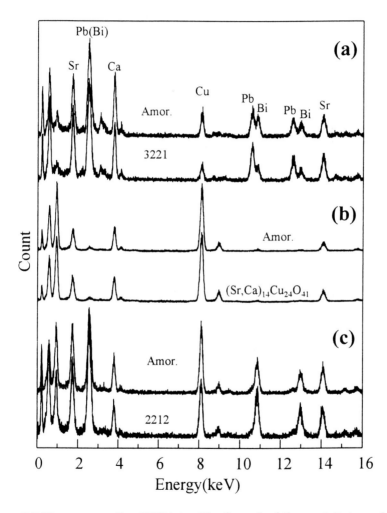

Figure 9.7 The corresponding EDX intensities for each of the precipitates and the corresponding amorphous regions in Fig.9.6 are shown.

Table 9.2 The compositions of the Pb containing intermediate precipitates.

Quenching Temperature	Pb	Bi	Ca	Sr	Cu
820°C	13.70	4.45	28.27	34.22	14.79
840°C	17.70	5.11	29.75	34.57	12.88
840°C after 20 h	19.89	2.42	52.11	8.90	6.70

indicated substantial "chemical" activities well below the generally acknowledged reaction temperature [Xu et al. 1994, Xu et al. 1995, Polonka et al. 1991]. However, it is difficult to directly associate these observations with the formation of liquid since conventional in situ X-ray diffraction technique is not very effective in detecting a small amount of liquid in a composite tape.

Above 800°C, a significant improvement in the alignment is noted in the Bi-2212 platelets from the X-ray diffraction data in Fig.9.1. However, no obvious alignment of the platelets parallel to the tape face was seen in TEM, although some local alignment among platelets was observed [see Fig.9.6(d)]. Also, the intermediate phase 3221 has grown considerably in size in this temperature range such that it was possible to perform more precise chemical analysis of this precipitate using a wavelength dispersive technique on a polished surface with SEM. The results of the measurements are shown in Table 9.2 for the specimens quenched from 820°C and 840°C, and after 20h at 840°C. (Compositions are determined by averaging the measurements from 5-6 precipitates.) The composition of the precipitates varied with temperature and heat treatment duration as seen in the table. Thus, it was difficult to assign a specific fractional composition by using whole numbers (as 3221 or 451). The specimen heated for 20h at 840°C reveals a dramatic change in composition due to the fact that at this temperature the phase becomes unstable with respect to formation of $(Ca,Sr)_2PbO_4$ and eventually disappears (discussed below). Also, this variation in the composition of the "3221" phase is consistent with the reported insensitivity of the positions of the X-ray diffraction lines to both the ratio and the combined amount of Pb and Bi with respect to the other elements [Luo et al. 1994]. However, a detailed electron diffraction analysis re-

9.3. Evolution at Early Stages of Heat Treatment

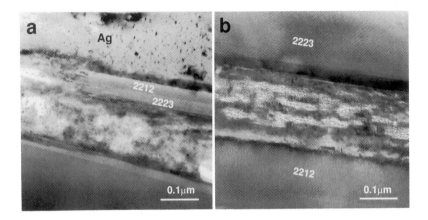

Figure 9.8 (a) An example of the amorphous region between the Ag matrix and Bi-2212 is shown in the specimen Tape C quenched from $840^{\circ}C$, and (b) The river patterned amorphous area between Bi-2212 platelets for a specimen quenched after 12h at $840^{\circ}C$.

veals that, depending on the composition, the crystal structure can be hexagonal or c-centered monoclinic [Dou et al. 1991].

vi. at $840^{\circ}C$ for 4, 12, 20, and 30h

In addition to the rapid grain growth of Bi(2212,2223), the most prominent change at this stage of the heat treatment is the change in the morphology of the amorphous areas which are trapped between the Bi-2212/ Bi-2223 platelets. The regions such as those shown in Fig.9.6(c) transformed to give the appearance of river patterns [light and dark contrast shown in Fig.9.8(b)]. Within the amorphous area, segregation of Ca causes the river-like patterns. With annealing times up to 30h at $840^{\circ}C$ the areas covered by such amorphous areas increased in volume. Occasionally, there are areas where the river patterns and ring shaped amorphous areas are in contact, but separated by a distinct boundary rather than by a smeared one. Thus, there appears to be very little interaction between the ring shape amorphous and the river patterned areas.

The growth of Bi-2212 platelets appears to take place into the amorphous regions for soak times less than 12h at $840^{\circ}C$, but it was difficult to assess whether the platelets are growing from or being dissolved into the liquid. Since, in this tape, as shown by the transmission X-ray rocking curve measurements in Fig.9.1, the alignment of Bi-2212 and Bi-2223 continued after the temperature had reached 840°C, one is tempted to conclude that the alignment of the platelets is assisted by the presence of a substantial amount of liquid which enhances the growth of the platelets. However, after 30h at 840°C, the alignment of the platelets is essentially completed, yet there were a significant number of amorphous areas trapped between the Bi(2212,2223) platelets. Thus, the presence of the river patterned liquid is not the sole cause nor even a major contributing factor for the alignment process. Thus, one has to conclude that the alignment of the platelets in these tapes is accomplished by a combination of factors, which include the general physical confinement imposed by the Ag matrix, the very highly anisotropic growth rate of the a-b plane with respect to that for the c-axis direction, and the presence of liquid which enhances the anisotropic growth rate. Previously, Merchant et al. stressed the importance of the formation of Bi-2223 in the alignment process as well as of the planar confinement of the growth of the cuprates by the Ag matrix [Merchant et al. 1994]. However, we believe that the formation of Bi-2223 is coincidental to the alignment. As shown in Fig.9.1, a major fraction of the alignment takes place even before Bi-2223 begins to form. Although in Tape C, the alignment continues to improve while Bi-2223 is being formed, this is likely to be due to the highly anisotropic growth of Bi-2223 platelets rather than the formation of Bi-2223 itself.

As the time at this temperature increases, regions of the Bi-2223 phases are increasingly apparent. We will defer this subject to a later section where the formation mechanisms will be discussed in detail.

Tape B

The alignment characteristic for Tape B is very different from that for Tape C. Thus, for comparison, Tape B was heat treated with the same heat treatment schedule used for Tape C. Below are the results for Tape B of a TEM study of microstructural changes during the early stages of heat treatment which are compared with those for Tape C.

    i. as rolled

In comparison with the precursor powder for Tape C, the powder

9.3. Evolution at Early Stages of Heat Treatment

in Tape B was only lightly deformed by the rolling process as seen in Fig.9.9(a). Thus, one concludes that the deformation of Bi-2212 platelets seen in Fig.9.3 for Tape C is due to the mechanical milling of the precursor powders. The important difference in the condition of the initial powders is that the major secondary phases are CaO and CuO in this tape instead of Ca_2CuO and CuO in Tape C. An area where CaO is heavily segregated is shown in Fig.9.9(a) to emphasize the presence of CaO. The presence of CaO and CuO in the precursor powders produced by a spray pyrolysis method is also noted by others using a standard x-ray diffraction measurement.

ii. 780, 800, and 820°C

Examples of the micrographs taken for the specimens quenched from 780 and 800°C are shown in Figs.9.9(b) and 9.9(c). A number of fine new precipitates are found but they are too small to be identified by composition at 780°C. As in the case of Tape C, Pb from the Bi-2212 platelets in this tape had dissolved out at 780°C. As temperature increases, the precipitates grow in size sufficiently to be identified. Remarkably, the secondary precipitate assemblage in this tape is totally different from that in Tape C. The precipitates for Tape C in this temperature range were, listed roughly in decreasing order in quantity, $(Ca_{0.8}Sr_{0.2})CuO_2$, CuO, 3221, SrO, (14,24), and $(Ca,Sr)_2CuO_3$. Interestingly, all of these precipitates except for 3221 are those found in the CuO SrO pseudo- binary phase diagram if one recognizes the fact that Ca and Sr substitute quite easily. Although it is difficult to estimate the relative phase fractions from TEM studies, $(Ca_{0.8}Sr_{0.2})CuO_2$ was prominent since the precipitates are often quite large (1-5 mm) while the (14,24) and Ca_2CuO_3 are far less visible. Finding $(Ca_{0.8}Sr_{0.2})CuO_2$ as the major secondary phase is surprising since this phase was not detected at all in Tape C. It also turns out that this phase has an ordered structure as indicated by the superstructure reflections in the electron diffraction pattern. This is consistent with an earlier report on this phase by X-ray diffraction [Hwang *et al.* 1990].

Another notable difference observed between Tapes B and C, is the fact that for Tape B the amorphous area was relatively small, even when quenched from 820°C. This observation is very surprising and implies that liquid does not play an important role in the alignment process since the FWHM for this tape is reduced nearly to the mini-

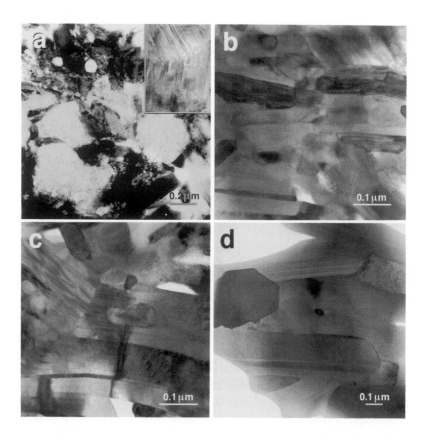

Figure 9.9 Examples of the microstructures illustrating the morphology of Tape B during the ramp up portion of the heat treatment showing Bi- 2212 and precipitates. (a) as rolled, (b) at $780^{o}C$, (c) at $800^{o}C$ and (d) at $840^{o}C$. For the as-rolled sample, an area with a high concentration of CaO is selected to illustrate the presence of this phase before heat treatment.

9.3. Evolution at Early Stages of Heat Treatment

mum value by the time the temperature had reached 840°C. Thus, it appears that the alignment process is primarily controlled by the highly anisotropic growth of the Bi-cuprate in the physically confined space imposed by the Ag matrix. It is also noted that the Ca contents in Bi-2212 platelets in Tape C are generally higher than those in Tape B. This may be attributed to the fact that Ca in the precursor powder for Tape C is higher than that for Tape B, but it is more likely that a significant amount of Ca is required for the formation of $(Ca_{0.8}Sr_{0.2})CuO_2$ precipitates to deprive Ca from Bi-2212 in Tape B. However, whether a high Ca in Bi-2212 in Tape C promotes formation of liquid is not clear.

iii. 840°C for 0, 20, and 40h

In addition to the continued enlargement of Bi-2212 platelets [See Fig.9.9(d)], the major change at this stage for the heat treatment of Tape B is the disappearance of the $(Ca,Sr)CuO_2$ precipitate particles and increased $(Ca,Sr)_2CuO_3$ as well as the (14,24) precipitates. Another very important difference in this tape from that for Tape C is that even at this stage of the heat treatment amorphous regions are less abundant as shown in Fig.9.9. The importance of this observation relative to the formation of Bi-2223 will be discussed later.

Tape A Heat Treated at 845°C

Previously, the measurements of the HWHM and the fractional conversion of Bi-2212 to Bi-2223 for the air-quenched Tape A specimens were made by the ex situ transmission X-ray diffraction techniques. The results from these measurements are shown in Fig.9.10. As reported earlier, one of the important observations from these measurements is that the alignment process is essentially completed prior to the formation of Bi-2223 [Wang et al. 1996b]. These results strongly suggest that the processes for the alignment and phase conversions are independent in this tape. Also, interestingly, a specimen, which was oil- quenched form 845°C after 20h soaking, exhibited a large volume of amorphous regions compared with Tape B. This difference may originate in the difference in the secondary phases in the precursors for these tapes as shown in Tape B.

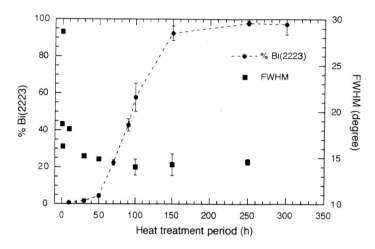

Figure 9.10 The results of the *ex-situ* transmission X-ray measurements of the FWHM of the [200] peak rocking curves and the fraction of Bi-2223 phase relative to the sum of Bi-2212 and Bi-2223 are shown for Tape A as a function of heat treatment period at $845^\circ C$ in air.

9.3.3 The Formation of Bi-2223

As mentioned above, the most detailed microstructural characterization relative to the formation of Bi-2223 was performed on those air quenched specimens from Tape A which were heat treated at $845^\circ C$ in air. Since this set of specimens exhibited a very slow growth rate for Bi-2223, two other tapes (Tapes B and C) with much faster rates of formation were examined for certain characteristics of the reaction mechanism [e.g. the spatial distribution of the Bi-2223 phase within a grain and the formation of Bi-2223 adjacent to the liquid]. Their results are compared with the earlier observations for Tape A. These specimens were prepared using different starting precursor powders (both Tape B and C). The comparative examination of Tapes B and C is of interest since, as discussed above, the alignment process continues in Tape C when the Bi-2223 begins to form while in Tape B, the process is completed before the initiation of the Bi-2223 formation. First, we will summarize the main observation regarding the formation of Bi-2223 in

9.3. Evolution at Early Stages of Heat Treatment

Tape A since a brief description of the study has already been given elsewhere [Wang et al. 1996b]. Then, the new findings on the other tapes will be given and contrasted with those of Tape A.

Heat Treatment of Tape A at 845°C in Air

In order to determine the microstructural origin for the phase conversion process, the cross sections of each of the specimens, which were used for transmission XRD previously, were examined by a lattice imaging technique of TEM with the electron-beam direction parallel to the $a-b$ planes. This technique allows us to easily differentiate the different superconducting phases of the Bi cuprates by observing the sizes of the half unit cells along the c-axis. Initially, we sought supporting evidence for a standard nucleation and growth model for the Bi-2223 phase, for example, the observation of the regions where Bi-2212 and Bi-2223 grains are separated by a thin amorphous layer or secondary phases. However, in spite of our exhaustive search for microstructural evidence for such a reaction, we found none. [Note that we have observed some evidence supporting the nucleation and the growth of Bi-2223 in Tape C which will be discussed in the later section.] Rather, we found in all of the Tape A specimens that each grain consists of a mixture of the Bi-2212 and Bi-2223 of varying proportion. Examples of grains with such a mixture of the phases are often shown, e.g. Fig.3 of [Wang et al. 1996b]. Here, we show a similar example from Tape B in Fig.9.11. It was generally found that there were regions of finely mixed phases as well as other regions of relatively pure Bi-2212 and Bi-2223. Occasionally, ordered regions of alternating layers of one half cell each of Bi-2212 and Bi-2223 were found; these were identified as the Bi(4435) phase, an intermediate between Bi-2212 and Bi-2223 [Bian et al. 1995]. Based on the above observations and combined with our failure to observe solid evidence supporting the nucleation and growth model, we propose that the Bi-2212 to Bi-2223 phase conversion is likely to be via the insertion of the additional Ca-Cu-O bi-layer into the existing Bi-2212 at least in this tape [Bian et al. 1995, Cai et al. 1995, Cai and Zhu 1997, Wang et al. 1996b].

If the insertion of planes is the mechanism for phase conversion, the numerical fraction of half unit cells of Bi-2223 as observed by TEM

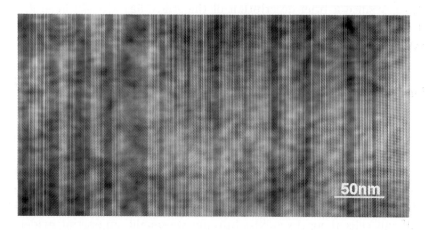

Figure 9.11 A TEM lattice image of an area with a fine mixture of Bi-2212 and Bi-2223, viewed along the [200] direction, showing a partially intercalated grain by Cu-Ca-O bi-layers. The image was taken from Tape B after a heat treatment for 40h at 840^oC in air. In this image, each paired light and dark line is a half unit cell along the c-axis. When a single Ca-Cu-O bi-layer insertion is made, the width of the pair increases from ~ 1.54 to ~ 1.85 nm corresponding to half of c-axis lattice parameter of Bi-2212 and Bi-2223, respectively.

should be quantitatively related to the percentage of Bi-2223 determined by transmission x-ray diffraction. In order to confirm this, the fraction of the wider spaced dark and light line pairs [corresponding to 1/2 of the unit cell of Bi-2223 along the c-direction] in a number (20-50) of the micrographs for each specimen were determined. The results are shown in Fig.9.12. In the figure, each horizontal bar corresponds to a percentage of Bi-2223 in a TEM image taken from a grain or grains. The number of pairs counted varied from 150 to 400 for each image, i.e. for each bar. Also, shown in the figure with solid circles are the average values of these measurements for each heat treatment. This is compared with the results of the transmission x-ray diffraction measurements for the percent conversion from Fig.9.10 (shown by open circles). These results are in very good agreement. There are small discrepancies between these two averages at the onset of the Bi- 2223 formation and at very long heat treatment times. These differences are believed to be due to the fact that the x-ray measurements are less sensitive to indi-

9.3. Evolution at Early Stages of Heat Treatment

vidual or very small clusters of intercalated layers. Initially, there are some single or groups of insertions which are not dense enough to contribute to the x-ray diffraction intensities; similarly, at the end of the heat treatment, there are some individual un-intercalated layers, i.e. Bi- 2212, which also cannot be seen in x-ray diffraction. Thus, the intensities inferred from x-ray measurements for the early and late stages of the heat treatments may be less sensitive than those determined by the TEM technique.

If a "uniformly random" intercalation of the Cu-Ca-O planes into the Bi-2212 lattice occurs, one would expect both a broadening and a shift in the XRD lines for Bi-2212 and Bi-2223 due to the change in lattice constants [Manaila et al. 1995, Tarascon et al. 1988b, Kramer et al. 1993]. However, neither is this observed in these tapes, nor has it been seen by others. (For example, see [Osamura et al. 1997]) As illustrated in Fig.9.12, the process of the conversion is not uniform but statistically random, i.e. the degree of the conversion varies extensively from one grain to another as well as within a grain. The intercalation is confined to a limited area at a given times. The areas, where the intercalation is initiated, convert relatively quickly to the Bi-2223 phase while other areas remain as Bi-2212. Thus, the x-ray diffraction pattern essentially presents a mixture of these two phases rather than a single intermediate phase with line broadening or the line shifting as expected for a uniformly random intercalation process.

Also, if the insertion of the Cu-Ca-O bi-layers is the mechanism for conversion, we might expect to catch the insertion process "in progress" when the specimen is quenched. Careful examination of the micrographs does show some of these single intercalating planes. Since the end of an inserting plane is an edge dislocation , the tip of the plane is surrounded by a stress field which exhibits a dark circular area. Also, occasionally, at a very early stage of the conversion, a group of insertion planes is observed, as shown in Fig.9.13(a). This suggests that one intercalating plane may have a co-operative effect on the nucleation and growth of the neighboring planes. Moreover, if the insertion of the Ca-Cu-O bi-layers is nucleated, it occurs at the grain boundaries. The nucleation and growth of a pair of inserting planes from a grain boundary are observed as shown in Fig.9.13(b). In this figure, the planes are tilted with respect to the beam direction and thus, the

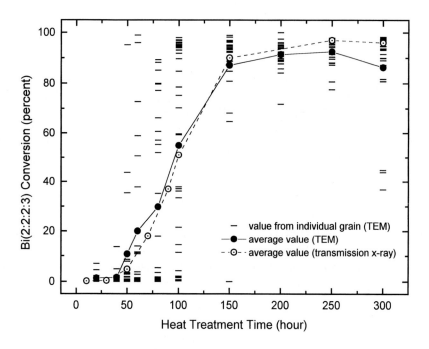

Figure 9.12 The fractional conversion of Bi-2212 to Bi-2223 in Tape A is shown as determined by TEM taken from the same specimens shown in Fig.9.10. The fractions are determined by counting the number of the insertions of the Cu-Ca-O layers in the lattice image micrographs. Each bar represents a data from an image for the given condition. The solid circles are the averaged value of the bars for each heating duration while the open circles are the fraction which were determined by the X-ray measurements in Fig.9.10.

9.3. Evolution at Early Stages of Heat Treatment

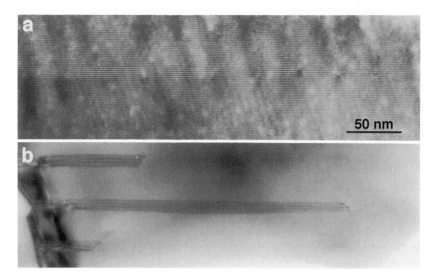

Figure 9.13 (a) An example of the Cu-Ca-O insertions being made into Bi-2212 by a group of the planes (an arrow points to an end of one of the intercalating planes), and (b) a pair of the inserting planes from a grain boundary. Here, the direction of the electron beam is tilted with respect to the a-b planes of the crystal, and thus, with respect to inserting planes. This results in the stacking-fault like images for the insertions. These images are from Tape A after 20h at 845^oC.

images of the planes appear very similar to a pair of stacking faults. In addition, Figs.9.14(a) and 9.14(b) show another example of an intercalation nucleating at a grain boundary in Tape C. Here, a number of the insertions into Bi-2212 grains are simultaneously initiated from a grain boundary.

These results certainly support the idea that the insertion of the Ca-Cu-O bi-layers into the existing Bi-2212 grains is the mechanism for the formation of Bi-2223 in the composite tapes. Furthermore, such an intercalation process can be accomplished by the pipe diffusion of the participating elements through the edge dislocation cores at the growth fronts of the layers [Bian et al. 1995, Cai et al. 1995, Cai and Zhu 1997]. Earlier, an intercalation mechanism was also invoked for the phase conversion of Bi-2201 to Bi-2212 from an amorphous precursor [Kramer et al. 1993]. For these highly anisotropic materials, it appears reasonable that layer insertion mechanism is the mechanism

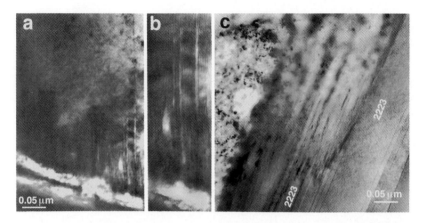

Figure 9.14 (a and b) Another examples of the nucleating Cu-Ca-O bi-layer insertions into Bi-2212 grain from a grain boundary in Tape C after 30h at $840^{\circ}C$. An arrow points to one of the intercalating planes. (c) An example of the phases which exist in the regions adjacent to the amorphous areas in Tape C after 30h at $840^{\circ}C$. An amorphous area is surrounded by Bi-2223 platelets with a Bi-2223 step which is growing into the amorphous area along the existing Bi-2223.

for the phase conversion. Previously, the Avrami relationship [Hulbert 1969] between the fraction of the Bi-2223 and the heat treating period was employed to infer the reaction mechanism. Based on this, some have concluded that the conversion mechanism is two dimensional diffusion [Hu et al. 1995]. One of the difficulties with such an analysis is that the model is based on the conventional nucleation and growth mechanism of phase conversion and does not take into account the precise microstructural details for the kinetics of the phase changes. As seen above, it is necessary to determine microscopically how the changes are taking place to guide us to the correct mechanism for the formation of Bi-2223 in composite wire.

9.3. Evolution at Early Stages of Heat Treatment

Heat Treatment of Tape B at 840°C in Air

As described in Sec. 9.2, in this tape, there are only small areas of amorphous regions for the entire heat treatment schedule in contrast to Tape C which contains a considerable amorphous region above 800°C. As shown in Fig.9.11, the limited presence of liquid in this tape appears to produce cuprate platelets which are finely intercalated. This observation strongly supports intercalation of the Ca-Cu-O layer discussed above as a primary phase conversion mechanism in this tape.

Heat Treatment of Tape C at 840°C in Air

In order to further demonstrate that the above observations about the formation process of the Bi-2223 phase are generally applicable and not specific to the precursor powder used in Tapes A and B (aerosol pyrolysis powders), the phase conversion in Tape C was investigated. Earlier, it was shown that the alignment process continues during the formation of Bi-2223 in Tape C after the reaction temperature is reached. [In Tapes A and B, the alignment is completed well before the initiation of Bi-2223 growth.] Thus, here, we focus on the possibility of a different conversion mechanism being manifested in Tape C, due to possibly the different precursor powder in this tape from those in Tapes A and B.

As mentioned in Sec.9.2, those specimens of Tape C quenched from 840°C contained significant amorphous areas and the amount increased with longer soak periods. Also, if the nucleation and growth [Grasso et al. 1995, Merchant et al. 1994] or the layer-by-layer epitaxial deposition of the Bi-2223 phase in the core [Morgan et al. 1991 1992, Morgan et al. 1994] is the mechanism for the phase conversion in the core, then such process is most likely to occur at or adjacent to the amorphous areas. Thus, we carefully examined those areas in the specimens quenched from 840°C after 12, 20, and 30h, soak periods. Very interestingly, the platelets adjacent to the amorphous regions are in most cases either pure Bi-2212 or Bi-2223 without or with very little intercalating layers, in contrast to the fine mixture of the two phases which was seen for Tapes A and B at intermediate reaction stages. An example is shown in Fig.9.14(c). Based on these observations, one may argue in support of the layer-by-layer deposition mechanism for the conversion process, sometimes called a "solution reprecipitation" mechanism, However, we often found a large step of Bi-2223 on the face of the straight Bi-2223 or Bi-2212 plate that was shown in Fig.9.14(c), next to an amorphous region. Hence, it appears that Bi-2223 platelets can grow out of the liquid as a block after nucleating in or near the liquid. Another example of the nucleation and the growth of Bi-2223 is shown in Fig.9.8(a). Here, a platelet, which consists of Bi-2212 and Bi-2223, is growing side by side into liquid. In this case, both Bi-2212 and Bi-2223 can grow together from their respective nuclei. The fraction of Bi-2223 platelets facing the liquid increases with the soak time as expected.

Occasionally we observed very thin layers of Bi-2201, which were attached to either Bi-2212 or Bi-2223 plates facing the amorphous region. Also, segments of Bi-2201 were found in the amorphous region. It is not clear whether the Bi-2201 phase formed during the quench or was actually present at the reaction temperature, although we have observed similar Bi-2201 in the specimens which were quenched ~ 10 time faster than the standard oil quench. Also, it is possible that these Bi-2201 fragments are a precursor to the growth of Bi-2212 or Bi-2223, or that Bi-2201 is part of a dissolution process of Bi-2212 into the liquid. There is not enough information to clarify this issue to our satisfaction at this time.

In contrast to the cuprate platelets surrounding the liquid, the platelets in the areas away from the liquid regions were found to consist of a fine mixture of two Bi superconductor phases or were heavily faulted by the insertions. The morphology of the grains in these areas is very similar to that in Tapes A and B as shown in Fig.9.11. Furthermore, in these grains, the insertion of the Cu-Ca-O bi-layers is accomplished at grain boundaries, as for the tapes made from spray pyrolysis powders. Such an example is shown in Figs.9.14(a) and 9.14(b), taken from a specimen with a soak duration of 30h at 840°C. Here, a number of the insertion planes are nucleated at a grain boundary. Thus, it appears that in the tapes with this particular precursor powders, two possible Bi-2223 formation mechanisms may be operating in parallel, depending on whether or not a given location is in contact with a liquid.

9.4 Dislocation and Bismuth 2212-to-2223 Transformation

9.4.1 Introduction

Several attempts have been made [Zhu and Nicholson 1992, Luo et al. 1993a, Guo et al. 1994] to analyze the kinetics of Bi-2223 formation using the Avrami equation

$$\ln\left(\frac{1}{1-C}\right) = K_0 \exp(-E/RT)t^\alpha \tag{9.1}$$

where C is the fraction of Bi-2223 transformed at time t, T is the temperature, E is the activation energy, $K_0 = 1.71 \times 10^{-22}$ is the rate constant, R is the universal gas constant and α is the Avrami exponent. It is generally agreed upon that the Avrami exponent obtained from experimental data can provide some insight into the reaction mechanism. Interestingly, although almost all the above-mentioned authors conclude that their data supports a diffusion-controlled, two-dimensional transformation, the Avrami exponents they obtained vary greatly, ranging from 0.5 [Zhu and Nicholson 1992] to close to $1 - 1.5$ [Luo et al. 1993a, Guo et al. 1994], which seem to be more consistent with one-dimensional diffusion-controlled transformation mechanism with a varying nucleation rate [Hulbert 1969].

As shown in Figs.9.12 and 9.13, during the annealing process, the Bi-2212/Bi-2223 system consists of fast-growing intercalating Ca/CuO_2 bi-layers instead of compact Bi-2223 domains. Unlike the conventional nucleation-and-growth mechanism, where a reactant diffuses from grain boundaries into the interior of the bulk materials, leading to a compact advancing "front" of the products, the transformation from Bi-2212 to Bi-2223 appears to be accomplished via the layer-by-layer intercalation of the extra Ca/CuO_2 planes into the Bi-2212 matrix. Fig.9.15 shows the [002] lattice image from a sample annealed at 825°C for 30 hours and 150 hours, respectively (the experimental detail is described by Bian et al. [Bian et al. 1995]) From the lattice image of the system at the very early stage of the transformation (Fig.9.15(a)) to the stage where the transformation is almost completed (Fig.9.15(b)), we did not observe any edge-dislocation inside the Bi-2212 grain formed by inserting the extra Ca/CuO_2 planes. It is clear that the fast intercalation

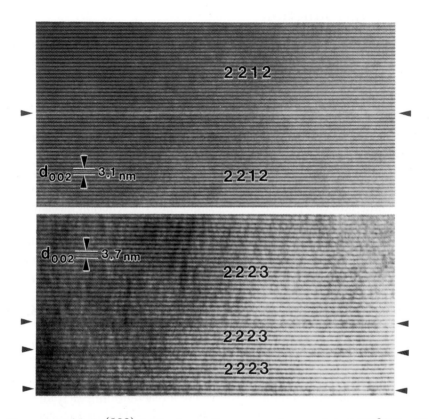

Figure 9.15 (a) A (002) lattice image from a sample annealed at 825°C for 30 hours. The crystal matrix was dominated by the Bi-2212 phase. A pair of arrowhead indicates the insertion of a Ca/CuO$_2$ bi-layer which forms a half unit-cell of the Bi-2223 phase. (b) A (002) lattice image from a sample annealed at 825°C for 150 hours. The majority of the crystals had completed the transformation from the Bi-2212 phase to the Bi-2223 phase. The remaining Bi-2212 phase are marked by the arrow heads.

9.4. Dislocation and Bismuth 2212-to-2223 Transformation

of individual Ca/CuO$_2$ plane into the Bi-2212 matrix, rather than the nucleation and growth of compact Bi-2223 domain, is the path for Bi-2212 to Bi-2223 transformation, at least for the samples we studied in the previous sections.

The growth of the intercalant Ca/CuO$_2$ planes is much faster than the bulk cation diffusion rate seems to allow, which lead us to propose that a different diffusion mechanism is at work in the Bi-2212/Bi-2223 system. We believe that the mechanism of cation diffusion is by the way of pipe diffusion through the cylindrical void created by the edge-dislocation which accompanies the insertion of a Ca/CuO$_2$ plane. These voids are located at the "interfaces" between Bi2212 and Bi-2223 where the reaction takes place. As the transformation progresses, these voids move with the Bi-2212/Bi-2223 "interface"; thus, the reaction progresses uninhibited. It should be noted that the term "interface" here denotes the edge-dislocation caused by the intercalation of extra Ca/CuO$_2$ planes in Bi-2212 matrix. Since the Bi-2223 is formed by the intercalation of individual Ca/CuO$_2$ plane rather than the nucleation and growth of compact 2223 domain, there is no Bi-2212/Bi-2223 interface *plane* in the traditional sense. In this sectionwe will show that pipe-diffusion via voids created by edge-dislocations in Bi-2212/Bi-2223 is indeed plausible. We will calculate the size of the voids using a layer-rigidity model [Thorpe 1989, Cai *et al.* 1990] modified for Bi-2212/Bi-2223 systems, with its parameters calculated from a shell model simulation and show the results of our model are consistent with the existing experimental data for the time dependence of the volume fraction of Bi-2212 converted to Bi- 2223.

9.4.2 Edge-dislocations as Channels for Fast Ion Diffusion

The transformation from Bi-2212 to Bi-2223 can be regarded as a chemical reaction between the Bi-2212 precursor and the secondary phases such as $CaPbO_3$ and CuO, to provide the necessary extra Ca and Cu. As shown in Fig.9.16, the only structural difference between Bi-2212 and Bi-2223 is a pair of extra Ca/CuO_2 planes inserted in the case of Bi-2223 and the associated lattice expansion. It is therefore convenient to divide the layered structure of Bi-2212 into two parts: the block layers (also called "host layers" below), which consist of BiO, SrO and the Ca/CuO_2 plane in Bi-2212, and the "gallery layers" into which the extra Ca/CuO_2 planes are inserted in the case Bi-2223. In this context, we can regard the Bi-2223 as a stacking fault of Bi-2212. The interfaces in a transforming system between Bi-2212 and Bi-2223 layers can then be regarded as an edge-dislocation. The approximately cylindrical void created at the extra half plane of this dislocation is a line of vacancies which can be an easy path for the extra Ca, Cu and oxygen ions to diffuse from the surface (or grain boundary) into the bulk material provided that the size of the void is large enough. The size of the voids will depend on the rigidity of the block layer as well as the compressibility of the Ca/CuO_2 plane along c-axis. Obviously the more rigid the layer is, the large the void will be. In the limit of infinite layer rigidity, the size of the void will be infinite. The size of the void in Bi-2212/Bi-2223 system will be evaluated in the next section. We will see that the size of the void is about twice the size of the biggest ions (oxygen), thus indicating that the dislocations at the Bi-2212/Bi-2223 interface can be easy paths for the reactant ions to diffuse into the bulk.

At the annealing temperature ($\sim 830^oC$) both $CaPbO_3$ and CuO are liquid. The grinding and mixing prior to the annealing make sure that this liquid phase is evenly coated around each Bi-2212 grain. As a Ca/CuO_2 plane nucleates near the surface of the Bi-2212 grain (grain boundary), the void created by the partially inserted Ca/CuO_2 plane then opens up a channel for the reactant ions to diffuse into the bulk, thus permitting the edge-dislocation to climb and the reaction to proceed rapidly at this location. Since the void moves into the bulk as the edge-dislocation climbs, the reactants are always in contact with the

Bi-2212/Bi-2223 "interface". The growth of the Ca/CuO$_2$ plane can therefore be very fast. The transformation from Bi-2212 to Bi-2223 is likely then to be limited by the nucleation rate of the Ca/CuO$_2$ plane near the 2212 grain boundary and the diffusion of reactant ions along the moving dislocation lines. This mechanism will give rise to the intercalation of individual Ca/CuO$_2$ planes in the Bi-2212 matrix, as we observed in TEM experiments, [Bian et al. 1995] instead of compact Bi-2223 domains one would expect from conventional nucleation and growth theory.

9.4.3 The Layer Rigidity Model

To address the question of the volume expansion at the cores of the edge-dislocations (extra atomic half-planes) produced during the Bi-2212 to Bi-2223 transformation, it is important to consider the large anisotropy in physical properties in these layered systems. It is reasonable to assume that the major expansion takes place along the direction perpendicular to the layers, denoted as c-axis. In fact, it has been shown by diffraction experiments that the lattice parameters along the a- or b-axis change very little during Bi-2212 to Bi-2223 transformation, while the lattice parameter along c-axis increases from 30.9Å to 37.8Å.

Since the only structural difference between Bi-2212 and Bi-2223 compounds is the extra layers of Ca and CuO$_2$ and the associated c-axis expansion, it is reasonable to regard the extra Ca and CuO$_2$ layer in Bi-2223 phase as "intercalants" and the rest of the structure as "host layers". Therefore we can regard the Bi-2212/Bi-2223 compound as a type of intercalation compound in the form of $A_{1-x}B_xL$, with $0 \leq x \leq 1$, where B is the intercalant (extra Ca and CuO$_2$ layer in Bi-2223 phase), A is a vacant (but collapsed) layer in the Bi-2212 phase (see Fig.9.16), which will be regarded as an intercalant of a smaller size, and L denotes the host layer which represents the rest of the structure. Therefore the single phase Bi-2212 (AL) and Bi-2223 (BL) can be regarded as the limiting cases for $x = 0$ and $x = 1$, respectively. The dislocation at the interface of Bi-2212/Bi-2223 can be modeled as an intercalation compound with B occupying the semi-infinite plane, as shown in Fig.9.17.

In the past, several attempts have been made to study the c-axis ex-

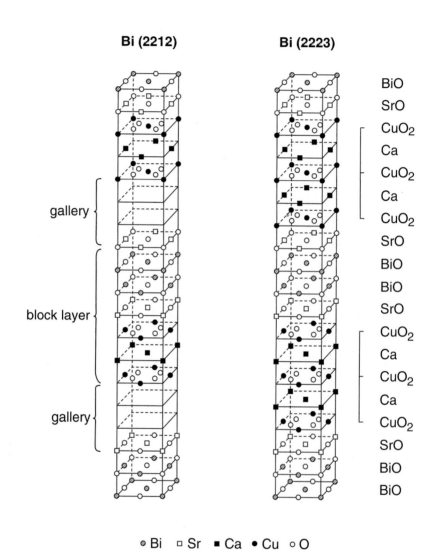

Figure 9.16 The unit cells of Bi-2212 and Bi-2223. The layered structure of Bi-2212 can be divided into block layers and galleries for intercalation of the extra Ca/CuO$_2$ plane to form Bi-2223. The "gallery height" in the 2212 structure is shown here expanded from its normal value for clarity of comparison of the two structures.

9.4. Dislocation and Bismuth 2212-to-2223 Transformation

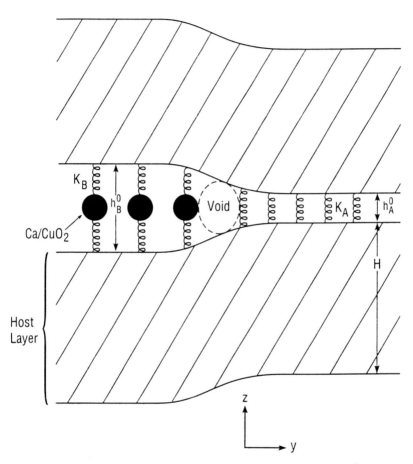

Figure 9.17 Side view of the intercalants in the transforming Bi-2212/Bi-2223 system. The extra Ca/CuO_2 plane is shown schematically as large intercalant "atoms". The rest of the structural features are shown schematically as the "host layer". The separation between the host layers are given by the local gallery height h_i.

pansion of intercalation compounds within the framework of two quite distinct types of models [Thorpe 1989, Cai et al. 1990]. It is assumed in these models that the host layers have finite transverse as well as bending rigidity while the intercalants have finite compressibility as well as different sizes. The first model [Thorpe 1989], referred to as bilayer model, assumes that the compressibility of the host layer is much smaller than that of the intercalant, so that the correlation between different galleries can be ignored. The second model [Cai et al. 1990], referred to as the multilayer model, assumes that the compressibility of the host layer is much larger than that of the intercalant, so that the correlation between different galleries can be mapped into an Ising-type model. In the case of Bi-2212/Bi-2223 compound, we believe that the first type of model is more suitable. Since the compressibility of the host layer is inversely proportional to its thickness, and in the case of Bi-2212/Bi-2223 system, the host layer thickness is much larger than the average gallery height, we find the compressibility of the host layer is about 10% of that of the intercalant. In this section we set up a spring model that describes both the layer rigidity and the size and stiffness of the intercalant species. The model can be solved analytically if the compressibility of the intercalants A and B is the same.

Consider a layered system with composition $A_{1-x}B_xL$ where L represents the host layer of thickness H which consists of BiO, SrO and Ca/CuO_2 plane in the Bi-2212 phase. A and B are two different types of intercalants which are assumed to occupy a set of well defined lattice sites. The total energy of the system can be approximated by a sum of two major contributions; one associated with the interaction between the intercalants and the host atoms and the other between host atoms themselves. We assume the intercalants to be frozen into the ordered structure (i.e. the time it takes to nucleate the Bi-2223 phase is much shorter than the time it takes for the Ca, Cu and O ions to diffuse from the grain boundary to the bulk). Therefore the direct interaction energy between the intercalants does not play any role in the layer distortion. As discussed in the introduction, the compressibility of the host layer is much smaller than the intercalant, so that the interlayer correlation is mostly related to the various thickness of the host layer instead of the correlation between intercalants of different galleries. Since the variation of the host layer thickness ΔH is very small compared to H, we can

9.4. Dislocation and Bismuth 2212-to-2223 Transformation

neglect the interlayer correlation and consider a single gallery bounded by two host layers with intercalants occupying the lattice sites inside the gallery.

The total energy of the host layer-intercalant system can be written as

$$E = \frac{1}{2}\sum_i K_i(h_i - h_i^0)^2 + \frac{1}{2}K_T \sum_{<i,\delta>} (h_i - h_{i+\delta})^2$$
$$+ \frac{1}{2}K_F \sum_i \left[\sum_\delta (h_i - h_{i+\delta})\right]^2, \quad (9.2)$$

where h_i is the local gallery height at the site i where an intercalant either (A or B) sits. h_i^0 is the gallery height for the pure system (AL or BL). It can be obtained from the diffraction measurements that for Bi-2212 (AL) $h_A^0 = 2.21\text{Å}$, for Bi-2223 (BL) $h_B^0 = 6.18\text{Å}$. K_i is the spring constant representing the compressibility of the local intercalant i. The terms involving the spring constants K_T and K_F describe respectively the transverse and bending rigidity of the host layers [Thorpe 1989]. K_T and K_F are related to the elastic constants of the Bi-2212 compound by the following relations:

$$K_T = \frac{H}{a_0} c_{44} \quad (9.3)$$

$$K_F = \frac{H^3}{12a_0^3} \left[\frac{(c_{11} + c_{33})c_{33} - 2c_{13}}{c_{33}}\right] \quad (9.4)$$

where $a_0 = 3.8\text{Å}$ is the lattice constant in the ab plane and $H = 14.35\text{Å}$ is the thickness of the host layer.

The compressibility $K_A = K_i(A)$ of the intercalant A (the vacancy) can also be obtained from the elastic constants given the relationship

$$K_A = \frac{a_0}{h_A^0} c_{33}. \quad (9.5)$$

The compressibility $K_B = K_i(B)$ of the intercalant B (the Ca and CuO_2 plane) cannot be calculated due to the lack of data for the elastic properties of Bi-2223. However it is reasonable to assume that $K_B > K_A$. Here we will calculate the gallery height for the case of $K_B = K_A$ so as to estimate the minimum size of the void.

The relevant elastic constants for Bi-2212 were calculated by Baetzold [Baetzold 1995] using a suitable shell model potential. They are found to be

$$c_{11} = 24.08 \times 10^{11} \text{ dyne/cm}^2, \quad c_{13} = 6.71 \times 10^{11} \text{ dyne/cm}^2,$$
$$c_{33} = 14.59 \times 10^{11} \text{ dyne/cm}^2, \quad c_{44} = 6.18 \times 10^{11} \text{ dyne/cm}^2. \quad (9.6)$$

Minimizing E in Eqn.(9.2) with respect to the h_i, we find

$$\mathbf{Mh} = \mathbf{\Phi} \quad (9.7)$$

where $\mathbf{M}$ is a tridiagonal matrix with

$$\begin{aligned} M_{ii} &= K_i + 2K_T + 6K_F, \\ M_{i,i\pm 1} &= -K_T - 4K_F, \\ M_{i,i\pm 2} &= K_F, \end{aligned} \quad (9.8)$$

and $\mathbf{h}$ and $\mathbf{\Phi}$ are two column vectors

$$\mathbf{h} = \begin{pmatrix} h_1 \\ h_2 \\ \vdots \\ h_N \end{pmatrix}, \quad \mathbf{\Phi} = \begin{pmatrix} K_1 h_1^0 \\ K_2 h_2^0 \\ \vdots \\ K_N h_N^0 \end{pmatrix}. \quad (9.9)$$

One can then obtain the h_i's by diagonalizing the $\mathbf{M}$ matrix. In the case of $K_A = K_B = K$, the dispersion relation for the system is given by

$$\lambda_q = K + 2K_T[1 - \cos(qa_0)] + 4K_F[1 - \cos(qa_0)]^2 \quad (9.10)$$

and the expanded form of $(\mathbf{M}^{-1})_{nm}$ can be expressed as

$$(\mathbf{M}^{-1})_{nm} = \frac{1}{N} \sum_q \frac{\exp[iqa_0(n-m)]}{\lambda_q} \quad (9.11)$$

with $q = \frac{2\pi r}{Na_0}$ where $r = 1, 2, ..., N$ and λ_q being the eigen value of the matrix $\mathbf{M}$. Then we have

$$h_n = \frac{1}{N} \sum_m (\mathbf{M}^{-1})_{nm} \Phi_m = \frac{1}{N} \sum_q \frac{K}{\lambda_q} \sum_m e^{iqa_0(n-m)} h_m^0. \quad (9.12)$$

9.4. Dislocation and Bismuth 2212-to-2223 Transformation

Taking a system with half of the $y-z$ plane occupied by intercalant B (i.e. Bi-2223 phase), we assume the dislocation is at $y = z = 0$ and the equilibrium height h_n^0 is defined by

$$h_n^0 = \begin{cases} h_B^0, & \text{for } n = -\frac{N}{2}+1, -\frac{N}{2}+2, ..., -1 \\ h_A^0, & \text{for } n = 1, ..., \frac{N}{2}-1, \frac{N}{2} \end{cases} \quad (9.13)$$

Taking the limit of $N \to \infty$ and changing the summation over q to integral, we find that

$$h(y) = \frac{h_A^0 + h_B^0}{2} + \frac{h_A^0 - h_B^0}{4\pi} \int_0^\pi \frac{K \sin[\theta(\frac{y}{a_0} - \frac{1}{2})]}{\lambda_\theta \sin(\theta/2)} d\theta \quad (9.14)$$

where $\theta = qa_0$ and $\lambda_\theta = K + 2K_T(1-\cos\theta) + 4K_F(1-\cos\theta)^2$ as defined in Eqn.(9.10).

The solid line in Fig.9.18 shows the profile of h_n for $K_A = K_B$ and the values of parameters defined in Eqs.(9.3)–(9.5). Since no high-resolution TEM image of the stacking fault of Bi-2212/Bi-2223 at a growth front is currently available, we performed calculation on a similar system to verify the accuracy of the results we obtained using this model. A space profile of the gallery height h_n is measured from the high-resolution TEM image of the stacking fault near the columnar defect produced by heavy-ion irradiation in $YBa_2Cu_3O_7$ [Zhu et al. 1993b]. The stacking fault consists of an extra CuO plane in the $YBa_2Cu_3O_7$ matrix. The theoretical profile of h_n is calculated using the elastic constants obtained from the atomistic simulation of $YBa_2Cu_3O_7$ using the shell model potential suitable for this material. [Baetzold 1990] The calculated as well as measured h_n for the stacking fault of $YBa_2Cu_3O_7$ system are plotted along with the calculated h_n for Bi-2212/Bi-2223 system in Fig.9.18. We can see that the calculated results agree very well with the experimental data, indicating that our assumption of $K_A = K_B$ is reasonable in this type of systems.

Of the three types of reactant ions (Ca^{2+}, Cu^{2+} and O^{2-}), the largest is O^{2-}, which has a diameter of about 2.9Å. We can see from Fig.9.18 that the lateral size of the void is about twice as large as the oxygen ion diameter, if we consider the the void to be the region where $h_n \geq 3$Å for $y \geq 0$. Therefore we have confirmed that the dislocation line is indeed likely to be an easy path for reactant ions to diffuse from the

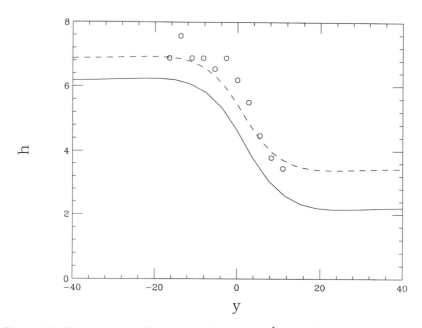

Figure 9.18 The spatial profile of the gallery height h Solid line: Bi-2212/Bi-2223 systems for $K_A = K_B = K$; Dashed line: YBa$_2$Cu$_3$O$_7$/YBa$_2$Cu$_4$O$_8$ system for $K_A = K_B = K$; open circle: Experimental data for the YBa$_2$Cu$_3$O$_7$/YBa$_2$Cu$_4$O$_8$ system. Note the units of both h and y are anstrom.

liquid phase at the 2212 grain boundaries into the bulk, causing the edge-dislocation (the extra Ca/CuO$_2$ plane) to rapidly climb into Bi-2212.

9.4.4 Kinetics of Bi-2212 Conversion to Bi-2223

Since the reactant is always in contact with the Bi-2212/Bi-2223 interface, the reaction rate of Bi-2212 to Bi-2223 perpendicular to the dislocation line is independent of the volume fraction of Bi-2223 and the growth in this direction is very fast. We therefore need to consider only the diffusion along the dislocation line and assume that the reactants are consumed immediately upon reaching the Bi-2212/Bi-2223

9.4. Dislocation and Bismuth 2212-to-2223 Transformation

interface, since the diffusion is probably a much slower process than any local rearrangements of atoms which might be required. Thus the rate of growth of the product layer along the dislocation line is given by

$$dl/dt = D/l \tag{9.15}$$

where l is the distance from the grain boundary measured along the dislocation line and D is the effective diffusion coefficient of the reactant ions. Since that the diffusion coefficient and the size of the void (see the previous section) is independent of the volume fraction of Bi-2223, integrating the above equation yields

$$l = \sqrt{2Dt} \tag{9.16}$$

The volume of the reactant ions consumed by the formation of the product phase is given by

$$R = S\sqrt{2D}t^{1/2} \tag{9.17}$$

where S is the size of the void. For a Bi-2212 grain, the number of Ca/CuO_2 planes nucleated at the grain boundary in the time interval between t' and $(t' + dt')$ is $I(t')Vdt'$ where $I(t')$ is the nucleation rate per unit area of the grain boundary. Therefore the total volume of reactant ions consumed at time t (assuming the nucleation can occur everywhere at the grain boundary, including the transformed region) is given by

$$V_e^C = A\int_0^t I(t')R(t-t')dt' = S\sqrt{2D}A\int_0^t I(t')(t-t')^{1/2}dt' \tag{9.18}$$

where A is the total area of the grain boundary. It should be pointed out that V_e^C is different from the actual volume of reactant consumed since it assumes that the nucleation can occur in the transformed region and it treats all regions as though they continued growing irrespective of other regions. To correct this problem, we assume the volume of the transformed region is V^C and work out the relationship between V^C and V_e^C. Consider a small region of which a fraction $[1 - (V^C/V)]$ remains untransformed (where V is the total volume of the grain). During the time interval dt, of the total transformed volume dV_e^C, a fraction $[1 - (V^C/V)]$ on the average will lie in previously unreacted material, and thus contribute to dV^C while the reminder of the dV_e^C will be in already transformed material. Therefore we obtain

$$dV^c = \left(1 - \frac{V^C}{V}\right) dV_e^C, \qquad (9.19)$$

or

$$V_e^C = -V \ln\left(1 - \frac{V^C}{V}\right). \qquad (9.20)$$

Let $[1 - (V^C/V)] = 1 - C$ where C is the volume fraction of the Bi-2223 and substitute Eqn.(9.20) into Eqn.(9.18), we have

$$\ln\left(\frac{1}{1-C}\right) = \frac{S\sqrt{2D}A}{V} \int_0^t I(t')(t-t')^{1/2} dt'. \qquad (9.21)$$

The volume fraction of Bi-2223 depends sensitively on the nucleation rate $I(t)$. It is often assumed that the time-dependence of the number of nucleation sites is a classical first-order rate process, i.e.

$$dN(t)/dt = -fN(t) \qquad (9.22)$$

where f is the frequency of an empty gallery in Bi-2212 turns into a nucleation site for Bi-2223 and $N(t)$ is the number of such nucleation sites at time t. Integrate the above equation yields

$$I(t) = fN(t) = fN_0 \exp(-ft) \qquad (9.23)$$

where N_0 is the number of empty galleries in the Bi-2212 grain at $t = 0$. Substitute Eqn.(9.23) into Eqn.(9.21) yields

$$\ln\left(\frac{1}{1-C}\right) = \frac{S\sqrt{2D}N_0 A}{V}\left[t^{1/2} + \frac{i}{2}\sqrt{\frac{\pi}{f}} \exp(-ft)\mathrm{erf}(i\sqrt{ft})\right] \qquad (9.24)$$

where erf(z) is the error function with complex variable z. We can then obtain the Avrami exponents for various values of f.

In the limiting case that ft is very small, i.e. the time it takes to produce a Bi-2223 nuclei is much longer than the time it takes to grow the Ca/CuO$_2$ across the sample, we can expand Eqn.(9.24) in terms of ft

$$\ln\left(\frac{1}{1-C}\right) = \frac{S\sqrt{2D}N_0 A}{V}\left(\frac{2}{3}ft^{3/2} - \frac{4}{15}f^2 t^{5/2} + ...\right). \qquad (9.25)$$

9.4. Dislocation and Bismuth 2212-to-2223 Transformation

compare the above equation with Eqn.(9.1) we have $\alpha \approx 1.5$ which agrees well with the experimental results for the preheated sample [Luo et al. 1993a, Guo et al. 1994]. It should be pointed out the assumption in this limiting case is valid only if the reactants (Ca, Cu and O) are evenly coated around the Bi-2212 grain and the Bi-2212 grain are small enough. The prediction of our model is therefore more consistent with experiments with small precursor powder which are heat-treated before annealing to ensure the even distribution of the reactants.

On the other hand, if ft is very large, i.e. i.e. the time it takes to produce a Bi-2223 nuclei is much shorter than the time it takes to grow the Ca/CuO_2 plane across the sample, We can consider that all the empty galleries in Bi-2212 have turned into nucleation sites for Bi-2223 before the Ca/CuO_2 planes begin to grow. In that case we find

$$\ln\left(\frac{1}{1-C}\right) = \frac{N_0 A R}{V} = \frac{S\sqrt{2D}N_0 A}{V} t^{1/2}, \qquad (9.26)$$

i.e. $\alpha = 0.5$. We can see that this will be the case if there is local shortage of reactant ions which will leads to the slow growth of Ca/CuO_2 plane. Experiments in [Zhu and Nicholson 1992] apparently falls into this situation. As we can see from Fig.4 in [Zhu and Nicholson 1992], The Avrami exponent α increases as the temperature rises, from 0.5 at $840^\circ C$ to 0.79 at $870^\circ C$. The small value of Avrami exponents in [Zhu and Nicholson 1992] is either due to the large Bi-2212 grain size as indicated by the high porosity ($\sim 40\%$) or the uneven distribution of the reactant ions [Zhu and Nicholson 1992]. When the temperature rises, the growth of the Ca/CuO_2 plane accelerates due to the increase of ion diffusion constant D and the reactant ions distribute more evenly around the Bi-2212 grain due to the decreased viscosity of the ionic liquid of Ca, Cu and oxygen.

It should be noted that Avrami exponents obtained previously are from X-ray diffraction data, which implicitly assumes that the product phase Bi-2223 form compact domains as X-ray diffraction cannot detect randomly intercalated Bi-2223 structure as shown in Fig.9.15. Thus if our model of random intercalation of Ca/CuO_2 bi-layer in Bi-2212 matrix is correct, then X-ray diffraction data underestimates the volume fraction of Bi-2223 phase at the early stages of the reaction and overestimate the volume fraction of Bi-2223 phase when the reaction is nearly

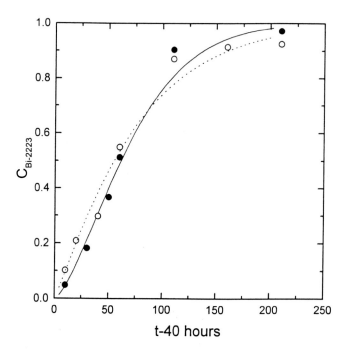

Figure 9.19 The time dependence of the volume fraction of Bi-2223 for samples prepared at 845^oC in air. Solid circle X-ray diffraction data; open circle: TEM data. The solid line and the dashed line are Avrami fit for X-ray and TEM data, respectively.

complete. We have measured the volume fraction of Bi-2223 of over one hundred quenched samples at various stages of transformation directly from lattice image and compared it with X-ray diffraction data. We found indeed that X-ray data significantly underestimated the volume fraction of Bi-2223 at the beginning of the transformation, while underestimate the volume fraction of Bi-2223 near the completion of the transformation, as shown in Fig.9.19. The Avrami exponent obtained using Eqn.(9.1) from TEM data is thus much lower than that from X-ray diffraction data, as shown in Fig.9.19. The Avrami exponent obtained from TEM data is 1.12 ± 0.08 while X-ray diffraction data gives 1.7 ± 0.1 for the same sample. The Avrami exponents obtained are also very sensitive to the time that the Bi-2212 to Bi-2223 transofmration takes place. It has been shown that when the sample was heated above 800^oC, Bi-2212 first went through a partial melting and recrystalization transformation, which is accompanied by the rapid alignment of of Bi-2212 platelets. The experiments indicate that for the initial ≈ 40 h the Bi-2223 phase was not formed [Wang et al. 1996a].Therefore the Bi-2212 to B-2223 transformation did not begin until 40 hours after the sample was heated to 845^oC. We thus choose the starting time of the transformation to be 40 hours. The value of the Avrami exponents we obtained from experiments are consistent with the one dimensional nucleation and growth model we proposed.

9.5 Summary

In conclusion we have studied the grain alignment and phase evolution mechanism during the Bi-2212 to Bi-2223 transformation using various theoretical as well as experimental techniques.

The main conclusions are:

(1) The temporal characteristics of the c-axis alignment of the Bi-cuprate platelets depend strongly on the phase assemblage and the fabrication method of the precursor powders. If the major secondary phase is Ca_2CuO_3 and the powder is produced by a sintering and milling process, the alignment is only partially completed before the initiation of the formation of Bi-2223 and the remainder occurred soon thereafter as Bi-2223 is formed. On the other hand, if the powder is produced by an aerosol spray pyrolysis method, regardless of the precursor secondary phase assemblage, the alignment is completed before the initiation of the Bi-2223 formation. However, irrespective of the starting precursor for a given tape geometry, the final degree of alignment is primarily dependent on the temperature for the Bi-2223 formation.

(2) The alignment is accomplished by the planar confinement provided by the Ag matrix in combination with the very large anisotropic growth rates in the a-b plane relative to the c-axis direction. The presence of liquid appears to have little effect on the alignment process. Tape B exhibits a small amount of liquid while it attains the same degree of the alignment as Tape C which contains a large amount of liquid (see Sec.9.3).

(3) The primary formation mechanism for Bi-2223 is the Ca-Cu-O bi-layer intercalation into Bi-2212. This is particularly true for the tape in which very little liquid is formed during heat treatment. In the case where a large amount of liquid is present, nucleation and growth of blocks of Bi-2223 appears to be the dominant mechanism in the regions adjacent to the liquid. The first is the primary mechanism for the tape with the spray pyrolysis powder and both mechanisms were operative in the tape with the mechanically milled powder.

(4) A microscopic kinetics and diffusion mechanism for the Bi-2212 to Bi-2223 transformation was proposed. The layer-rigidity model yields the local gallery height which is in good agreement with experimental observations. We propose that the Bi-2212 to Bi-2223 transformation is a one dimensional diffusion controlled nucleation and growth process with edge-dislocations at the Bi-2212/Bi-2223 interface acting as channels for fast ion diffusion. The Avrami exponent depends sensitively on the nucleation rate of the Bi-2223 (the extra Ca/Cu_2 plane) in the Bi-2212 matrix and varies between 0.5 to 1.5.

(4) It is also confirmed that Pb dissolves out of Bi-2212 in the precursor powder to form an intermediate phase, 3221. Subsequently, this phase decomposes as Pb is reabsorbed back into Bi-2212. Some Pb remains outside the Bi superconductors as Ca_2PbO_4. This process is independent of composition and secondary phase assemblage of the precursor powder.

(5) The secondary phase assemblage at the intermediate stage of the heat treatment is determined by the initial phase assemblage of the precursor powder. If the powder contains primarily CaO and CuO in addition to Bi,Pb(2212), the primary secondary phase formed in the ramp-up portion of the heat treatment is $(Ca,Sr)CuO_2$, and this phase becomes unstable against the formation of $(Ca,Sr)_2CuO_3$ at higher temperatures. If the initial powder contains primarily $(Ca,Sr)_2CuO_3$ after Bi,Pb(2212), this phase is retained as primary precipitates at the intermediate stage and the $(Ca,Sr)CuO_2$ phase is not formed at all.

Chapter 10

Charge Distribution

10.1 Introduction

The charge carriers in high temperature superconductors are the electron holes confined to the CuO_2-plane [Pickett 1989, Nücker et al. 1989], and thus, the distribution of charge plays a key role in determining their superconducting properties. Several groups of researchers have calculated the electronic structure of different superconducting oxides [Pickett 1989, Krakauer et al. 1988, Oles and Grzelka 1991], and core-level spectroscopic studies are plentiful, both emission spectroscopy, and absorption spectroscopy with incident electrons and incident x-rays. In absorption spectroscopy, attention has focused on the near-edge structure of the K- and L-edge of copper, and, in particular, the K-edge of oxygen which exhibits clear signatures of the electron holes that are responsible for superconductivity [Fink et al. 1994, Nücker et al. 1995]. On the other hand, there are few experimental studies of the spatial distribution of the electron charge in these superconductors.

In high-temperature superconductors, the density of electron holes is typically considerably less than 1% of the total density of electrons. However, the electron diffraction patterns and images of these superconductors, with their high local concentration of charge, was expected to be strongly influenced by the charge distribution. One reason for this expectation is the large crystal unit cell resulting in reflections at small angles which are very sensitive to the charge. We realize this from the classical picture of the scattering of fast electrons by an atom. Charged particles interact with the electrostatic potential, and thus, for small scattering angles, which correspond to large impact parameters, the incident particle sees a nuclei that is screened by the electron cloud. Thus, the scattering amplitude is mainly determined by the net charge of the ion at small scattering angles, θ. In mathematical terms the scattering amplitude of incident electrons, f_e, is given by the Mott formula:

$$f_e(\theta) = \frac{8\pi^2 m_0 e^2}{h^2} \left(\frac{\lambda}{\sin\theta}\right)^2 (Z - f_x) \qquad (10.1)$$

where Z is the charge of the nucleus, λ the wavelength of the incident electrons, m_0 the rest mass of electron, h Planck's constant, and e the charge on the electron. The scattering amplitude for incident x-rays, f_x, is determined by the spatial distribution of electrons around the atom. Near the forward direction, where f_x is close to Z, small changes in f_x may drastically alter f_e. For example, for a Ba atom $Z^{Ba} = 56$, when $\sin\theta/\lambda \approx 0$ the x-ray scattering amplitude for an ionized Ba$^+$ and Ba^{++} atom is $f_x^{Ba^+} = 55$ and $f_x^{Ba^{++}} = 54$, respectively. The difference in their scattering amplitude is $\Delta f_x = (f_x^{Ba^{++}} - f_x^{Ba^+})/f_x^{Ba^+} = 1.8$. In contrast, for electrons, $\Delta f_e = [(Z^{Ba} - f_x^{Ba^{++}}) - (Z^{Ba} - f_x^{Ba^+})]/(Z - f_x^{Ba^+}) = 100\%$. Thus, electron diffraction has a greater sensitivity than x-ray diffraction in addressing the valence electron distribution in crystals with Bragg reflections present at small reciprocal distances i.e. at small scattering angles [Cowley 1953, Anstis et al. 1973, Zhu and Tafto 1996a].

Due to the screening at low angles an electron hole, and a bond electron after reversal of sign, has the scattering amplitude of a stripped hydrogen atom, H$^+$. This scattering amplitude is shown in Fig.10.1 where it is compared with the scattering amplitude of a neutral Bi atom. Note that the scattering amplitude of the hole is larger than that of the high Z neutral Bi atom at small angles. It also follows from these considerations that the electron scattering amplitude of any charged ion approaches that of H$^+$ multiplied by the excess charge, or valency, when the scattering angle becomes small.

In the following sections, we discuss our studies on charge distribution in $Bi_2Sr_2CaCu_2O_8$ and $YBa_2Cu_3O_7$ high temperature superconductors using advanced transmission electron microscopy. In the former, we consider structural deviation due to the charge modulation

10.1. Introduction

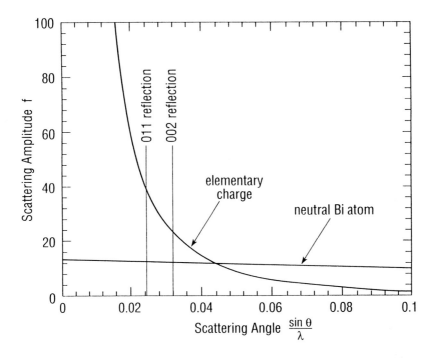

Figure 10.1 Electron scattering amplitude of an electron hole and a Bi atom at small scattering angles. Also indicated is also the angle at which the innermost reflections, the superstructure reflections 110, and the fundamental reflections 002 of $Bi_2Sr_2CaCu_2O_8$ are located on the angular scale.

based on computer simulations of experimental images and diffraction patterns; in the latter, we use a more quantitative approach to address the charge distribution by determining the 001 structure factor using a novel diffraction technique.

10.2 $Bi_2Sr_2CaCu_2O_8$

10.2.1 A General Nano-scale Description: Difference Structure

The major structural feature in $Bi_2Sr_2CaCu_2O_8$ (hereafter denoted as Bi-2212) is the structural modulation which results in superlattice reflections. To address such modulation, we use a standard method of crystallography of adding and subtracting matter to describe the deviation from the average structure (see, e.g., [Cowley 1976]). Let us start with an average structure with total scattering amplitude $\bar\psi$ and let the actual structure ψ deviate from the average structure by a perturbation $\Delta\psi$. The scattering amplitude of the crystal in the kinematical case at position **s** in reciprocal space is thus:

$$\psi(\mathbf{s}) = \bar\psi + \Delta\psi \qquad (10.2)$$

and the intensity:

$$I = \psi\psi^* = (\bar\psi + \Delta\psi)(\bar\psi + \Delta\psi)^* = \bar\psi\bar\psi^* + \bar\psi\Delta\psi^* + \bar\psi^*\Delta\psi + \Delta\psi\Delta\psi^*. \qquad (10.3)$$

Because $\bar\psi$ is periodic, it has values other than zero only at the Bragg position of what we define here as the fundamental reflections, and thus only the last term contributes to the intensity outside the fundamental reflections. Thus, it suffices to consider $\Delta\psi$ to assess what is outside the fundamental reflections, i.e., the diffuse scattering and superstructure reflections.

The deviation from the average structure that gives rise to $\Delta\psi$, and thus, the superstructure reflections, can be described in terms of the removal and addition of ions in the average structure in such a way that charge balance is maintained. Displacement of an atom, for example, is to remove it from its position in the average structure and add it at its position in the actual structure. Sometimes it is convenient to distinguish between modifications of the average structure that leaves the composition of the material unchanged, and those that do not. To the first category belongs displacement, interchange of atoms between two sites, and charge transfer. We restrict ourselves to the situation where the deviation from the average structure can be described by

10.2. $Bi_2Sr_2CaCu_2O_8$

identical clusters [Zhu and Tafto 1996b] of scattering amplitude ΔF at the position $\mathbf{r_j}$. Thus:

$$\Delta \psi = \sum_j \Delta F \exp(2\pi i \mathbf{s} \cdot \mathbf{r_j}). \tag{10.4}$$

Here $\Delta F = F_I - F_R$, where we have removed scattering matter of scattering amplitude F_R and inserted scattering matter of amplitude F_I at the position $\mathbf{r}_j$. Moving a single atom or electron A from position $\mathbf{r_j}$ to $\mathbf{r_j} + \Delta \mathbf{r}$ gives:

$$\begin{aligned}\Delta F &= f_A \sum_j [\exp(2\pi i \mathbf{s} \cdot (\mathbf{r_j} + \Delta \mathbf{r})) - \exp(2\pi i \mathbf{s} \cdot \mathbf{r_j})] \\ &\approx f_A 2\pi i \mathbf{s} \cdot \Delta \mathbf{r} \sum_j \exp(2\pi i \mathbf{s} \cdot \mathbf{r_j})\end{aligned} \tag{10.5}$$

This approximation applies at small $\mathbf{s} \cdot \Delta \mathbf{r}$. For modifications that change the composition:

$$\Delta F = (f_A - f_B) \sum_j \exp(2\pi i \mathbf{s} \cdot \mathbf{r_j}). \tag{10.6}$$

Here f_A and f_B are the scattering amplitudes of the species added and removed from position $\mathbf{r_j}$. In the case of introducing a vacancy or interstitial, either f_A or f_B is zero.

We conclude that displacements over small distances contribute negligibly to the electron diffraction pattern at small angles (Fig.10.2(a)) because of the short s-vectors. Displacement of atoms over larger distances, of the order of half the unit cell, and substitution may contribute equally in all parts of reciprocal space apart from the general fall-off of the scattering amplitude with angle. Charge transfer gives major contributions only close to the origin of reciprocal space as is apparent from the curve of Fig.10.1 and also from Fig.10.2(b) which shows the scattering amplitude from the superlattice in Bi-2212 due to the charge. We also note here the complimentary of the displacement of atoms over short distances and charge transfer over larger distances, a feature unique to electron diffraction. Charge transfer shows up at reflections at small angles while displacement of atoms occurs at large angles.

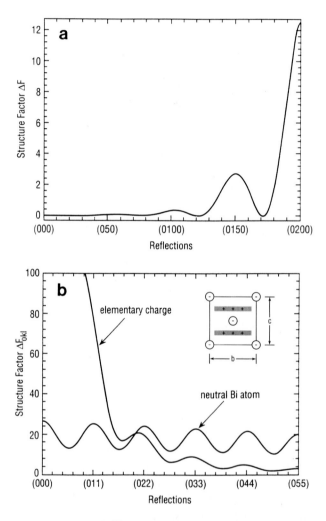

Figure 10.2 Structure factor ΔF describing the deviation from an average structure in $Bi_2Sr_2CaCu_2O_8$. (a) deviation due to the displacement of atoms, (b) due to the charge. Note, the charge transfer contributes to reflections at small angle while the lattice displacement contributes at large angle.

10.2.2 Direct Imaging of Charge Modulation

Experimental electron-diffraction patterns and electron-microscope images are usually compared to calculations made under the assumption that the atoms are neutral, and for small crystal unit-cells with interplanar spacings less than 5 Å this is a good approximation in most cases. Decades ago, Cowley [Cowley 1953] pointed out that electron diffraction should be able to reveal ionicity in crystals, and effects of deviation from neutral atoms have been retrieved from convergent-beam electron diffraction and used to study charge distribution [Anstis *et al.* 1973, Zuo *et al.* 1989, Gjonnes *et al.* 1993]. These studies required interpretation procedures involving many parameters to extract interesting information about charge distribution, while the system we study enables us to directly image the charge distribution with a resolution of about 1 nm [Zhu and Tafto 1996b].

Bi-2212 has the idealized crystal unit cell shown in Fig.10.3(b), i.e., when the mixing of superstructure periods, that leads to incommensurability, is disregarded. The unit cell can be described as an average structure with structure factor, F_g, and a deviation from this, ΔF_g. It is useful to think of the average structure as consisting of five identical units along the b-axis with lattice parameters $a = 0.538$ nm, $b \approx 5a = 2.69$ nm and $c = 3.06$ nm, giving rise to the fundamental reflections. The large crystal unit-cell results in reflections at small angles, and such reflections may change their intensities by orders of magnitude by charge transfer over distances that are a considerable fraction of the crystal unit cell.

It was established by x-ray [Imai *et al.* 1988], neutron [Gao *et al.* 1993a], and electron diffraction, and by high- resolution electron microscopy (HREM) [Horiuchi *et al.* 1988] that the deviation from the average structure of Bi-2212 is reasonably well described by the introduction of a displacement field. However, in electron diffraction superstructure reflections are seen also near the forward direction, Fig.10.3(a) and they should not be there in a kinematical diffraction pattern unless the displacements are extremely large. Unfortunately electron diffraction is prone to multiple scattering. Thus, we can not rule out the possibility that the superstructure reflections near the center of these diffraction patterns are caused by the other reflections further out in reciprocal space; these may act as new incident beams bringing super-

296   Chapter 10. Charge Distribution

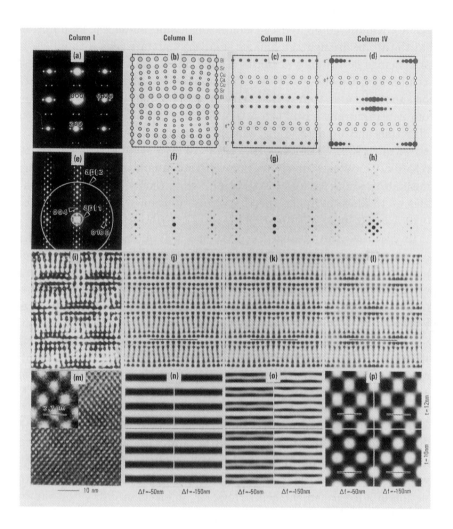

10.2. $Bi_2Sr_2CaCu_2O_8$

Figure 10.3 Electron diffraction patterns and images. The experimental ones are shown in Column I, and the calculated ones in Columns II, III, and IV, based on three different models of the charge distribution. All the figures refer to the (100) projection, except for (a) to the (001) projection. The bar in the experimental ((I) and inset of (m)) and calculated ((j), (k), (l) and (p)) images indicates the size of the crystal unit cell along the b-axis. Column II: based on a model with neutral atoms with atomic positions from [Horiuchi *et al.* 1988], as indicated in (b). Column III: based on a model the same as in (b) except for the addition of charge modulation along the c-direction as calculated in [Gupta and Gupta 1994] but with an evenly distributed charge in the b direction, and indicated in (c). Column IV: same as in (b) but with additional modulation of the charge in the b-direction as well as in the c-direction as indicated in (d). (e) Experimental diffraction pattern with the size of the objective apertures, apt 2 and apt 1, for the calculated high-resolution electron microscope image of the same row and lower resolution image of the row below, respectively. (f), (g), and (h) are calculations for their different charge distribution. (i) Experimental HREM image, the corresponding HREM image calculations in (j), (k), and (l) are for a crystal thickness of 27Å, other parameters used are also the same as in [Horiuchi *et al.* 1988]. (m) Experimental lower resolution image with corresponding calculations for the different models at different thickness and different defocuses in (n), (o), and (p). The inset in (m) is magnified for comparison with the calculated images (n), (o) and (p).

structure spots caused by displacement into the central region of the diffraction pattern.

To further address the question of displacement of atoms versus charge transfer, we performed dynamical calculations of electron diffraction and images using a standard multislice program. We made the program to include the scattering amplitudes of holes and electrons, and performed calculations in the (100) projection of Bi-2212 for three different models of the crystal. In Fig.10.3(b) and Column II, the model is based on neutral atoms using the displacement parameters reported by Horiuchi et al. [Horiuchi et al. 1988], similar parameters were reported in other work [Imai et al. 1988, Gao et al. 1993a]. In the model of Column III, we introduced charged atoms and used the charge distribution from the electron structure calculations of Gupta and Gupta [Gupta and Gupta 1994]. They found a hole concentration in the CuO_2 layer of 0.37 per Cu (also Bi) atom in the material, and an excess electron concentration of the same magnitude in the BiO layer. Their charge distribution is relative to atoms of the following valance: O^{2-}, Ca^{2+}, Sr^{2+}, Cu^{2+}, and Bi^{3+}. Thus, relative to neutral atoms, which are the frame of reference in our calculations, the excess charge is that shown in Fig.10.4. We added these charges in the model which was based on neutral atoms, and assumed the charge is evenly distributed in CuO_2 and BiO planes to arrive at the calculated results of Column III. In the last model (Fig.10.3(d)) and column IV, we maintained the average concentration of charge in each CuO_2 plane and BiO plane at the same level as in the previous model, but we modulated the charge in the BiO layer along the b-axis. With reference to formal valence, this model shows a pileup of electron charge in the region of the BiO double layer where the atomic planes expand along the b-direction; this agrees with the notion that electron doping increases BiO distance [Pham et al. 1993].

In our comparisons, the diffraction pattern of the last model with the pronounced (011) reflections (Fig.10.3(h)) shows the best agreement with observations, Fig.10.3(e). However, these calculations are for very thin crystals, while the maximum thickness of the illuminated region under our parallel beam diffraction experiments could be much larger. A notable difference in the calculated HREM images is the dark, fairly large elliptic regions for the model with charge modulation in the

10.2. $Bi_2Sr_2CaCu_2O_8$

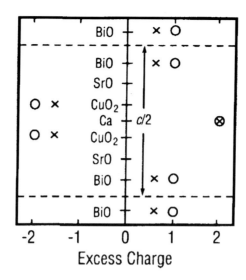

Figure 10.4 Deviation from the charge of neutral atoms on the different planes along the c-direction. ○ for formal valence, and × after creation of the holes as calculated in [Gupta and Gupta 1994].

BiO layer. Very similar features are seen in the high resolution images, Fig.10.3(i). Similar HREM images were observed by others previously [Matsui et al. 1988a, Matsui et al. 1988b], which were pointed out as being a striking feature [Matsui et al. 1988b]. In a subsequent calculation [Horiuchi et al. 1988] these workers substituted 40% of the atoms of some of the Bi sites by Sr thereby improving the agreement with experiments. We obtain excellent agreement with our experimental images (Fig.10.5) without interchanging atoms. Nevertheless, we tested the possibility that substitution could be the source of the inner reflections by making calculations with a gradual change in the occupancy across the unit cell. The image of Fig.10.3(m) was obtained by using a much smaller objective aperture so that, in addition to the direct beam, only the four innermost superstructure spots and the two fundamental reflections (00 ± 2) were inside the aperture (apt.1), as seen in Fig.10.3(e). Primarily, modulation along the c direction is seen in the calculated images for different thicknesses and defocuses using the models without charge modulation in the b-direction, Figs.10.3(n)

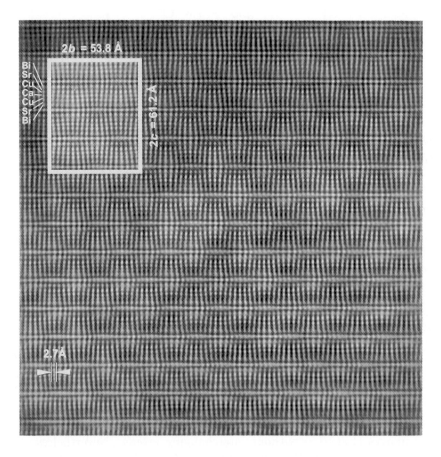

Figure 10.5 A digitized experimental HREM image (after noise reduction) from an area with a slight increase in crystal thickness from top to bottom. The embedded image consisting of four unit-cells is a calculated one based on the displacive and charge modulation as shown in Fig.10.3(b) and Fig.10.3(d). Parameters used for the simulation are the same as Fig.10.3(l). Good agreement between the experimental and calculation confirms the validity of the model.

and 10.3(o). This change is what we expect since the dominating beams now are (002) and (000). When charge modulation along the b-direction is introduced, good agreement with the observed image is achieved. A noteworthy point is that contrary to high-resolution images (Fig.10.3(i)-(l)) obtained by using a large objective aperture (apt.2), these images (Fig.10.3(p)) are very robust, meaning that they are relatively insensitive to small changes in thickness, and extremely insensitive to changes in imaging conditions. By further decreasing the objective aperture to exclude the fundamental reflections (00 ± 2), the image (Fig.10.6(a)) is almost exclusively caused by the charge modulation, though admittedly, the spatial resolution is poor. An aperture that includes one or two fundamental reflections is probably the optimum if this approach is to be developed into a quantitative technique. Then, if the structure as determined by x-ray or neutron diffraction, for example, is well known, the fundamental reflection may be used as a reference beam in an electron interferometric approach to study charge distribution. In Fig.10.6(b), we show that images obtained with a very small objective aperture are very powerful in mapping the long-range disorder of the charge modulation and this also demonstrates the incommensurability of the charge distribution. Presently, the technique is not quantitative and we cannot make a clear choice between the rather simplistic picture of charge transfer presented here and more detailed models based on electronic-structure calculations of the spatial distribution of valence electrons, such as those we used for $YBa_2Cu_3O_7$ (see Sec.10.3.2). What appears clear, however, is that there is a considerable modulation of the valence electron density in the BiO double layer, and this may be caused by additional oxygen in this layer, at least in part.

10.3 $YBa_2Cu_3O_7$

10.3.1 A Novel Diffraction Method: Parallel Recording of Electron Diffraction Intensity as a Function of Thickness

Since $YBa_2Cu_3O_7$ does not have structure modulation which generates superlattice reflections, the method we used for Bi-2212 cannot be di-

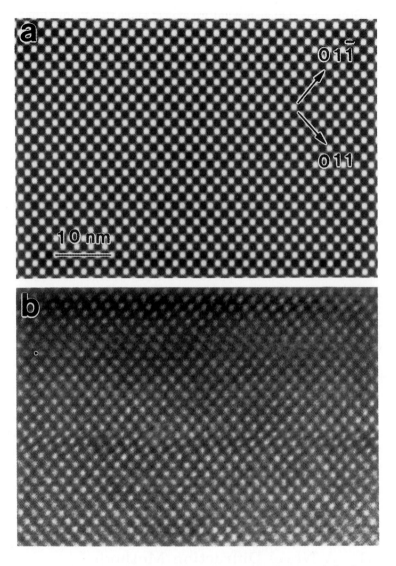

Figure 10.6 a) Calculated image with only the incident beam and the four superstructure spots of type (011) inside the objective aperture. b) a similar experimental image where disorder of the image feature caused by the charge modulation is present, both incommensurability and out of phase boundaries.

10.3. $YBa_2Cu_3O_7$

rectly applied to this material. However, since $YBa_2Cu_3O_7$ also has relatively large lattice parameter along the c-axis, we can study the charge distribution in the crystal by determining the amplitude and phase of Bragg reflections at low angles. As discussed above, the spatial distribution of electrons in crystals can be addressed by x-ray and by electron diffraction. When the amplitudes and phases of the x-ray structure factors are known, the spatial distribution of electrons in the crystal, $\rho(\mathbf{r})$, can be obtained by Fourier transform. Similarly, the electrostatic potential in the crystal, $U(\mathbf{r})$, can be extracted by Fourier transform of ideal electron-diffraction data. When isolated atoms are brought together to form a solid, it is mainly the outer valence electrons that are rearranged; this necessitates very accurate x-ray diffraction experiments to register such minute differences, $\rho(\mathbf{r}) - \rho_a(\mathbf{r})$. Here, $\rho_a(\mathbf{r})$ is the electron distribution to be expected in the crystal when the electron clouds around the atoms are unperturbed by the fact that the atoms are in the vicinity of other atoms, i.e. using the scattering amplitudes of isolated atoms to calculate the structure factors of the crystal. In electron diffraction, since the interaction takes place between the incident electrons and the electrostatic potential of the crystal, the difference between $U(\mathbf{r})$ and $U_a(\mathbf{r})$ can be very large.

Fig.10.7 illustrates the great potential of low-angle reflections in addressing the charge distribution in a crystal, where the Fourier components of the electrostatic potential (for short, Fourier potential) for inner reflections of $YBa_2Cu_3O_7$ are calculated by assuming ionic bonding and different distributions of electron holes in the crystal unit cell. We start with an ionic model for the charge distribution by delocalizing the transferred electrons around the atoms so that the scattering amplitudes of Y^{3+}, Ba^{2+} and Cu^{2+} [Doyle and Turner 1968], and O^{2-} [Rez et al. 1994] are in agreement with tabulated values converted from x-ray scattering amplitudes. We used the atomic positions determined by neutron diffraction [Jorgensen et al. 1990] to calculate these structure factors of the orthorhombic unit cell of size $a = 0.382$ nm, $b = 0.388$ nm and $c = 1.169$ nm. We note, in Fig.10.7, the huge variation in the value of the (001) Fourier potential. With the origin chosen at the Cu atom in the CuO chain, this value changes from a large positive to a large negative one when the holes are moved from the CuO chains to the CuO_2 planes. Thus, if the kinematical intensities and the phases

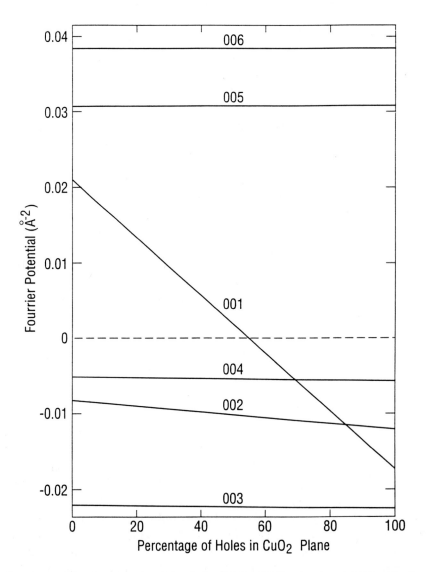

Figure 10.7 The Fourier potentials of the (001) - (006) reflections of $YBa_2Cu_3O_7$ assuming ionic bonding, for different distributions of holes in the CuO_2 plane.

10.3. $YBa_2Cu_3O_7$

of these reflections could be retrieved from electron diffraction experiments, we would be well armed to determine the charge distribution in large unit cell crystals.

However, a major obstacle in electron diffraction has been the difficulty of accurately measuring intensity. In special cases, this problem was circumvented by observing in relatively thick regions of crystal, typically 100 nm, features in the diffraction patterns associated with the strong dynamical coupling between the Bragg beams [Terasaki et al. 1979, Gjonnes and Hoier 1971] and, more recently, by many-beam simulations of the whole convergent beam electron diffraction (CBED) patterns [Tsuda and Tanaka 1995, Vincent et al. 1984] until there is a good fit with the experimental data. These procedures were quite successful in determining structure factors in small unit cell crystals [Zuo et al. 1989], but are not suited for the reflections at small scattering angles of crystals with large unit cell. A somewhat different approach is to use a small convergent beam angle to study large unit cell crystals, and very thin regions of the crystal to reduce the dynamical effects [Olsen et al. 1985]. This was the technique we applied at first in attempting to quantitate charge distributions in high-temperature superconductors by operating our transmission electron microscope at 200keV. As was pointed out by Anstis et. al [Anstis et al. 1973], we found that strong reflections at small angles cause strong coupling between the Bragg beams and thus, very rapid oscillations of the beam intensities with thickness, so that the diffraction pattern was greatly altered by small changes in thickness, as shown in Figs.10.8(a) and (b). We could not control the thickness to within the required about 1 nm for our crystals because they were wedge-shaped at the edge, rather than being platelets a few tens of a nm thick. During the work, we realized that by focusing the electron probe above (or below) the specimen, as shown in Fig.10.9, we could obtain thickness profiles, or pendellosung plots, of many reflections simultaneously, starting from zero thickness up to a maximum that could range from 10 nm to several hundred nm, depending on the distance from the specimen to the cross-over [Zhu and Tafto 1997]. Figs.10.10(a) and (b) show such patterns of the $(00l)$ row of $YBa_2Cu_3O_7$ at a low- and high-magnification, i.e. the probe farther from, and closer to, the specimen.

The technique used in this study does not significantly differ in prin-

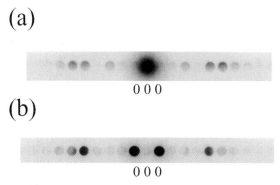

Figure 10.8 Electron diffraction patterns of the $(00l)^*$ reciprocal row in $YBa_2Cu_3O_7$. (a) and (b) show diffraction patterns at estimated thicknesses of 10 nm and 45 nm, respectively.

ciple from using the thickness fringes in dark-field images that were previously used to determine structure factors in small unit cell crystals [Ichimiya and Uyeda 1977]. The great advantage of our approach is that many reflections in large unit cell crystals can be recorded simultaneously, thereby ensuring that the exposure and the crystallographic direction of the incident beam are exactly the same for all reflections. A further advantage, compared with other techniques of convergent-beam electron diffraction, is that the increase in intensity can be followed at very small thickness where the relationship between the structure factors and the intensities is simple, and where the problems of normal and anomalous absorption are minimized. This technique can be improved further by preparing a perfect crystal wedge, thus obtaining a linear relationship between image and thickness. With imaging plates or a CCD (charge couple device) camera, the intensity as a function of thickness of many reflections can be recorded digitally. We expect that this will improve the spatial resolution of the charge distribution by allowing accurate determinations to be made of the structure factors of several reflections.

10.3. $YBa_2Cu_3O_7$

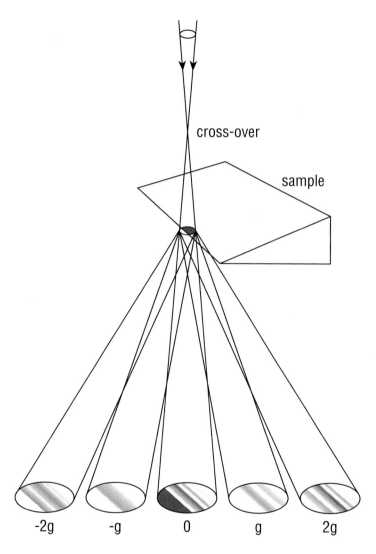

Figure 10.9 Schematics of the experimental configuration used to obtain the diffraction patterns with parallel recording of diffraction intensities as a function of thickness.

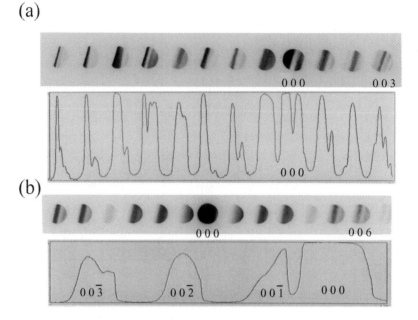

Figure 10.10 Examples of diffraction patterns and line scans of the intensity profiles of different reflections as a function of thickness using the diffraction condition shown in Fig.10.9. The probe is farther from the specimen in (a) than in (b) resulting in lower magnification of the thickness oscillations. The line scan in (b) is from $(00\bar{3})$ to (000) only. Note the onset of each intensity profile.

10.3.2 Distribution of Electron Holes

To determine the amplitude of the inner reflections, we focus attention on the onset of increase in intensity near zero thickness where the kinematical approach holds, and the relationship between the intensities, I_g, of the different reflections, g, the Fourier potentials, U_g, and the thickness, z, is:

$$I_g = |\psi_g \psi_g^*| \propto |U_g|^2 z^2. \tag{10.7}$$

Thus, in the very thin region, less than 10 nm, the amplitudes of the Fourier potentials can be extracted. From Fig.10.7, we note that the (003) reflection is very little influenced by the charge distribution, and, thus, can be used as reference to determine the absolute value of the

10.3. $YBa_2Cu_3O_7$

(001) and (002) Fourier potentials. By analyzing several diffraction patterns using relatively high spatial resolution, i.e. with a small probe close to the specimen as is the case in Fig.10.10(b), we found that the Fourier potential of the (001) and (002) reflection both are close to half the value of the (003). Assuming fully ionic bonding and holes only in the CuO chains and CuO_2 planes, this estimated absolute value of the (001) structure factor corresponds to 80% of the holes being in the CuO_2 plane for the negative sign, and 28% for the positive one (see Fig.10.7). In the thicker regions of the specimen, where the diffraction is dynamical due to coupling between many beams, there is phase information. We used a multislice program to perform the dynamical calculations and included 47 beams along the (001) reciprocal row. With 28% of the holes in the CuO_2 planes, there was poor agreement with the experimental thickness profiles from thicker regions, while the agreement was good with 80 % of the holes in the CuO_2 planes. Calculations for hole fractions of 76% and 56% in the CuO_2 plane, which are shown in Figs.10.11(a) and (b), demonstrate the great sensitivity of electron diffraction to charge distribution. Through a quantitative refinement procedure we developed, as shown in Fig.10.12, the best fit was with a hole fraction of $76 \pm 8\%$ in the CuO_2 plane. The diffraction pattern shown in Fig.10.8(b), with the electron probe focused on the specimen, is consistent with this charge distribution if the thickness is 45 nm. This remarkable pattern, with virtually no intensity in the direct beam (000) and the (002), (003), and (004) reflections, could not be reproduced in calculations in which the hole concentration in the CuO_2 plane deviated by more than 8 % from 76%. From bond valence calculations, using the bond distance determined by Jorgensen et al. [Jorgensen et al. 1990], Brown [Brown 1991] concluded that 2/3 of the holes were in the CuO_2 planes. Hole distributions also have been calculated from first principles [Oles and Grzelka 1991] and determined by core-level absorption spectroscopy [Nücker et al. 1995]. In those studies, some holes were found in the BaO planes. We tried models with holes in the BaO plane in addition to the CuO chain and the CuO_2 plane, but achieved reasonable agreement only with the majority of the holes in the CuO_2 planes. Thus, whether the remaining small fraction of the holes were in the BaO or CuO planes had little influence on the calculated diffraction pattern.

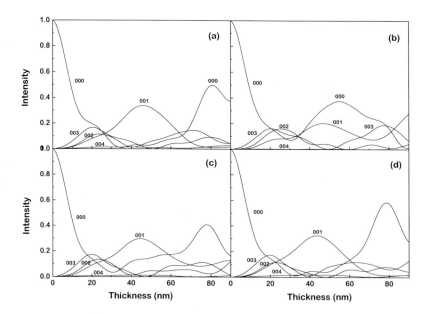

Figure 10.11 Dynamical calculations of the thickness profiles of the electron diffraction intensities for different hole distributions assuming fully ionic bonding in (a) and (b) and with 76 and 56 of the holes in the CuO_2 plane, respectively. In (c) there is partly covalent bonding assigning the charge +3, +2, +1.62, and -1.69 to the Y, Ba, Cu, and O ions, respectively (from [Krakauer et al. 1988]), and with an additional charge transfer of 0.08 holes per unit cell from the chains to the planes in (d).

In these calculations, We assumed fully ionic bonding, but other charge distributions also may be consistent with our observations. Introducing some covalent character by assigning the valences +3, +2, +1.62, and -1.69 to the Y, Ba, Cu, and O atoms, respectively, as suggested from electronic structure calculations using the local density approximation [Anstis et al. 1973], did not give a good fit to the observations, Fig.10.11(c). However, with the small correction of moving 0.08 holes per unit cell from the CuO chain to the CuO_2 plane, the agreement became good, Fig.10.11(d). This demonstrates the limitation of knowing, with sufficient accuracy for assigning charge distribution, only a few structure factors. At the same time, it once again demonstrates

10.3. $YBa_2Cu_3O_7$

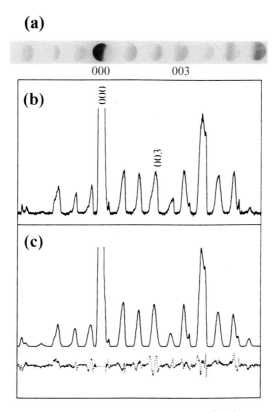

Figure 10.12 (a) An electron diffraction pattern of the $(00l)$ systematic row of in YBCO. (b) Intensity scan of the diffraction pattern of (a). Note the $00\bar{2}$ peak is shadowed due to overlap with the high intensity outside the specimen in the 000 disc. (c) The calculated intensity profiles using the charge assignment described in the text. The difference (dotted-line) between the observations and calculations is included. To evaluate the goodness-of-fit and estimate the error range of the measurement, we modified the R factor traditionally used in crystallography. For detail, see [Wu *et al.* 1997b].

the great sensitivity of electron diffraction to charge transfer in crystals with large unit cells.

In this case we determined only the (001) structure factor with sufficient accuracy to give valuable information about charge transfer. The value of this structure factor is -0.45 ± 0.15 nm. The determined value of the structure factor of the (001) reflection, which is very sensitive to the distribution of valence electron in the crystal unit cell, corresponds in the purely ionic model to $76 \pm 8\%$ of the electron holes being located in the CuO_2 planes of the *fully oxygenated samples*. Note that the effect of moving the Ba atoms only 0.0005 nm would influence the 001 *x-ray structure factor* more than an inaccuracy of this magnitude. Also, in a previous electron diffraction study that relied on the coupling between low- and high-angle reflections, the positions of the atoms and their Debye-Waller factors had to be known accurately [Gjonnes et al. 1993]. However, with our approach, the reflections at low scattering angles of the thin areas are much more dominated by charge distribution than changes in the positions of the atoms.

10.4 Conclusion

While it has been long known that, in principle, electron diffraction can provide useful information on ionicity in crystals, we have experimentally verified this and demonstrated that electron microscope-images on a nanometer scale and electron diffraction patterns (amplitudes and phases of Bragg reflections at low angles) are very sensitive to the distribution of valence electrons and charge transfer over a few tenths of a nanometer. Although in this chapter we discussed only the charge distribution in $YBa_2Cu_3O_7$ and $Bi_2Sr_2Ca_{1-n}Cu_nO_{2n+4+\delta}$ (n=2) cuprates, the methods described can be applied to the other structurally complicated Bi-based superconductors, n=1 (Bi-2201) and n=3 (Bi-2223), and most likely to the closely related Tl-based superconductors, as well as other crystals with large unit cells.

Chapter 11

Experimental Techniques for Grain Boundary Studies

Transmission electron microscopy has a unique capability in studying grain boundaries and interfaces. With an advanced field-emission microscope, we can directly not only examine grain boundary structure on an atomic scale, but also determine the crystallographic orientation of a boundary and measure its chemical composition by forming an electron-probe smaller than 1nm. In the following sections, we discuss some difficult, but most important techniques in investigating grain boundaries. We discuss how to measure grain-boundary geometry, i.e., the five degree of freedom (boundary misorientation and boundary plan-normal), and how to compare the experimentally determined boundary geometry with the observed boundary dislocations, in the framework of Coincidence-Site-Lattice theory, by determining the Burgers vectors, line directions and spacings of the dislocations. We also discuss how to quantitatively measure the oxygen/hole density across and along the boundary using electron energy-loss spectroscopy, and local lattice parameters near the boundary using convergent beam electron-diffraction. The techniques we discuss here can be applied to any other crystal systems.

11.1 Analyses of Grain Boundary Crystallography

11.1.1 Determination of Grain Boundary Geometry

In order to determine the crystallography of a grain boundary we first determine the orientation of grains separated by the grain boundary. A widely used technique in orientation determination is the Kikuchi patterns [Kikuchi 1928] in convergent beam electron diffraction. A Kikuchi pattern, formed by incoherently scattered electron diffracted by lattice planes in a relatively thick region of a crystal, consists of a large number of pairs of Kikuchi lines. For any pair of Kikuchi lines, one line corresponds to θ_B and the other to $-\theta_B$ Fig.11.1, where θ is the scattering angle. Thus, the geometry of the Kinkuchi pattern is very sensitive to the crystal orientation. To determine arbitrary orientations of a polycrystal sample, a complete Kikuchi map, generated by a computer by applying Bragg's law to the crystal, can be very useful. For an orthorhombic system, the map is composed of six standard triangles covering all stereographic projections. Usually, a knowledge of the positions (not including intensities) of the Kikuchi lines is sufficient for indexing a simple crystal. However, for a complex crystal, such as $YBa_2Cu_3O_{7-\delta}$, the Kikuchi patterns are so complicated that most of the observed Kikuchi lines [Zhu et al. 1991e], as shown in Fig.11.1(c), can not be matched with the geometrically calculated Kikuchi maps for a relatively small value of N [see Fig.11.1(a), $N = h^2+k^2+l^2/(c/a)^2 \leq 12$, centered at [101] pole, camera length L=300mm, E=200KV]. In order to match the observed Kikuchi pattern, it is essential that Kikuchi lines for higher indices be generated and the weak and kinematically forbidden Kikuchi lines eliminated. After calculating the intensity of Kikuchi lines and setting an appropriate range of intensity value for printing the calculated Kikuchi lines, all of the observed Kikuchi patterns can be matched quite easily and accurately. Fig.11.1(b) demonstrates an example of a kinematically calculated Kikuchi pattern. All the parameters used are the same as Fig.11.1(a) except for the value of N [$N = h^2 + k^2 + l^2/(c/a)^2 < 40$]. The calculated Kikuchi lines match the experimental pattern [compare Fig.11.1(b) and (c), and note that

11.1. Analyses of Grain Boundary Crystallography

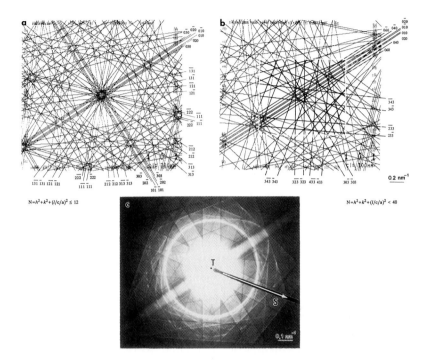

Figure 11.1 (a) A geometrically calculated Kikuchi pattern centered at the $[101]$ pole. The parameters used for the calculation are $a = b = 3.86$Å, $c = 11.79$Å, $N = h^2 + k^2 + l^2/(c/a)^2 \leq 12$, $E = 200$kV, and camera length= 300 mm. (b) A calculated Kikuchi pattern of the $[101]$ pole with $N = h^2 + k^2 + l^2/(c/a)^2 < 40$. Other parameters used are the same as (a). To match the calculated Kikuchi patterns with the experimental ones, we eliminated weak and forbidden pairs of Kikuchi lines based on kinematical intensity calculations of the Kikuchi patterns [compare (a) and (b)]. (c) An experimentally observed Kikuchi pattern of the $[101]$ pole in textured $YBa_2Cu_3O_{7-\delta}$. T and S represent the incident beam direction (beam out) and the beam stopper direction, respectively. The Kikuchi lines with strong intensity seen in (c) were drawn as thick lines in the center square of (b). Note that the scale of the pattern in (c) is double that of (a) and (b). The pattern matches (b) very well but cannot match (a).

the scale in Fig.11.1(a) and (b) is half of that in Fig.11.1(c)]. The lines with strong intensity seen in Fig.11.1(c) were drawn as thick solid lines in the center square of Fig.11.1(b). The structure factor F_g used in the calculation is expressed by

$$F_g = \sum_j f_{xj} \exp(-i2\pi \mathbf{g} \cdot \mathbf{r}) \qquad j = 1, 2, ...13$$
$$f_x = \sum_{i=1,4} a_i \exp[-b_i (\sin\theta/\lambda)^2] \qquad (11.1)$$

where j represents the jth atom site in the crystal unit cell, f_x is an atomic scattering factor, a_i and b_i are parameters determined by curve fitting procedures. The parameter a_i can be found in [Doyle and Turner 1968]. Experimentally, the Kikuchi patterns were recorded for different camera lengths: 200mm for locating the zone axis and 600mm for precisely matching the observed Kikuchi lines with the calculated ones. By changing the lattice parameters, the Kikuchi map may be used for crystals with different oxygen stoichiometry.

To determine the orientation of a grain, at least two crystallographic directions for each grain must be measured. We choose an incident beam direction **T** (beam out) and a direction along the shadow of the beam stopper **S** (which is perpendicular to the incident beam) as the two reference directions. These two directions were recorded in a Kikuchi pattern with the beam stopper pointing to the 000 center spot [see Fig.11.1(c)]. Sharp Kikuchi patterns were obtained by focusing the electron beam with a small convergent angle and tilting the crystal away from a low index zone axis. The observed Kikuchi pattern was superimposed onto a calculated Kikuchi map to locate the beam direction. The beam direction with an accuracy better than 0.03° was then indexed on a calculated Kikuchi pattern on a computer screen by moving the cursor to the exact position of the center spot. The second reference direction can be indexed by rotating a simulated beam stopper about the center spot on the screen to its exact position. Experimentally, the error of the indexing from the measurement is much smaller than the local bending of the crystal itself. In general, for determining the misorientation, only one pair of sharp Kikuchi patterns, one pattern from each of the grains across the boundary, are necessary.

11.1. Analyses of Grain Boundary Crystallography

Grain Boundary Misorientation

The misorientation of two adjacent grains can be described by rotation angles and rotation axes. Based on a method similar to that developed by Young, Steel and Lytton [Young et al. 1973] and Chen and King [Chen and King 1987a], the misorientation matrix **R**(orth), which refers to an orthonormal coordinate system, can be written as

$$\mathbf{R}(\text{orth}) = [\mathbf{t1}, \mathbf{s1}, \mathbf{t1} \times \mathbf{s1}][\mathbf{t2}, \mathbf{s2}, \mathbf{t2} \times \mathbf{s2}]^{-1} \quad (11.2)$$

where $[\mathbf{t1}, \mathbf{s1}, \mathbf{t1} \times \mathbf{s1}]$ and $[\mathbf{t2}, \mathbf{s2}, \mathbf{t2} \times \mathbf{s2}]$ are two 3×3 matrices formed by six column vectors. **t** and **s** are two reference directions, in the present study are the beam direction and the beam stopper direction, respectively; the subscript refers to the grain. According to the properties of rotation matrix [Bollmann 1982], the rotation angle α can be expressed by

$$\alpha = \frac{1}{2} \arccos(r_{11} + r_{22} + r_{33} - 1) \quad (11.3)$$

where r_{ij} are the elements of the rotation matrix. The rotation axis [c1 c2 c3] is given by

$$c1 : c2 : c3 = (r_{32} - r_{23}) : (r_{13} - r_{31}) : (r_{21} - r_{12}). \quad (11.4)$$

All the calculations were carried out in orthonormal coordinates, which must be transformed from the crystal system, say tragonal for Bi-based oxides. For a vector **X** in tetragonal coordinates

$$\mathbf{X}(\text{orth}) = \mathbf{S}\mathbf{X}(\text{tet}) \quad (11.5)$$

where **S** is a structure matrix for a tetragonal crystal. For a matrix **R**

$$\mathbf{R}(\text{orth}) = \mathbf{S}\mathbf{R}(\text{tet})\mathbf{S}^{-1}, \quad (11.6)$$

where **R** denotes a matrix in tetragonal coordinates. Furthermore, for a tetragonal symmetry (P4/mmm), there exist 8 symmetric operations corresponding to rotations about [001], [100], [010], [110] and [-110] axes, respectively. Mathematically, we have

$$\mathbf{R}_{ij}(\text{tet}) = \mathbf{E}_i \mathbf{R}(\text{tet}) \mathbf{E}_j \qquad i, j = 1, 2, ...8 \quad (11.7)$$

where $\mathbf{E}_i$, $\mathbf{E}_j$ are the symmetric matrix. In total, 64 equivalent rotation matrices, and thus 64 pairs of rotation axis and rotation angle can be found. Among them, the minimum rotation angle about an axis with positive sense is usually chosen as a description of the misorientation, the so-called disorientation. The disorientation (hereinafter denoted as misorientation) is then compared with the theoretical values of exact CSLs. In general, the experimentally determined (or observed) structure usually deviates from the exact (or ideal) CSL structure. The deviation can be evaluated by O2 lattice, which is a second order O-lattice (see section 5.2) generated by two O-lattices instead of two crystal lattices. We have

$$\mathbf{R}(O2) = \mathbf{R}(obs)\mathbf{R}(ideal) - 1 \qquad (11.8)$$

where $\mathbf{R}(obs)$ and $\mathbf{R}(ideal)$ are rotation matrices of observed and ideal coincidence rotation, respectively. $\mathbf{R}(O2)$ is a transformation matrix which describes the deviation of an observed grain boundary from the exact coincidence. The misorientation angle obtained from $\mathbf{R}(O2)$ is an important parameter which represents the deviation from an ideal CSL. When the misorientation angle is smaller than a certain critical value, the boundary is considered as a coincidence site lattice boundary.

O-lattice and b-net

Bollmann's O-lattice theory is derived from the earlier concept of CSL [Bollmann 1970]. It is a geometrical approach to characterizing interface structure, and describes the matching and mismatching of the misoriented lattices at an interface. If two misoriented crystal lattices are allowed to inter-penetrate, there will be a periodic set of points in crystal space (not necessarily the lattice points of either lattice) where the two lattices share coincidence sites, known as the O-elements. These sites are the locations of the best match, or equivalently, sites of minimum strain. The O-elements can be separated from one another by the so-called O-cell-wall, i.e., planes bisecting the connection between two O-elements. In this way, a cell structure representing the area of the maximum disregistry between the two lattices is introduced into the crystal space. Consequently, the intersection of a grain boundary with the cell walls is the dislocation network of the boundary.

The position of the O-elements (defined by vector $\mathbf{X}^{(o)}$) and the

11.1. Analyses of Grain Boundary Crystallography

Burgers vector of the dislocations (defined by vector $\mathbf{b}^{(L)}$) can be determined by the O-lattice equation [Bollmann 1970]:

$$\mathbf{X}^{(o)} = (\mathbf{I} - \mathbf{A}^{-1})^{-1}\mathbf{b}^{(L)}$$

which is the same as Eqn.(5.4). We can simplify it by considering a pure rotation about the z-axis. The transformation matrix $\mathbf{A}$ is then given by:

$$\begin{pmatrix} \cos\theta & -\sin\theta & 0 \\ \sin\theta & \cos\theta & 0 \\ 0 & 0 & 1 \end{pmatrix} \quad (11.9)$$

when θ is small, $\sin\theta \approx 0$, and $\cos\theta \approx 1$, Eqn.(5.4) becomes:

$$\begin{pmatrix} b_1^{(L)} \\ b_2^{(L)} \\ b_3^{(L)} \end{pmatrix} = \begin{pmatrix} 0 & -\theta & 0 \\ \theta & 0 & 0 \\ 0 & 0 & 0 \end{pmatrix} \begin{pmatrix} x_1^{(0)} \\ x_2^{(0)} \\ x_3^{(0)} \end{pmatrix}. \quad (11.10)$$

Equation 11.10 suggests that the vectors $\mathbf{b}^{(L)}$, which define a two-dimensional lattice of Burgers vector of the boundary dislocations along the rotation axis, known as the b-net, can be obtained directly from the O-lattice vectors $\mathbf{X}^{(o)}$ clockwise $90°$ along the same axis, with the magnitude reduced by a factor of θ. With recent advances in computer graphics, it is feasible to draw a projected (two-dimensional) O-lattice accurately, and hence, the corresponding b-net. This approach is very useful for predicting the possible Burgers vectors of dislocation arrays for a given boundary misorientation (see the case study). Such an approach also can be applied to a large-angle grain boundary; however, θ is then a small angular deviation from a certain coincidence orientation, rather than the deviation from a perfect crystal, and the rotation is between two O-lattices rather than two crystal lattices. Eqn.(5.4) should be modified as

$$\mathbf{X}^{(o)} = (\mathbf{I} - \mathbf{A}_c\mathbf{E}^{-1}\mathbf{A})^{-1}\mathbf{b}^{(L)} \quad (11.11)$$

where $\mathbf{A}_c$ is the rotation matrix of the corresponding coincidence boundary, and $\mathbf{E}$ is the constraint matrix for non-cubic crystals [Shin and King 1989, Zhu et al. 1993c].

320   Chapter 11. Experimental Techniques for Grain Boundary Studies

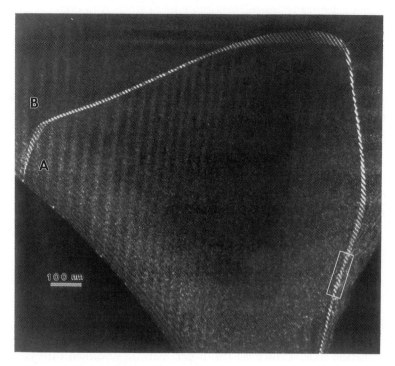

Figure 11.2 A dark-field image of dislocation network located at an entire grain boundary observed in melt-textured bulk $YBa_2Cu_3O_{7-\delta}$.

11.1.2   Case Study of $YBa_2Cu_3O_{7-\delta}$

In this section we illustrate the technique discussed by applying it to $YBa_2Cu_3O_{7-\delta}$ compounds [Zhu 1994]. Fig.11.2 is a dark-field image showing the periodic line contrast at an entire grain boundary observed in melt-textured bulk $YBa_2Cu_3O_{7-\delta}$ (at the bottom left and right of the figure is vacuum). These linear features are not moiré fringes (because the direction of the fringes does not change with the diffraction conditions), but are arrays of grain boundary dislocations. The grain boundary is curved, and has various boundary plane normals. The spacing and the line direction of dislocations at the grain boundary vary with the boundary plane normal. In diffraction, a selected-area-diffraction pattern along the $[3\bar{1}1]$ axis showed a clear lattice rotation between the adjacent crystals. Because there were no segments

11.1. Analyses of Grain Boundary Crystallography

of the boundary plane either parallel to or perpendicular to the rotation axis, this is an arbitrary grain boundary. Such boundaries are frequently seen in bulk $YBa_2Cu_3O_{7-\delta}$. Experimental determination of the Burgers vector of a dislocation array at a grain boundary using the $\mathbf{g} \cdot \mathbf{b} = 0$ criterion is more complicated than that of one within the interior of a grain. For a grain boundary, there are three possible two-beam diffraction conditions:(1) operating within only one grain, (2) operating simultaneously in both grains with the same diffraction vector, or (3) operating simultaneously in both grains but with different diffracting vectors in each grain. It was reported that the $\mathbf{g} \cdot \mathbf{b}$ criterion is not valid for condition 3, unreliable for condition 1, and is most suitable for condition 2 [Humble and Forwood 1975, Forwood and Humble 1975]. In the present case, the Burgers vector of the dislocations was determined through three steps: (1) the possible Burger vectors were first derived by constructing the O-lattice and b-net; (2) the $\mathbf{g} \cdot \mathbf{b} = 0$ criterion was applied using several diffracting vectors from either crystal, or common to both crystals for all possible b's, including those that underwent dislocation reactions; (3) the Burgers vector of the dislocations was finally determined by comparing the measured and calculated values of their line direction and line spacing.

The Grain Boundary Crystallography

The exact misorientation of the two crystals was determined using the method described in the previous section. The best-fit rotation matrix representing the rotation of lattice 1 to lattice 2 (using Cartesian coordinates denoted by a subscript c) was:

$$R = \begin{pmatrix} 0.9978 & -0.0627 & -0.0197 \\ 0.0614 & 0.9962 & -0.0609 \\ 0.0234 & 0.0596 & 0.9979 \end{pmatrix}_c$$

which is a rotation of $5.163°$ about the [0.6702 -0.2414 0.2398] axis in the orthorhombic coordinates.

The rotation axis is very close to $[3\bar{1}1]$ ($\approx [3\bar{1}3]_c$) and the rotation plane is very close to $(3\bar{1}9)$ ($\approx (3\bar{1}3)_c$). The crystallographic orientation of the orthorhombic crystal, the observed rotation axis and the rotation plane are shown in Fig.11.3. Figure 11.4 shows a superpo-

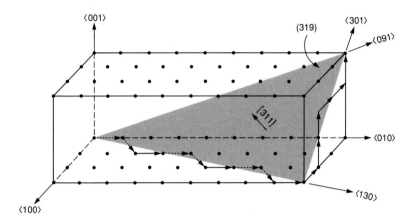

Figure 11.3 The crystallography of the observed grain boundary (rotation angle, 5.163°; rotation axis, [0.6702, -0.2414, 0.2398] in orthorhombic coordinates). The rotation axis is very close to $[3\bar{1}1]$ ($\approx [3\bar{1}3]_c$), and the rotation plan is very close to $(3\bar{1}9)$ ($\approx (3\bar{1}3)_c$).

sition of the two $(3\bar{1}9)$ lattice plane generated accurately by rotating one lattice 5.163° in relation to another about the $[3\bar{1}1]$ axis. It is a two-dimensional periodic pattern. The O-elements (coincidence sites), denoted as black dots, represent the best-fit locations, while the positions between the O-elements represent the worst-fit locations. The O-elements form the O-lattice. By bisecting the O-elements (for a very small rotation, $\theta \approx 0$, the bisection should be perpendicular), we generated the O-cell-wall along the rotation axis. A lattice of Burgers vectors (b-net) then was assigned to each segment of the O-cell-wall with their coordinates rotating 90° from the O-lattice. As illustrated in Fig.11.4, the two base vectors of the b-net are very close to [130] and $[\bar{3}01]$. We note that to demonstrate the lattice mismatch in this figure, we ignored the large difference in scale between the O-lattice and the b-net, and for convenience, chose lattice sites as O-element sites. Although the exact $\mathbf{X}^{(O)}$ and $\mathbf{b}^{(L)}$ can be calculated numerically, in practice, they can be easily determined by the corresponding regions of the best fit and the worst fit through the visual periodicity of the superimposed lattice pattern. Thus, based on the geometry (rotation axis and rotation angle) of an arbitrary boundary, we can construct the corresponding

11.1. Analyses of Grain Boundary Crystallography

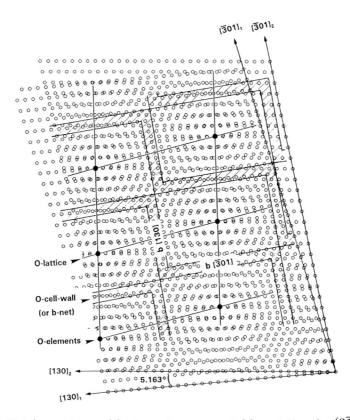

Figure 11.4 A superimposed lattice pattern generated by rotating the $(3\bar{1}9)$ lattice $5.163°$ with respect one to another about the $[3\bar{1}1]$ axis. The O-elements (coincident sites), denoted as full circles, represent the best-fit locations, while the position between the O-elements represent the worst-fit locations. By bisecting the O-elements, we generated the O-cell wall and b-net. The Burgers vector of the dislocations are assigned to be $[130]$ and $[\bar{3}01]$, with a $90°$ rotation from the O-lattice.

O-lattice and b-net; such constructions enable us to derive the possible Burgers vectors of the boundary dislocations without tedious O-lattice calculations.

The Burgers Vectors of the Grain Boundary dislocations

According to Bollmann's O-lattice theory, the misfit of the lattice at the boundary is accommodated by arrays of dislocations whose net Burgers vectors must lie in a plane with its plane normal parallel to the rotation axis. Thus, two set of dislocations with a Burgers vector perpendicular to the rotation axis ([130] and [$\bar{3}$01] in Fig.11.4) would be sufficient to accommodate the boundary misorientation. Nevertheless, the dislocations predicated from the O-lattice construction are based on the geometry of the boundary, and they may not have low energy. Since the elastic strain energy of a dislocation is proportional to $|\mathbf{b}|^2$, a dislocation $\mathbf{b}_1$ would dissociate into $\mathbf{b}_2$ and $\mathbf{b}_3$, if $\mathbf{b}_1^2 < \mathbf{b}_2^2 + \mathbf{b}_3^2$. Thus, in the present case, the following reactions will be energetically favorable:

$$a[130] \Rightarrow a[110] + a[010] + a[010],$$

$$a[010] \Rightarrow a/2[010] + a/2[010],$$

$$a/3[301] \Rightarrow a[100] + a/3[001].$$

Taking these dislocation reactions into account, we then compare these four types of dislocation arrays with Burgers vectors [110], [010], [$\bar{1}$00], and 1/3[001] (as demonstrated in Fig.11.3), with experimentally observed ones.

Figure 11.5 (a)-(e) show the enlarged images of dislocation arrays from a segment of the boundary marked as the rectangular box in Fig.11.2. All the dislocations were imaged at the same magnification under two-beam conditions: $\mathbf{g}_A = 200$ (a), $\mathbf{g}_A = 020$ (b), $\mathbf{g}_B = 1\bar{1}0$ (c), $\mathbf{g}_A = \mathbf{g}_B = 10\bar{1}$ (d), $\mathbf{g}_A = \mathbf{g}_B = 0\bar{2}0$ (e), where d_{110} and d_{010} are denoted as the spacings of the [110] and [010] type dislocations (with Burgers vectors [110] and [010]), respectively, and d_1 and d_2 are denoted, respectively, as $\frac{1}{2}$[010] type, and [$\bar{1}$00] type of dislocations (and/or $\frac{1}{3}$[001] dislocations, because [$\bar{1}$00] and $\frac{1}{3}$[001] dislocations have a very close line direction and a line spacing). The Burgers vectors were determined by the method described in section 11.2.3. To illustrate the configuration

11.1. Analyses of Grain Boundary Crystallography

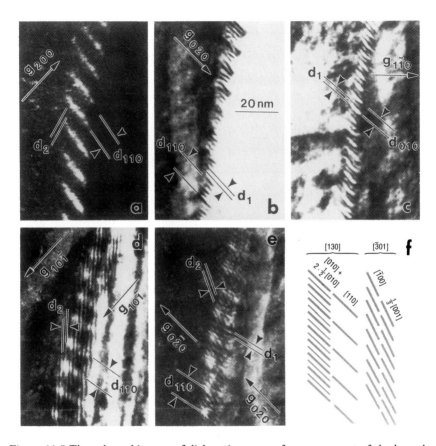

Figure 11.5 The enlarged images of dislocation arrays from a segment of the boundary marker as box in Fig.11.1. All the images were recorded at the same magnification but under different two-beam conditions: (a) $\mathbf{g}_A = 200$; (b) $\mathbf{g}_A = 020$; (c) $\mathbf{g}_B = 1\bar{1}0$; (d) $\mathbf{g}_A = \mathbf{g}_B = 10\bar{1}$; (e) $\mathbf{g}_A = \mathbf{g}_B = 0\bar{2}0$. (d) was imaged away from the $[1\bar{1}1]$ zone axis, while the others were imaged away from the $[001]$ zone axis, d_{110} and d_{010} are the spacings of the $[110]$ (and/or $\frac{1}{3}[001]$) dislocation spacings respectively. (f) A sketch of two groups of the dislocation arrays.

Table 11.1 The $\mathbf{g} \cdot \mathbf{b}$ values of the observed arrays of grain-boundary dislocations.

g	$\mathbf{g} \cdot \mathbf{b}$ for following $\mathbf{b}$				
	$[\bar{1}00]$	$\frac{1}{2}[010]$	$[010]$	$[110]$	$\frac{1}{3}[001]$
$[200]$	-2	0	0	2	0
$[020]$	0	1	2	2	0
$[10\bar{1}]$	-1	0	0	1	$\frac{1}{3}$
$[1\bar{1}0]$	-1	$-\frac{1}{2}$	-1	0	0
$[110]$	-1	$\frac{1}{2}$	1	2	0
$[\bar{1}12]$	1	$\frac{1}{2}$	1	0	$\frac{2}{3}$
$[011]$	0	$\frac{1}{2}$	1	1	$\frac{1}{3}$
$[111]$	-1	$\frac{1}{2}$	1	2	$\frac{1}{3}$

of the dislocations, we have sketched the observed dislocation arrays (Fig.11.5(f)); the dislocations are $\frac{1}{2}[010]$, $[010]$, $[110]$, $[100]$, and $\frac{1}{3}[001]$ with mixed edge and screw character. Under nine different diffraction conditions, we observed two groups of dislocation arrays with $\frac{1}{2}[010]$, $[010]$ and $[110]$ in one group ($[130]$ group), and $[\bar{1}00]$ and $\frac{1}{3}[001]$ in the other ($[\bar{3}01]$ group). The line directions of the different types of dislocation arrays in each group are nearly parallel. Under some diffraction conditions, the arrays of dislocation were not distinguishable; however, under others, they showed a different contrast, which enabled us to define them. For example, we observed $\frac{1}{2}[010]$ type dislocations denoted as d_1 in Fig.11.5(b), (c) and (e); however, we can clearly differentiate $[010]$ from $\frac{1}{2}[010]$ in Fig.11.5(c), and both are out of contrast in Fig.11.5(a) and (d). In Fig.11.5(e) $[\bar{1}00]$ dislocations show residual contrast because the value of $\mathbf{g} \cdot (\mathbf{b} \times \xi)$ is very large. According to Hirsch, Howie, Nigholson, Pashley, and Whelan [Hirsch et al. 1965], dislocations with mixed edge and screw characters will be out of contrast only when both grains are under the two-beam condition with $\mathbf{g} \cdot \mathbf{b} = 0$ and $\mathbf{g} \cdot (\mathbf{b} \times \xi) \leq 0.64$. Table 11.1 lists the $\mathbf{g} \cdot \mathbf{b}$ values of these dislocations for the nine different diffractions showing the validity of the $\mathbf{g} \cdot \mathbf{b} = 0$ extinction criterion in $YBa_2Cu_3O_{7-\delta}$ ($\delta \approx 0.0$) when the two-beam condition is fulfilled in either one grain or in both grains.

11.1. Analyses of Grain Boundary Crystallography

The Line Directions and Line Spacings of the Grain Boundary Dislocations

The dislocation Burgers vectors were finally examined from several different projections by comparing the calculated line direction and line spacing of the dislocation arrays (Eqs.5.1, 5.6, 5.7) with the measured ones:

$$\xi = (\mathbf{B}_1 \times \mathbf{L}_{1p}) \times (\mathbf{B}_2 \times \mathbf{L}_{2p}) \tag{11.12}$$

where $\mathbf{L}_{1p}$ and $\mathbf{L}_{2p}$ are the projections of the dislocation line directions under beam directions $\mathbf{B}_1$ and $\mathbf{B}_2$, respectively. If two sets of boundary dislocations can be observed, then the grain boundary plane normal, $\mathbf{n}$, can be accurately determined by:

$$\mathbf{n} = \xi_1 \times \xi_2. \tag{11.13}$$

The spacing of the dislocation arrays, D, at the grain boundary was determined by measuring the projected spacing of dislocation D_p that was recorded along the beam direction $\mathbf{B}$:

$$D = \frac{D_p}{(\xi \times \mathbf{n}) \cdot \mathbf{B}/(|\xi \times \mathbf{n}||\mathbf{B}|)}. \tag{11.14}$$

Table 11.2 lists one example measured and calculated under the beam direction [-0.019, -0.062 0.327] (image projection direction corresponding to the image shown in Fig.11.5(e)). The second to fifth rows are the comparisons of the calculated and measured line directions of the dislocations, and the sixth to ninth rows are the comparisons of the calculated and measured line spacings of the dislocations. For the line directions, the angular deviation between calculated values and measured ones is about 12° for the [130] group, while it is less than 3° for the [$\bar{3}$01] group. The larger deviation from the [130] group may be attributed to a larger error in the measurement, because the projected length of the dislocations for this group is much shorter, only $1/2 \sim 1/4$ of that of the [$\bar{3}$01] group within an available $\pm 30^o$ tilt of the grain boundary. However, the calculated and measured values for both groups can be considered as being in fairly good agreement. For dislocation spacing, the match of the calculated values with the measured ones is even better. A deviation of less than 0.5 nm was within the range of measurement error, even when the spacing was measured

Table 11.2 Comparison of the measured and calculated dislocation line directions and line spacings projected along the beam direction of $[-0.019, -0.062, 0.327]$. Both grain A and grain B are under a two-beam condition ($\mathbf{g} = 0\bar{2}0$, as shown in Fig.11.5(e)). Orientations were indexed using Cartesian coordinates, denoted by the subscript C.

b	[110]	[020]		[$\bar{1}$00]	[00$\frac{1}{3}$]
		2[0$\frac{1}{2}$0]	[010]		
Calculated line direction	[-0.71 -0.07 -0.70]$_C$	[-0.69 -0.05 -0.72]$_C$		[-0.34 0.14 0.9]$_C$	[-0.36 0.13 -0.93]$_C$
Calculated line projection	[0.65 0.10 0.02]$_C$	[0.61 0.09 0.01]$_C$		[0.33 -0.07 0.00]$_C$	[0.34 -0.06 0.00]$_C$
Measured line projection	[0.94 0.35 0.04]$_C$	[0.94 0.35 0.04]$_C$		[0.98 -0.22 0.00]$_C$	[0.98 -0.22 0.00]$_C$
Deviation (degrees)	11.7	12.1		0.68	2.64
Calculated line Spacing (nm)	14.5	3.6	7.2	8.9	8.8
Calculated spacing projection (nm)	10.5	2.6	5.1	2.7	2.7
Measured spacing projection (nm)	10	2.5	5	2.5	2.5
Deviation (nm)	< 0.5	< 0.1	< 0.1	< 0.2	< 0.2

200,000 times. Thus, the crystallographic characteristics of the observed dislocation arrays were determined for assessing their role in accommodating the lattice misorientation at the boundary.

11.2 Electron Energy-Loss Spectroscopy of Oxygen K-edge

It is well known that the superconducting properties of high temperature cuprates strongly depend on the oxygen content of the crystal. In the case of $YBa_2Cu_3O_{7-\delta}$ ($0 \leq \delta \leq 1$), it can vary from an antiferromagnetic insulator at $\delta = 1$, to a 90K superconductor at $\delta = 0$. It has been shown that the local density of the unoccupied states (hole density near the Fermi level) in the oxygen-2p band are the relevant

11.2. Electron Energy-Loss Spectroscopy of Oxygen K-edge

superconducting-charge carriers, and is very sensitive to the oxygen sub-stoichiometry δ and to the fine structure of the oxygen 1s absorption K-edge, or more precisely, the pre-edge peak (hole peak). It is also well documented a one to one correspondence between the oxygen sub-stoichiometry δ and the integrated intensity of the pre-edge peak, which represents the transition of the electron from the oxygen 1s into unoccupied states. Investigations of the pre-peak of the oxygen K- edge as a function of oxygen sub-stoichiometry using techniques such as x-ray-absorption near-edge structure (XANES) [Takahashi et al. 1990, Yang et al. 1990] and electron energy-loss spectroscopy (EELS) [Nücker et al. 1988, Nücker et al. 1989] have been carried out extensively, and sheded light on our understanding of superconducting mechanism. Unfortunately, these techniques only give an overall electronic structural-information of a specimen and cannot provide information from particular local regions, such as from a defect or an interface, which is important in studying structure-sensitive properties of a material. EELS via an electron transmission microscope [Egerton 1986] (hereafter denote as TEM-EELS) is crucial in such studies. Fig.11.6 is a series of EELS spectra acquired from a grain boundary of a Bi-2212 sample.

11.2.1 Experimental Considerations of TEM-EELS

The TEM-EELS method offers the advantage of local information not accessible using the fluorescence technique. However, low resolution and high background, presumably due to the small probe-size, contribute to low quality of TEM-EELS data, especially for quantitative analysis. With a coherent field-emission source and a PEELS spectrometer, the quality of TEM-EELS has been significantly improved in recent years. The energy resolution of a spectrum from a state-of-the-art instrument, which is determined by the full-width at half-maximum of the zero loss peak, now can be better than 0.6eV that is essential to observe the oxygen pre-peak of Bi-based cuprates [Zhu and Suenaga 1995].

In TEM-EELS, there are image mode and diffraction mode to acquire a spectrum. Diffraction mode with a small camera length usually gives high intensity and good signal-to-noise ratio especially when a

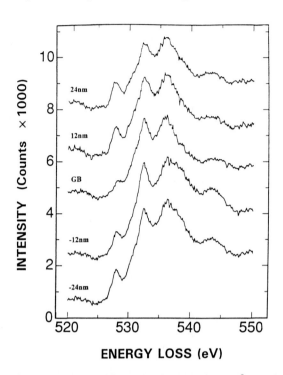

Figure 11.6 EELS spectra obtained from the vicinity of a 84.3° twist boundary (with a small tilt component). The labels show the distance from the grain boundary. The reduction of the oxygen pre-peak at the boundary is clear, suggesting the boundary was hole deficient. The region associated with hole depletion appears to be confined in a zone of less than 200Å.

small probe is required. A short acquisition time (less than 10 seconds) is also desirable, which may reduce any possible specimen (mechanical) and spectra (electronic) drifts, and minimized radiation damages to the area under study. The ideal spectra of the oxygen K-edge should have more than several hundreds counts per channel after the subtraction of the backgrounds.

Finding a suitable grain boundary in a bulk $YBa_2Cu_3O_{7-\delta}$ ($\delta \approx 0.0$) to perform an EELS study is not always trivial. We focused our attentions on "good" boundaries (clean, straight, structurally intact, and without grooving. An example of such a boundary is shown in Fig.11.7. Because EEL spectrum is very sensitive to the thickness of the speci-

11.2. Electron Energy-Loss Spectroscopy of Oxygen K-edge 331

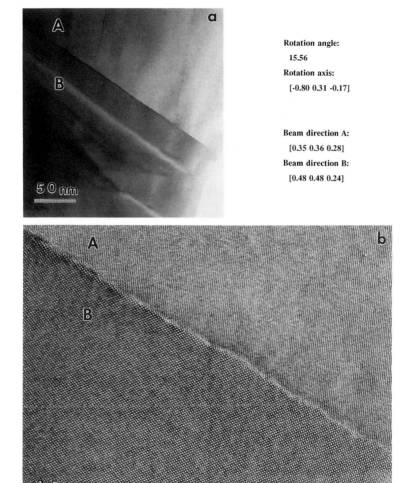

Figure 11.7 An example of a clean large-angle grain boundary in bulk $YBa_2Cu_3O_{7-\delta}$: (a) diffraction-contrast image; (b) phase-contrast image.

men, it has to be collected from an area of interest that appears to be of uniform thickness. This can be done by tilting the area and examining the trend of the thickness contour under a two-beam condition. Next, the oxygen absorption edge from the two abutting grains separated by a grain boundary have to be pre-examined under the boundary edge-on position while avoiding the orientations which result in strong scattering by either grain. If both sides of the grains show a clear oxygen pre-edge peak, a series of spectra were finally acquired across the boundary by scanning the electron probe across the boundary. It is important to note that unless the specimen is fully oxygenated throughout the grains, variations in the intensity of the oxygen pre-edge are often observed. Such specimens will make the interpretation of EELS at the grain boundary ambiguous.

The quality of a spectrum is often determined by the intensity, or the probe-size of the incident beam. Fig.11.8(a) shows two EELS spectra taken from a same spot from a $YBa_2Cu_3O_{7-\delta}$ crystal. The top trace was obtained using ~ 80Å electron probe while the bottom was with $15-20$Å probe. The spectrum collected by a larger probe clearly shows higher jump ratio. Although the use of the small beam results in a significant degradation in the quality of the shape of the oxygen pre-peak, it is required in the grain boundary study. The jump ratio of the oxygen pre-edge can be improved by artificially compensating the energy spread of the incident electron. This can be achieved using deconvolution of the original spectrum with a zero-loss spectrum. An example is shown in Fig.11.8(b), before and after the deconvolution. The deconvolution process clearly brings out a well developed pre-edge peak in the spectrum taken with small beam, even though the original spectrum only shows a bump at the pre-edge position.

In addition, one of the important characteristic features of EEL spectra from superconducting oxides is the orientational dependence of the spectra. However, in studying grain boundaries, because the boundary had to be tilted edge-on, the freedom to change the crystal orientation was very limited. Thus, understanding the orientational dependence of the oxygen K-edge and how to optimize the acquisition condition of EELS for the pre-peak becomes an important issue.

11.2. Electron Energy-Loss Spectroscopy of Oxygen K-edge 333

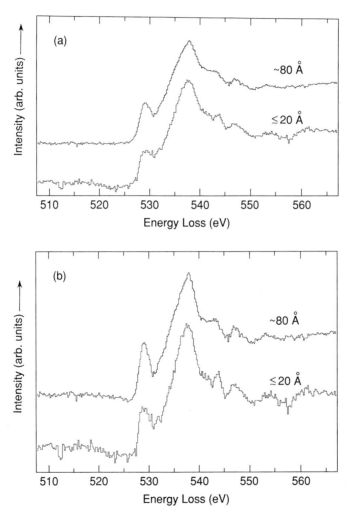

Figure 11.8 Comparison of the oxygen K edge pre-peak obtained from the same location with electron beams of different sizes. (a) As acquired, (b) after deconvolution of (a).

11.2.2 Orientational Dependence of the Oxygen Pre-peak

The momentum transfer is very anisotropic in high-temperature Superconductors. By changing the direction of a parallel incident beam of electrons or photons, it was found that the fine structure of oxygen K-edge was different when the incident beam is parallel or perpendicular to the ab-plane [Batson and Chisolm 1988, Nücker et al. 1988, Eibl et al. 1991, Nücker et al. 1989, Saini et al. 1994, Faiz et al. 1994, Fink et al. 1994]. These experiments on the orientation dependency of electron momentum transfer showed that the hole was mainly located at oxygen in 2-dimensional Cu-O planes with O-$2p_{x,y}$ symmetry. In the case of Bi-2212, for the $\mathbf{q}\|\mathbf{c}$ spectrum, in which O$2p_z$ states are probed, there is no spectral weight between 527.5 and 529.3 eV, while for $\mathbf{q}\|ab$, where O$2p_{x,y}$ states are probed, a well pronounced pre-peak is observed. In contrast, for YBa$_2$Cu$_3$O$_{7-\delta}$ ($\delta \approx 0$), the intensity ratio of the pre-peaks for $\mathbf{q}\|c$ with respect to that for $\mathbf{q}\|ab$ is about 0.48 [Fink et al. 1994]. This higher anisotropy in Bi-2212 causes complications in observing the oxygen pre-peak in the system, and thus, makes measuring the hole distribution at boundary regions very difficult, in particular, when $\mathbf{q}$ is nearly parallel to the c-axis. In an ideal case, a grain boundary plane will be positioned edge-on to the beam for a nano-probe high-resolution measurement at the boundary. However, the orientations of the adjacent grains may not be at optimum for observing the pre-peak. Thus, in reality, some compromises may need to be made with regard to the spatial resolution in the interest of maximizing the intensity of the pre-peak.

The intensity distribution of the core-edge with an energy between E and $E+dE$ in EELS is known to be proportional to the double-differential-cross-section $d^2\sigma/(dEd\theta)$ integrated over all finite scattering angles, θ [Egerton 1986]. This cross-section is given by

$$\frac{d^2\sigma}{dEd\theta} = \frac{8\pi \sin\theta \gamma^2 R}{Eq^2} \frac{k_1}{k_0} \frac{df(q,E)}{dE}$$

where $f(q,E)$ is the generalized oscillator strength, k_0 and k_1 are wave vectors of the fast electron before and after scattering, $R = 13.6$ eV is the Rydberg energy, γ is the relativistic factor, and q is the electron momentum transfer which is dependent on the scattering angle.

11.2. Electron Energy-Loss Spectroscopy of Oxygen K-edge

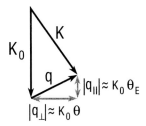

Figure 11.9 A sketch of the experimental conditions of the orientation dependency of the momentum transfer **q**, where **k₀** an **k** represent the wave vectors of incident electrons and scattered electrons, respectively.

When $\theta \ll 1$ rad and $E \ll E_0$, where E_0 is the incident-beam energy, **q** can be broken down into a component $q_\parallel$ (parallel to $\mathbf{k}_0$) and $q_\perp$ (perpendicular to $\mathbf{k}_0$) with $q^2 = q_\parallel^2 + q_\perp^2 \approx k_0^2 \theta_E^2 + k_0^2 \theta^2$, as sketched in the inset of Fig.11.9. Here, θ_E is the characteristic scattering angle at which $q_\parallel \approx q_\perp$ [Egerton 1986]. For oxygen 1s, θ_E is about 2.6 mrad for 100kV electrons, and is about 1.32mrad and 0.88mrad for 200kV and 300kV electrons, respectively. Using the generalized oscillator strength in a hydrogen-like-model [Egerton 1986], the ratio $q_\parallel//q$ and the double-differential cross-section $d^2\sigma/(dEd\theta)$ are calculated and plotted as functions of the scattering angle for the oxygen 1s core loss [Fig.11.10(a)].

In TEM-EELS, we are more interested in the differential-cross-section that gives the fraction of incident electrons scattered into an angle range from 0 to β admitted by the beam-convergence and spectrometer-entrance-aperture, rather than at a particular scattering angle, θ. Figure 11.10(b) shows the calculated $d\sigma/dE$ after integrating $d^2\sigma/(dEd\theta)$, θ from 0 to β, and the ratio $\bar{q}_\perp/\bar{q}_\parallel$, where

$$\bar{q} = \frac{\int_0^\beta q(\frac{d^2\sigma}{dEd\theta})d\Omega}{\int_0^\beta (\frac{d^2\sigma}{dEd\theta})d\Omega}$$

as a function of the integrated angle β for the O-1s core loss, using the generalized oscillator strength in the hydrogen-like-model. As shown in the figure, at about 5.0 mrad, the amount of momentum transfer in the beam direction ($\bar{q}_\parallel$) is equal to that perpendicular to the

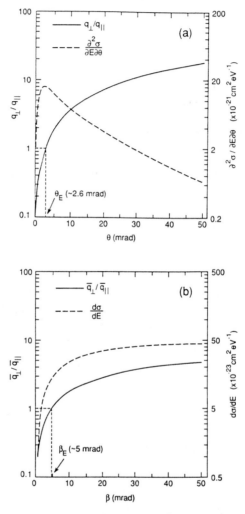

Figure 11.10 (a) Double differential cross-section $d^2\sigma/(dEd\theta)$ and the $q_\perp/q_\parallel$ ratio as functions of the scattering angle θ. (b) Differential cross-section $d\sigma/dE$ and $\bar{q}_\parallel/\bar{q}_\perp$ as functions of the collection angle β.

beam ($\bar{q}_\perp$) (i.e., in this condition, the spectra are independent of the crystal orientation). For a larger β, the component of $\bar{q}_\perp$ increases in the plane perpendicular to the electron beam. In a diffraction mode without an objective-lens aperture, β is determined by the size of the condenser-lens aperture, the entrance aperture of the spectrometer, and the camera length. By adjusting the beam convergent angle and/or spectrometer collection angle based on the crystal orientations at both sides of the boundary, the integrated intensity or the jump ratio of the oxygen pre-peak thus can be optimized. For example, for a crystal with its c-axis parallel to the beam, $\beta \gg 5\text{mrad}$ may be selected so that $\bar{q}_\perp$ is predominant. On the other hand, when the beam is parallel to the ab planes, in principle, one may choose a small β, however, as shown in Fig.11.10, the $d\sigma/dE$ may become too small to observe the pre-peak above the background noise. Thus, the crystal may have to be tilted to gain sufficient intensity of the pre-peak, and this may make the boundary plane no longer truly parallel to the beam.

The set-up of the scattering angle of the incident electron beam and the collection angle of the spectrometer in the EELS experiments is crucial to observing the oxygen pre-peak because of the high crystallographic anisotropy of the superconducting cuprates, particularly for Bi/2212 [Faiz et al. 1994, Zhu et al. 1993h, Chen et al. 1992, Pellegrin et al. 1993, Bianconi et al. 1992, Nücker et al. 1995, Saini et al. 1994, Eibl et al. 1991]. The set-up was adjusted for each crystal orientation by changing the size of the condenser and the spectrometer entrance aperture.

11.2.3 Quantitative Analysis of Oxygen/hole Concentration

The fine structure of the oxygen K-edge absorption spectra in high temperature superconductors is not only sensitive to the total oxygen concentration, but also to the density of mobile holes (charge carriers) [Nücker et al. 1988, Nücker et al. 1989]. To quantify our measurements of oxygen/ hole density across a grain boundary by TEM-EELS, and to obtain a one-to-one correspondence between the hole density and the shape of the oxygen pre-edge absorption peak, we have prepared a set of sintered $YBa_2Cu_3O_{7-\delta}$ with various oxygen contents by subjecting samples to an oxygen-reduction process (i.e., annealing them in air in a box furnace for several hours and quenching them into liquid nitrogen from different temperatures, [Tranquada et al. 1988]). The oxygen contents of these samples were then determined by thermal gravimetric analysis (TGA) and mass measurement (MM) techniques.

Figure 11.11 shows five spectra of the oxygen K-edge from these standard $YBa_2Cu_3O_{7-\delta}$ samples with oxygen contents ranged from $7 - \delta$ =6.3 to 7.0, together with the lattice parameters for each sample using high-resolution near-edge X-ray absorption fluorescence spectroscopy (NEXAFS) [Moodenbaugh and Fischer 1994, Zhu et al. 1994c]. The oxygen K absorption edge in $YBa_2Cu_3O_{7-\delta}$ exhibits two pre-peaks. One occurs near 528 eV which has been found to correspond to hole density and attains a maximum at $\delta = 0$ (optimum superconducting composition), while the other, near 530 eV, is associated with hybridized d state, and has maximum intensity in the insulating, non-superconducting regime. The high-resolution, low background, and good energy calibration provide a reliable basis for a quantitative analysis of these data. The oxygen K-edge was analyzed using the published procedures [Fischer et al. 1993] to yield pre-edge intensity and position for the two pre-peaks as a function of oxygen content (Fig.11.11). Although in TEM-EELS we usually can not resolve the two peaks we know to exist, we chose to use the fine structure of the pre-edge as determined in the fluorescence yield spectra as a framework for analyzing our EELS data. The EELS spectra were treated in the following manner. First, a portion of the pre-edge (background) region was fit with an exponential function by a standard AE^{-r} fitting procedure [Egerton 1986] (where A is a constant and E is the energy loss of the incident electrons) and subtracted. Then relative intensity and energy scales were imposed by matching peak shapes. Then the main-peak edge was subtracted from the pre-peak region (526-533 eV) using a combination of Gaussians. The remaining intensity was attributed to the combined contribution of the 528 eV hole peak and the 530 eV hybridization peak. The relative values of the remaining intensity was used to estimate oxygen content based on the known fluorescence prepeak intensity again oxygen content dependence.

11.2. Electron Energy-Loss Spectroscopy of Oxygen K-edge

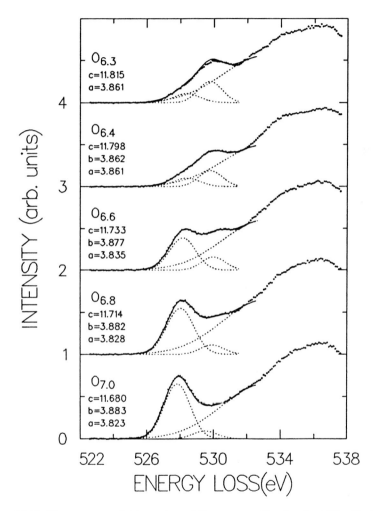

Figure 11.11 Fluorescence-yield spectra of the oxygen K edge for $YBa_2Cu_3O_{7-\delta}$ ($7-\delta$ =6.3, 6.4, 6.6, 6.8 and 7.0). The two pre-peaks on the K edge were fitted by two Gaussian functions. The oxygen content increases with the increase in the integrated intensity of the pre-peak (528eV).

11.3 Lattice Parameter from CBED Measurement of HOLZ Pattern

TEM confers the unique capability for studying local structural defects simultaneously by electron diffraction, real-space imaging, and chemical composition analysis. This quality is essential for studies of grain boundaries. One advantage of convergent beam electron diffraction (CBED) is its high spatial resolution. Typically, structural information can be gained from a region in a thin foil defined by the size of the incident probe. In CBED, lattice parameters can be determined by the positions of the Kikuchi line or high-order Laue zone (HOLZ) lines. The position of the HOLZ lines is more sensitive than the Kikuchi pattern to the change of the lattice parameters ($\Delta a/a$) because the high-order reflections involve ($\Delta a/a \propto g^{-2}$). In addition, HOLZ lines usually are intensity-deficiency lines in the central disk, formed by elastic scattering [Jones et al. 1977] and give a better signal-to-noise ratio. When the convergent angle is small, the HOLZ lines in the disk are straight lines. The position of the line can be expressed by a HOLZ-line equation:

$$K_y = -\frac{g_x}{g_y}K_x + \frac{g_z}{g_y}\left(K_z - \frac{\delta}{2g_z}\right) - \frac{g^2}{2g_y}. \qquad (11.15)$$

where the subscripts denote the x, y, and z components of wave-vector **K** of the incident beam, and the reciprocal lattice vector **g**. δ is a dynamic correction term due to the multiple scattering effect, and can be estimated by

$$\delta \approx -\sum_{h' \neq h, 0} \frac{|U_{h'}|^2}{2KS_{h'}}, \qquad (11.16)$$

where $U_{h'}$ is the dynamical structure factor, $S_{h'}$ is the excitation error, and h refers to all the zero order Laue zone (ZOLZ) reflections being considered. δ also can be obtained from many-beam calculations, and can be approximated as a constant in a small area of a CBED disk. Then, the effects of dynamical dispersion may be treated as a correction to the accelerating voltage, and accommodated by a change in the term

11.3. Lattice Parameter from CBED Measurement of HOLZ Pattern

K_z. When δ is zero, Eqn.(11.15) is the kinematic (geometric) HOLZ line equation.

For a given experimental HOLZ pattern, analysis of computer simulated HOLZ patterns was made using Eqn.(11.15), i.e., by adjusting the lattice parameters systematically until a convincing match with the experimental pattern was found. To achieve a highly accurate measurement of the lattice parameters, which are comparable to the x-ray and neutron methods, we developed a full nonlinear least-square error analysis. The comparison of the experimental and simulated pattern is refined by finding the minimum in a best-fit parameter, χ^2, as a function of the adjustable parameters [Zuo 1992]. We define

$$\chi^2 = \frac{1}{N-p} \sum_i \frac{1}{\sigma_i^2} (d_i^{cal} - d_i^{exp})^2. \tag{11.17}$$

where N is the total number of data points, and p is the number of parameters, d_i is the distance between two HOLZ line intersections, and σ_i^2 is the variance of the ith point. The subscript i distinguishes the distance between different intersections. In the present study, the process of finding the best fit, or the lowest χ^2, was automated with an optimization subroutine used to determine the lattice parameters from a set of measured distances between HOLZ line intersections. The values of the lattice parameters were refined by repeatedly comparing the distances between HOLZ line intersections and their variance until a minimum in χ^2 is found.

In this type of experiment, the selection of a zone axis is very important because this influences the error in the measurement caused by dynamical effects. Also, to determine the lattice parameters in a grain boundary region, the boundary must be planar and truly edge-on, so that the closest possible position to the boundary can be measured (a useful HOLZ pattern cannot be obtained exactly at or very close (< 10Å) to the boundary due to the overlapping of two diffraction patterns and severe lattice distortion, respectively). Thus, in our study, we first locate a grain which has a grain boundary with the c-axis as its boundary plane normal. We have chosen (230)* as a reference orientation. We then, tilt the crystal about the c-axis (the ab plane remains parallel to the beam) to an orientation a few degrees away from the [230] zone center for lattice parameter measurement. By doing this, we ensure that the grain boundary is edge-on and that the obtained HOLZ patterns are sharp with minimal dynamical effects. The probe diameter was less than 17Å with an unsaturated electron beam. We succeeded in measuring lattice parameters as close as 15 − 20Å to the boundary, and in recording a series of HOLZ patterns at 20Å±10Å intervals [Zhu et al. 1994c].

In the above fitting process, it is possible to obtain a convincing match to an experimental HOLZ pattern with all six cell parameters a, b, c, α, β, and γ as variables, provided that the HOLZ lines one uses are sensitive to each of those parameters. Otherwise, a large error could be introduced [Spence and Zuo 1992, Rozeveld et al. 1992]. At present, no optimization method can guarantee finding a global minimum in χ^2, and thus, a correct fit. Generally, using fewer variables tends to stabilize the refinement. In our study, we assumed that the strain near a grain boundary does not affect the cell angle of the crystal, i.e., $\alpha = \beta = \gamma = 90°$. We further note that when the crystal orientation is close to the $[230]^*$-axis, the component of **g** along this orientation is very sensitive to the variations in the lattice parameter along the c-axis, but less sensitive to the difference of the parameters along the a and b axes. Thus, in the refinement, we approximated $a = b$, and refined only lattice parameter a and c. Our experience shows that a two variables refinement is consistently successful in finding likely global minimum, while giving stability to the refinement not obtained when using more than two variables. In terms of the CCSL model, we are more interested in the change of the c/a ratio near the boundary rather than the absolute values of a, b, and c.

Figure 11.12(a)-(c) shows an example of experimental HOLZ line patterns obtained near a grain boundary at 100kV (Before measuring the lattice parameters, the nominal accelerating voltage was calibrated using many-beam calculations and a silicon reference sample). The HOLZ patterns were recorded near the $(230)^*$ zone axis at temperature of 90K with the probe diameter about 17Å (exposure time < 1 sec). Fig.11.12(d)-(f) show the calculated HOLZ patterns with the best fit to Fig.11.12(a)-(c), respectively. The arrow pairs in Fig.11.12(a)-(c) indicate three HOLZ lines, which correspond to three bold lines in Figs.11.12(d)-(f). The shift of the HOLZ lines caused by a small change in the lattice parameters (< 0.8% in Fig.11.12) is readily apparent

11.3. Lattice Parameter from CBED Measurement of HOLZ Pattern

when comparing the triangular area defined by these three lines. In the refinement process, 15 HOLZ lines and more than 300 intersections were used. The small values of χ^2 and the good match of the calculated and experimental patterns confirms the validity of the refinement. The lattice parameters were determined as (d) $c = 11.65(5)$Å, $a = b = 3.840(9)$Å, $\chi^2 = 1.3$; (e) $c = 11.64(1)$Å, $a = b = 3.848(5)$Å, $\chi^2 = 2.1$; and (f) $c = 11.73(6)$Å, $a = b = 3.879(3)$Å, $\chi^2 = 0.9$. The measurement has a precision of $\sim 0.02\%$ under an optimized condition.

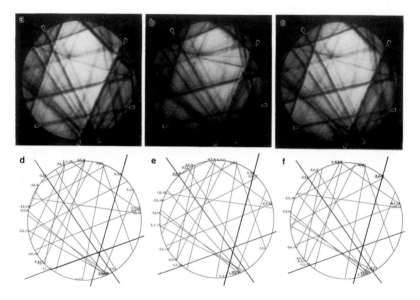

Figure 11.12 (a)-(c) Examples of consecutive experimental HOLZ line patterns recorded near a grain boundary at a nominal voltage 100kV close to the $[230]^*$ zone axis at temperature of 90K. The probe diameter is less than 17Å under an unsaturated condition. The distances from the boundary are (a) 100Å, (b) 80Å and (c) 60Å. (d)-(f) The simulated HOLZ line patterns of (a)-(c) respectively. The lattice parameters retrieved from the best fit to the HOLZ patterns and the associated χ^2 are as follows: (d) $c = 11.65(5)$Å, $a = b = 3.840(9)$Å, $\chi^2 = 1.3$; (e) $c = 11.64(1)$Å, $a = b = 3.848(5)$Å, $\chi^2 = 2.1$; (f) $c = 11.73(6)$Å, $a = b = 3.879(3)$Å, $\chi^2 = 0.9$.

Chapter 12

The Structure of Grain Boundaries

12.1 Introduction

In practical applications of polycrystalline metals and ceramics, the properties of the grain boundaries often determine their utility. This is unquestionably true for high temperature superconductors. One of the stumbling blocks in the practical application of the high T_c ceramic materials is their low critical current densities as compared to the typical values for single crystals and epitaxial thin films [Chaudhari 1987, Dinger et al. 1987]. It is generally believed that the predominant factor limiting the critical current is the disruption of the superconducting current at grain boundaries [Chaudhari et al. 1987b, Campo et al. 1987]. It is thus important to characterize the structure of the grain boundaries produced under various processing conditions and to understand the effect of grain boundaries on the superconducting properties of the sample.

Sections 12.2 and 12.3 of this chapter are devoted to the understanding of structure and properties of individual grain boundaries. We will use the techniques described in the previous chapter to examine in detail many aspects of the-grain boundary structure that are related to the superconducting properties of the high-T_c superconductors. As we will see, $YBa_2Cu_3O_{7-\delta}$ and Bi-2212 compounds have quite different grain boundary characteristics, while share many common attributes in crystal structure.

While it is important to characteristics of individual grain boundaries, it is the collection of grain boundaries with various sizes and shapes that determines the superconducting properties of the polycrystalline samples. We will describe a simple model in section 12.4 to study the overall superconducting properties of a polycrystalline sample using the statistical data obtained from experiments.

12.2 Grain Boundaries in $YBa_2Cu_3O_{7-\delta}$

The observations of higher critical current densities J_c measured in single crystals than in polycrystals of $YBa_2Cu_3O_{7-\delta}$ suggest that improving the superconducting characteristics of the grain boundary is essential in achieving high transport currents in the bulk materials. Earlier, it was reported that the values of J_c in thin films of $YBa_2Cu_3O_{7-\delta}$ are quite sensitive to the angular misorientation of the boundary and decreases drastically when the angle of the misorientation exceed $\sim$ 10° [Chaudhari et al. 1988, Dimos et al. 1988, Dimos et al. 1990]. However, more recently, it was also demonstrated that for polycrystalline thin films (the a/b- and c-axis orientated) [Hwang et al. 1990, Eom et al. 1991] and bicrystals [Babcock et al. 1990, Larbalestier et al. 1991] of $YBa_2Cu_3O_{7-\delta}$, large-angle ($\sim 90°$) grain boundaries transport high currents. An important issue to understand, and one that could form the basis for alternate processing approaches, is why certain grain boundaries behave as weak-link junctions in the cuprate superconductors. A discrete, and distinct, intergranular phase can be probably ruled out (unless the material is fabricated with considerable excess liquid phase) since TEM observations generally do not reveal such phase. An alternative possibility is that the superconducting electron pair potential at the grain boundary is suppressed relative to that in the adjacent grains because of the structure disorders, such as chemical segregation, hole depletion and strain-field, at the boundary. Such structural disorders likely vary with the misorientation of the boundary. In this section, we first present the angular distribution of the boundaries in textured bulk $YBa_2Cu_3O_{7-\delta}$. Then, we discuss how the grain boundary structures, i.e., grain boundary dislocations, oxygen/hole concentration and cation segregation at the boundary, and strain energy of the grain

12.2. Grain Boundaries in YBa$_2$Cu$_3$O$_{7-\delta}$

Figure 12.1 Examples of polarized optical micrographs of textured samples. (a) A well-textured sample, the microstructure consisting of colonies of aligned YBa$_2$Cu$_3$O$_{7-\delta}$ platelets containing randomly distributed black Y$_2$BaCuO$_5$ particles. (b) A poorly textured sample.

boundary, change with grain boundary misorientation in YBa$_2$Cu$_3$O$_{7-\delta}$.

12.2.1 Misorientation Distribution in Textured Bulks

Figure 12.1(a) and (b) show examples of polarized light micrographs of a well and a poorly textured sample, respectively. In the well-textured sample [Fig.12.1(a)], the microstructure consists of colonies (or bundles) of aligned and twinned YBa$_2$Cu$_3$O$_{7-\delta}$ platelets containing randomly distributed Y$_2$BaCuO$_5$ particles (which appear black). The grains are plate-like, primarily aligned in the a-b plane, with lengths in the range of 200-500μm. The bundles have an average size of 50μm in the direction perpendicular to the plane of the platelets. Qualitatively, using an optical microscope, we can see that the misorientation of the aligned grains within the colonies is quite small while the misorientation of grains between the colonies is comparatively large. The degree of texturing is sensitive to the processing parameters as well as to precursor materials. The samples textured from thermal sprayed deposits are usually aligned better than those textured from stoichiometric sintered pellets.

X-ray diffraction was performed on the sample surface (which is parallel to the surface of the TEM specimen used in the misorientation characterization). Fig.12.2 is a typical X-ray diffraction pattern show-

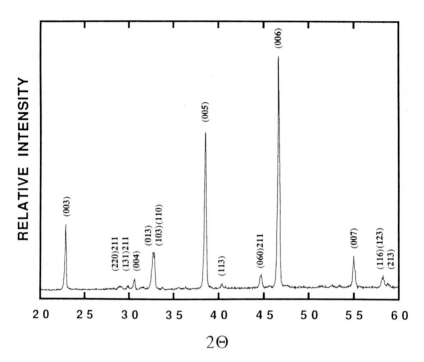

Figure 12.2 X-ray diffraction pattern from a textured bulk surface.

ing the presence of two phases, the major $YBa_2Cu_3O_{7-\delta}$ and the minor Y_2BaCuO_5. Strong intensities of $[00l]$ peaks of $YBa_2Cu_3O_{7-\delta}$ phase in the highly textured sample indicate that a large volume of ab-plane was aligned. This is consistent with the TEM observation.

In order to understand the nature of the grain boundaries of these specimens in detail, the relative crystallographic orientations of over 200 pairs of neighboring grains were characterized in terms of their disorientations, i.e., misorientation angles and rotation axes [Zhu et al. 1991e]. In Fig.12.3 a histogram is presented showing the distribution of the number of boundaries versus misorientation angle in the textured specimens. Preparations 1 and 2 represent the results from well-textured and poorly-textured samples, respectively. It is clear that the values of the angles are not randomly distributed. In the low angle region, a Gaussian-like distribution exists up to 20°. Beyond that

12.2. Grain Boundaries in $YBa_2Cu_3O_{7-\delta}$

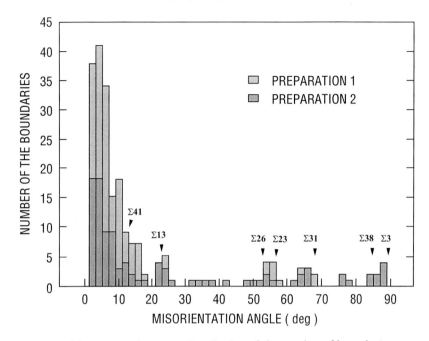

Figure 12.3 A histogram showing a distribution of the number of boundaries versus disorientation angle. Preparations 1 and 2 represent the results from well-textured and poorly textured samples, respectively.

several small peaks are observed approximately at $22°$, $55°$, $68°$ and $89°$. Our measurements indicate that more than 70% of the total 230 observed boundaries are small-angle boundaries (misorientation angle $\alpha < 15°$). Among these about 60% of the boundaries have misorientation angles smaller than $5°$ (42% of total). The observed large fraction of small-angle grain boundaries is simply due to the fact that these specimens are textured and, statistically, the probability of observing small-angle grain boundaries is large. These are primarily the grain boundaries within the colonies. For the large-angle boundaries, the observed preferred misorientation angles imply the possible existence of low-energy grain boundaries in this crystal, although the statistics for the high-angle boundaries may be insufficient to pinpoint low-energy misorientations.

The distribution of the rotation axes corresponding to these grain boundaries is plotted in a stereographic [100] projection and shown in Fig.12.4 (a) and (b) for the large-angle ($\alpha > 15^o$) and the small-angle ($0^o < \alpha < 15^o$) grain boundaries, respectively. The low-index poles are plotted as black dots. It is interesting to note that the rotation axis distribution appears to be not totally random. In fact, almost half of the rotation axes of adjacent grains in the both cases are located within a band $\pm 10^o$ perpendicular to the [001] axis, or lie in planes close to the ab basal plane. This implies that the boundaries with the rotation axis near the ab-plane may be the low energy boundaries, as new nuclei form on existing grains. This observation is not surprising for the small-angle boundaries within the colonies but is somewhat unexpected for large-angle boundaries. It is worth noting that the samples we used for studying misorientation distribution did not undergo a sufficiently slow cooling-process during grain growth. As a result, most of the grain boundaries might not be able to reach low-energy status.

Much higher occurrence of CLS related misorientation distribution of the [001] pure tilt- and twist-boundaries in $YBa_2Cu_3O_{7-\delta}$ was observed by Smith, Chisholm and Clabes [Smith et al. 1988] using naturally grown single-crystal tablets [Kaiser et al. 1987]. In contrast to our samples, their crystals had a very slow cooling-rate during crystal growth, the boundaries are likely to be able to rotate freely. The tablets habit a tetragonal symmetry and have a characteristic square or retangular tablet morphology with edges closely parallel to the a, b, and c edges of the crystal unit cell. It was found that most of the crystals are misoriented relative to their neighbors by a rotation about the c-axis. Fig.12.5 is the histogram for showing the misorientation angle θ between crystals using scanning electron micrographs. All the populated bins shown in Fig.12.1 can be corresponded to the well known CSL misorientations for cubic lattices. The most frequently observed boundary is $\Sigma = 17$ at 28.07^o, and then $\Sigma = 29$ at 43.69^o, and $\Sigma = 5$ at 36.87^o. The absence of some low Σ boundaries such as $\Sigma 25$ and $\Sigma 37$ was attributed to a consequence of the limitations of the measurements rather than a significant feature of the data. Related observations were made by Tietz et al. [Tietz et al. 1988] who found $\Sigma = 13, 17, 1$ and 29 coincidence boundaries in vapor deposited $YBa_2Cu_3O_{7-\delta}$.

12.2. Grain Boundaries in $YBa_2Cu_3O_{7-\delta}$

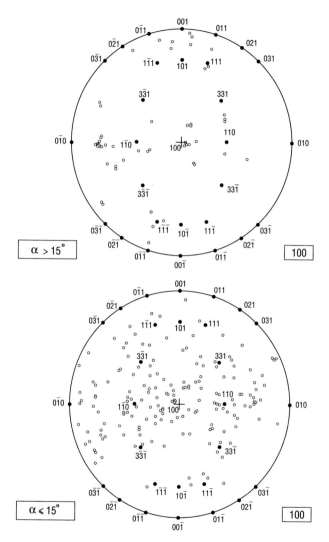

Figure 12.4 The rotation axis distribution in a stereograph of $[100]$ projection. (a) Large-angle grain boundaries ($\alpha > 15^o$); (b) small-angle grain boundaries ($\alpha \leq 15^o$).

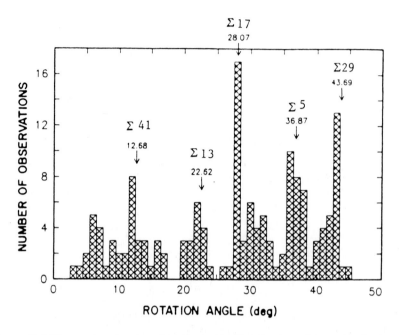

Figure 12.5 Historgram showing the orientation distribution of the individual crystals relative to their neighbors [Smith *et al.* 1988].

12.2.2 Grain Boundary Dislocations

Grain boundary dislocations (GBDs) are intrinsic feature of a grain boundary, and are observed in both small- and large-angle grain boundaries in $YBa_2Cu_3O_{7-\delta}$ [Babcock and Larbalestier 1990, Chisholm and Smith 1989]. These GBDs are often found to accommodate the deviation from the exact coincidence of a boundary. Observing GBDs is not always easy, this at least can be partly attributed to the difficulty of obtaining a proper two-beam condition to image the dislocations. Shown in Fig.12.6 are two sets of dislocation arrays at a small-angle grain boundary (misorientation angle: $4.01°$, rotation axis: [0.5 0.86 0.05]) in textured $YBa_2Cu_3O_{7-\delta}$. The micrographs were imaged from the same area under different diffraction conditions. Fig.12.6(a), (b), (d), (f) are bright-field images and (c), (e) are dark-field images. Since

12.2. Grain Boundaries in YBa$_2$Cu$_3$O$_{7-\delta}$

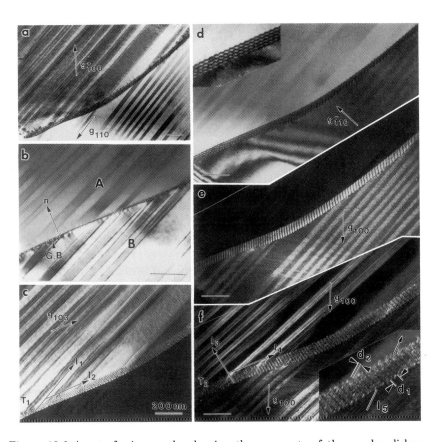

Figure 12.6 A set of micrographs showing the same sets of the regular dislocation arrays boserved at a single small-angle grain boundary with misorientation $4.01°/[0.5\ 0.86\ 0.05]$. (a) Two-beam condition $\mathbf{g}_A=[\bar{1}10]$, $\mathbf{g}_B=[110]$; (b) boundary is tilted edge-on and the direction n is the boundary plane normal; (c) $\mathbf{g}_A=[103]$; (d) $\mathbf{g} \cdot \mathbf{b} = 0$ condition, $\mathbf{g}_B=[\bar{1}10]$, and the dislocations are out of contrast (see inset); (e) $\mathbf{g}_B=[100]$; (f) $\mathbf{g}_A = \mathbf{g}_B = [100]$, and both sets of dislocation arrays can beclearly seen in the inset. Specimen positions (x,y): $+18°, +20°$ (c); $+4°, +1°$ (b); $-2°, -28°$ (f); $-22°, -28°$ (e). (c), (e) and (f) are dark-field images. All the scale bars are 200 nm. For definition of $\mathbf{T}$ and $\mathbf{I}$, see text.

this is a small-angle grain boundary, it is possible to tilt the specimen to positions where both sides of the crystal are under two-beam condition [(a): $\mathbf{g}_A = [\bar{1}00]$, $\mathbf{g}_B = [110]$, (f): $\mathbf{g}_A = \mathbf{g}_B = [100]$ (grain A is closer to the zone axis than grain B)]. Only under such conditions do the dislocations exhibit good contrast. From the inset of (f), two sets of dislocation are clearly visible. These are not Moiré patterns; these can be ruled out simply by changing the reflection condition (Moiré fringes should be perpendicular to $\Delta \mathbf{g} = \mathbf{g}_A - \mathbf{g}_B$). When $\mathbf{g} = [\bar{1}10]$ reflection is excited, the dislocation array contrast completely disappears [see inset of (d)]. We define these dislocations as [110] type ($\mathbf{b} = [110]$). In addition to determining the misorientation between the A and B grains and the Burgers vector of the dislocations, we analyzed the geometry of the boundary dislocation by tilting the specimens about 60° through a boundary edge-on position [Fig.12.6(b)]. The boundary plane normal and the two dislocation line directions were experimentally determined. $\mathbf{T}_1$ and $\mathbf{T}_2$ in Fig.12.6 are two incident beam directions (38° apart), $\mathbf{I}_1$ and $\mathbf{I}_2$ are the projections of the dislocation array 1 and the grain boundary under $\mathbf{T}_1$ incident beam [see Fig.12.6(c)]. $\mathbf{I}_3$ and $\mathbf{I}_5$ are two projections of the dislocation array 1 and array 2, respectively, under $\mathbf{T}_2$ incident beam, and $\mathbf{I}_4$ is the projection of the grain boundary also under $\mathbf{T}_2$ incident beam [see Fig.12.6(f)]. All the directions were determined by Kikuchi patterns. We found that the boundary plane normal $\mathbf{n}=[-0.44\ -0.89\ 0.11]$ [plane Miller index: (-0.30 -0.61 0.73)], and the line directions dislocation array 1 and 2 are [-0.36 0.78 0.51] and [0.85 0.52 0.09], respectively. The results indicate that the boundary is a mixture of tilt and twist in character and the boundary dislocations have both edge and screw components. The dislocation spacing for array 1 (d_1) is about 250Å, while for array 2 (d_2) is about 50Å.

Characterizing the crystallographic orientation of the dislocations using a two-beam brigt-field and dark-field technique in a complex crystal, such as a high T_c superconductor, can be very tedious. However, this is a necessary step to determine the structure of a general boundary (a mixture of tilt and twist characters). Special GBDs, such as [001] type dislocations in [100] tilt boundary and [100] type dislocations in [001] tilt boundary, have been observed in $YBa_2Cu_3O_{7-\delta}$ using HREM. The Burgers vectors were determined by directly measuring the closure failure of a Burgers circuit from images and confirmed by the

Read-Shockley formula $H = |\mathbf{b}|/\sin\theta$, where H is dislocation spacing, $\mathbf{b}$ dislocation Burgers vector and θ boundary tilt angle. However, one must bear in mind that HREM has its limitation for studying boundary dislocations. It is usually only good for observing edge dislocations seen edge-on in tilt boundaries. On the other hand, the conventional TEM technique of imaging and characterizing the nature of dislocations does not have such limitations and compliments the HREM technique.

12.2.3 Oxygen Content and CCSL Boundaries

A unique feature of the $YBa_2Cu_3O_{7-\delta}$ crystals is that its lattice parameters change with oxygen content, e.g., the c-constant increases systematically with the decrease of oxygen content from 11.69 for O_7 to 12.00 for $O_{6.2}$. This makes characterization of structural defects of the material more difficult. On the other hand, as a geometric approach, CSL was successfully used in studying grain boundaries in cubic systems. An important reason for the popularity of the CSL stems from the fact that the CSL description of a boundary structure can be verified by the experimentally determined grain boundary geometry and configuration of grain boundary dislocations using transmission electron microscopy [Balluffi 1979, Zhu 1994, Zhu et al. 1993h]. However, the difficulty for applying CSL to a non- cubic crystal is to define a CSL, or a reference interfacial structure.

As suggested by Bruggeman, Bishop, and Hartt [Bruggeman et al. 1972], a three-dimensional coincidence lattice can exist only when the lattice parameter ratios $a_2 : b_2 : c_2$ are rational. In a non-cubic system, $a_2 : b_2 : c_2$ are generally irrational, as they are for $YBa_2Cu_3O_{7-\delta}$. Thus, for a crystal other than cubic, the ratios of the lattice parameters at the boundary region must be constrained to the closest rational values to form a constrained CSL, or CCSL. In a hexagonal system, such constraint was demonstrated to be accommodated by grain boundary dislocations [Chen and King 1987, 1988]. For $YBa_2Cu_3O_{7-\delta}$, since the lattice parameters, particularly the c-lattice, can vary significantly with the oxygen substoichiometry δ [Jorgensen et al. 1990, Cava et al. 1990], the constraints for some misorientations may be realized readily by varying the oxygen content at the boundary region without introducing extra dislocations [Zhu et al. 1993h]. Furthermore,

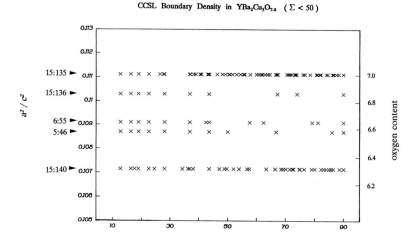

Figure 12.7 Distribution of the ideal CSL boundaries of $YBa_2Cu_3O_{7-\delta}$ ($\Sigma \leq 50$) as function of oxygen content in misorientation-angle space.

because of the small difference in the a and b lattice parameters and twinning in the grain interior (i.e., the symmetry of the entire grain is no longer orthorhombic but tetragonal), for a first-order approximation, it is reasonable to approximate $a = b$ for a grain boundary study in $YBa_2Cu_3O_{7-\delta}$. Thus, we need only consider constraining c^2/a^2 to be rational values. For $YBa_2Cu_3O_{7-\delta}$, the possible $a^2 : c^2$ are 15 : 135, 137.5, 138 and 140 for the oxygen contents of ~ 7.0 to ~ 6.3. For each ratio of $a^2 : c^2$, the possible CSL boundaries ($\Sigma \leq 50$) are plotted as a function of the misorientation angles in Fig.12.7. Even though many of the CSL boundaries in Fig.12.7 appear to be very close to each other on the misorientation angle space, some of these points in the figure are far apart if they are plotted in a space including a rotation-axis dimension. Also, it should be noted that the number of possible 15:135 boundaries is three times more than that for the 15:140 boundaries for $\Sigma \leq 50$, as shown in Fig.12.8. Note that very-low Σ-boundaries ($\Sigma \leq 10$) exist for 15 : 135. Since low Σ-boundaries generally have lower energies than high Σ-boundaries, the 15 : 135 boundaries may have a preference in forming. Furthermore, for boundaries with very large Σ values the CSL description tends to lose its physical significance.

12.2. Grain Boundaries in $YBa_2Cu_3O_{7-\delta}$

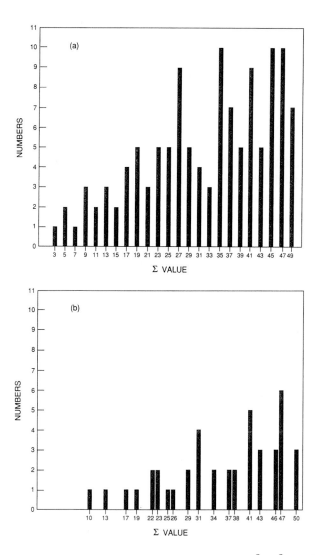

Figure 12.8 The number of ideal CCSL boundaries for (a) $a^2 : c^2 = 15 : 135$ and (b) $a^2 : c^2 = 15 : 140$ boundaries plotted as a function of Σ value.

Based on the average lattice parameters determined for $YBa_2Cu_3O_{7-\delta}$ ($0.0 < \delta < 0.7$), the most likely rational ratios of $c^2 : a^2$ that can form coincidence boundaries are 135 : 15 (corresponding to $\delta \approx 0.0$) and 140 : 15 (corresponding to $\delta \approx 0.7$). With the recent advance in technique for measuring lattice parameters using CBED, it is now possible to make simultaneous measurements of the local lattice parameters and the oxygen/hole contents at grain boundary regions. By doing so, we can verify the existence of the lattice constraint at the boundary and establish the origin of the hole depletion at the boundaries. More importantly, such measurements allow us to critically test the CSL model and provide us with a detailed knowledge of the grain-boundary characteristics. For these reasons, $YBa_2Cu_3O_{7-\delta}$ is an attractive model system for grain boundary studies of non-cubic crystals.

12.2.4 Variation of Oxygen/hole at the Boundaries

Figure 12.9 (a) and (b) are two spectra showing the oxygen K-edge fine structures collected across two large-angle grain boundaries, one is oxygen deficient and the other is fully oxygenated, from the same sample with a nominal composition of $YBa_2Cu_3O_7$. The horizontal axis is energy-loss of the electrons. Here, we focus only on the gross features of the spectra. For each spectrum, the oxygen absorption edge starts at ~ 527 eV, followed by a pre-edge peak at about 529 eV, and the main peak is located at about 537 eV. In the vertical axis, a distance scale indicates each beam position respect to the boundary. To compare the spectra and to observe the relative change of the fine structure of the oxygen K-edge across the boundary (the spectrum from the grain boundary position is marked by an arrow), we normalized the intensity of the oxygen main peak in each spectrum. The spectra are clearly shown that, respect to those away from the boundary, the spectrum from the oxygen deficient boundary shows reduced integrated intensity of the pre-peak, while it does not change for the fully oxygenated one.

Fig.12.10 plots the integrated intensity of the pre-peak and the equivalent oxygen content as a function of the distance from the $\Sigma 10$ boundary using the procedures in Sec.11.3. Although the oxygen pre-

12.2. Grain Boundaries in $YBa_2Cu_3O_{7-\delta}$

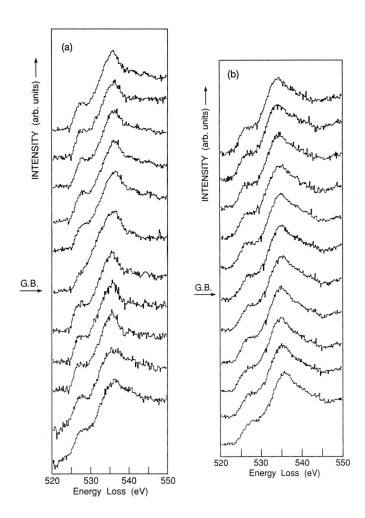

Figure 12.9 A series of electron-energy-loss spectra of the oxygen K edge collected at 50Å apart across (a) an oxygen-deficient grain boundary (GB) ($\Sigma = 31$; misorientation, $69.22°$; [310]; $a^2 : c^2 = 15 : 140$ (the spectrum acquired from the boundary is marked by an arrow) and (b) a fully oxygenated grain boundary ($\Sigma = 3$; misorientation, $83.56°$, [100]; $a^2 : c^2 = 15 : 135$) (no remarkable change in the spectra was observed across the boundary).

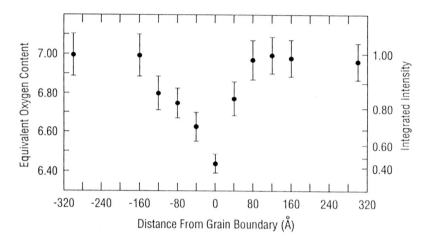

Figure 12.10 The integrated intensity of the oxygen K-edge prepeak and the equivalent oxygen content as a function of the distance from the $\Sigma = 10$ boundary. The hole density is lowest at the grain boundary and gradually increases before reaching a plateau characteristic of the grain interior. The uncertainty in position is ± 10Å.

peak is a direct measure of the hole density rather than oxygen content, for convenience, we use the latter to quantitatively describe the hole density at grain boundary by comparing the integrated intensity of the pre-peak to that of the calibrated ones (Fig.12.10). The oxygen content is lowest, $7 - \delta \approx 6.4$, at the boundary, and gradually increases before reaching a plateau ($7 - \delta \approx 7.0$) at the grain interior (The asymmetry of the hole density respect to distance from the grain boundary is expected, and can be caused by the asymmetric boundary geometry). The oxygen deficient region ($\delta > 0.2$) extends about $80 \sim 160$Å (half width); this is typical for oxygen-deficient boundaries in our study.

To examine the validity of the CCSL model for studying grain boundaries in $YBa_2Cu_3O_{7-\delta}$, we now examine, in more detail, the changes in the lattice parameter, or local strain, associated with the grain boundary misorientation. We examined a dozens of boundaries, a general trend we found is that a large c-lattice parameter is often associated with a low oxygen content, similar to those observed for the bulk $YBa_2Cu_3O_{7-\delta}$ by Jorgensen et al. [Jorgensen et al. 1990], and

12.2. Grain Boundaries in $YBa_2Cu_3O_{7-\delta}$

Cava et al. [Cava et al. 1990] using x-ray techniques.

As an example, we plot the oxygen/hole content, lattice parameters c and a, and the c/a ratio as a function of the distance from the boundary for two groups based on their ideal boundary constraints, or a^2/c^2 ratios ($a^2 : c^2 = 15 : 135$ in Fig.12.11, and $a^2 : c^2 = 15 : 140$ in Fig.12.12 [Zhu et al. 1994c]). Fig.12.11 shows two boundaries whose misorientations meet the constraint of the full oxygenation criterion, i.e., $c^2/a^2 = 135/15$; Boundary A (measured misorientation: $88.46°$ [0.9989, 0.0051,0.0035]; system $\Sigma 3$, $90°$ [100]) and Boundary B (measured misorientation: $61.65°$ [0.9011, 0.0051, 0.3018]; system: $\Sigma 9$, $70.53°$ [301]). Since these boundaries are satisfied with fully oxygenated boundary constraints, they might be expected to have either no or a very small oxygen deficiency nor lattice parameter variation at the boundary region compared with those in grain interior. This is true for Boundary A. No change in the oxygen K-edge EELS spectra was noted at the boundary region. The increase in the a and c values were small although notable, while the change of the c/a ratio was negligible considering the measurement error. In contrast, Boundary B exhibits very large increases in both a and c lattice parameter, at the position nearest to the boundary plane, and then the rate of the increases become small and similar to those for Boundary A. Associated with this large lattice expansion, a drastic drop in the oxygen/hole content at the boundary is observed, but this was limited to the very narrow boundary region ($< 40\text{Å}$).

A likely cause for the above differences in variation in the lattice parameters and the oxygen/hole contents between the two boundaries ($c^2/a^2 = 135/15$) can be attributed to the difference in their angular deviation from the ideal coincidence orientations. The measured deviation is less than $2°$ from $\Sigma 3$ for Boundary A while it is $\sim 9°$ from $\Sigma 9$ for Boundary B. Based on the CSL and O-lattice theories, such deviations have to be accommodated by arrays of secondary grain boundary dislocations. Thus, a large deviation such as seen in Boundary B results in closely spaced boundary dislocations which may contribute to a large change in local lattice parameter and cause the hole depletion at the boundary. To explore this possibility, we calculated the strain field of the array of grain-boundary dislocations, using the Frank's dislocation model for grain boundaries [Frank 1950], and the theory of

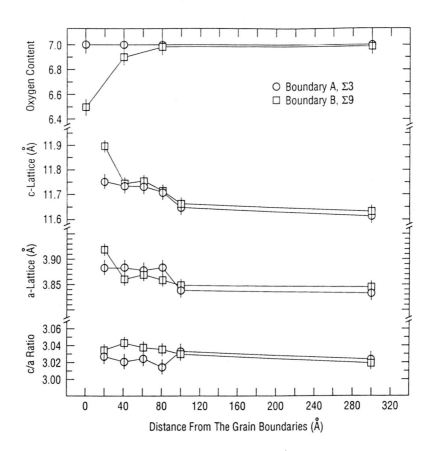

Figure 12.11 Lattice parameters c and a and the c/a ratio as functions of the distance from large-angle boundaries with $c^2/a^2 = 135/15$. The experimentally determined boundary misorientations are as follows: boundary A, $88.46°$, $[0.9989, 0.0051, 0.0035]$ (corresponding system, $\Sigma = 3$, $90°$, $[100]$, $\Delta\theta = 1.81°$); boundary B, $61.65°$, $[0.9011, 0.0051, 0.3018]$ (corresponding system, $\Sigma = 9$, $70.53°$, $[301]$, $\Delta\theta = 8.91°$). The closest position measured from the boundary was about $15 - 20$Å. The interval was 20Å within the distance 100Å from the boundaries. The uncertainty in position is ± 10Å.

12.2. Grain Boundaries in $YBa_2Cu_3O_{7-\delta}$

dislocation grouping [Li 1963]. We found that for a pure tilt or twist grain boundary there is no long-range stress, and the stress decreases exponentially to zero with increasing distance from the boundary. The stress at a position one dislocation spacing away from the boundary (or from the array of dislocations) is less than 3% of the stress at an equivalent position for a single dislocation. Thus, in general, the width of the discernible strain field of a grain boundary can be estimated by the spacing of the grain boundary dislocations. For most large-angle grain boundaries in $YBa_2Cu_3O_{7-\delta}$, this will be less than 20Å from the grain-boundary core in agreement with the observed lattice parameter variations in Boundary B. Thus, dense arrays of grain boundary dislocations associated with the large deviations give rise to a localized severe lattice distortion, and suppress the carrier density and thus the superconducting order-parameter.

Shown in Fig.12.12 are the variations in the lattice parameters and the oxygen contents as a function of the distance from the boundary for boundaries with $c^2/a^2 = 140 : 15$; Boundary C (measured misorientation: $64.10°$, $[0.9983, 0.0032, 0.0055]$; system: $\Sigma 10$, $66.42°$ $[100]$), Boundary D (measured misorientation: $83.78°$, $[0.6201, 0.6181, 0.0083]$; system: $\Sigma 13$, $85.59°$, $[110]$), and Boundary E (measured misorientation: $58.10°$, $[0.8433, 0.4215, 0.0015]$; system: $\Sigma 26$, $52.02°$ $[210]$). This is a group of the boundaries which is thought to be formed in a near equilibrium condition when the samples are sintered at high temperatures while oxygen deficient ($c^2/a^2 = 140 : 15$). Upon cooling, or during an oxygenation process, the interior of the grains become fully oxygenated and the c-lattice shrinks. However, due to their energetically favorable misorientations, these 140 : 15 boundaries are not likely to be able to reach an oxygenation state ($c^2/a^2 = 135 : 15$), i.e., remains oxygen deficiency. Systematic increases in the c-lattice parameters and the associated decreases in the oxygen content are evident for those oxygen deficient coincidence boundaries as the probe is moved towards the boundaries from the interior of the grains (Fig.12.12). At a location very close to the boundary planes ($\sim$ 20Å from the boundary), very large increases in the c-lattice parameters ($\sim 1.7 - 2.3\%$ greater than that of the matrix, even larger than that expected for $YBa_2Cu_3O_{6.3}$ samples). Such large increases, as seen in the case of Boundary B(about 2.2%), can be partially attributed to the lattice distortion caused by

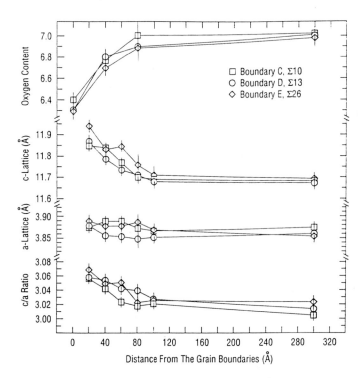

Figure 12.12 Lattice parameters c and a and the c/a ratio as functions of the distance from large-angle boundaries with $c^2/a^2 = 140/15$. The experimentally determined boundary misorientations are as follows: boundary C, $64.10°$, $[0.9983, 0.0032, 0.0055]$ (corresponding system, $\Sigma = 10$, $66.42°$, $[100]$, $\Delta\theta = 2.55°$); boundary D, $83.78°$, $[0.6201, 0.6181, 0.0083]$ (corresponding system, $\Sigma = 13$, $85.59°$, $[110]$, $\Delta\theta = 2.88°$); and boundary E, $58.10°$ $[0.8433, 0.4215, 0.0015]$ (corresponding system, $\Sigma = 26$, $52.02°$, $[210]$, $\Delta\theta = 6.09°$). The closest position measured from the boundary was about $15 - 20$Å. The interval was 20Å within the distance 100Å from the boundaries. The uncertainty in position is ± 10Å.

12.2. Grain Boundaries in $YBa_2Cu_3O_{7-\delta}$

the grain boundary dislocations, particularly for Boundary E, which has a substantially deviation ($\Delta\theta \approx 6.1°$) from the $\Sigma 26$ misorientation.

In contrast to the large variation in the c-lattice parameter, the insignificant change ($\sim 0.3\%$) in the a (or b) lattice parameter with distance from these boundaries could also be due to our assumption of $a = b$ in the lattice parameter refinement. However, the measured c/a (or $2c/(a+b)$) ratios from HOLZ patterns (~ 3.05), and the corresponding hole densities from EELS spectra at the boundaries are very close to the lattice parameter data shown in Fig.12.11 (3.06 for $\delta = 0.7$), and are consistent with the ratio predicated by the CCSL model ($c^2/a^2 = 140 : 15 = (3.055)^2$), i.e., the oxygen contents vary in accordance with the variation in the c/a ratio or the c-lattice parameter. The spatial range of the variation in oxygen for these 140 : 15 boundaries is nearly twice as large as that for Boundary B ($c^2/a^2 = 135 : 15$). The difference can be associated with the causes for the deficiencies. The former can be attributed primarily to the c-lattice expansion due to the grain boundary constraint, while the latter is due to the local lattice distortion originating from the grain boundary dislocations which accommodate the large angular deviation at the boundary.

The grain boundaries observed in bulk $YBa_2Cu_3O_{7-\delta}$ are mostly mixture of tilt and twist in character. It is important to note that grain boundary plane normal is equivalently important to boundary misorientation in determining the hole density or oxygen content at the boundary. Based on the Bollmann's theory, for a given boundary misorientation, the GBDs are defined by the intersection of the boundary plane and the planes with the worst lattice mismatch (O-cell-wall). Thus, a change in the boundary plane normal results in a change in the spacing and line direction of GBDs.

Figure 12.13 (imaged near the [100] axis for grain A, and the [001] axis for grain B, as shown by the insets of the selected-area-diffraction pattern and Kikuchi-pattern) shows an V-shape grain boundary with two different plane normals: [001]A/[010]B (boundary II) and [209]A/[230]B (boundary II). It is a near $\Sigma 3$ boundary with $c^2/a^2 = 135 : 15$. Although the entire V-shape grain boundary corresponds only one misorientation (both are arbitrary boundaries with mixed character of twist and tilt) careful EELS with a 2nm-probe showed that the boundary I has the same oxygen/hole concentration as that in grain

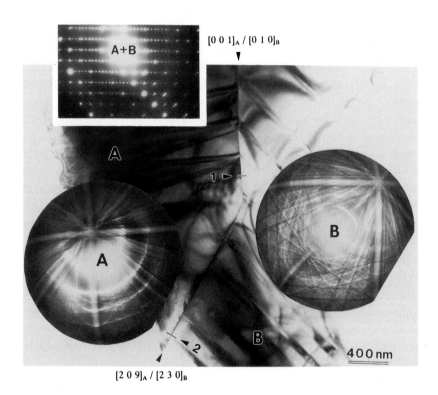

Figure 12.13 The V-shaped grain with two different boundary planes. The boundary was imaged with the beam nearly parallel to the [100] axis for grain A, and the [001] axis for grain B. The selected-area-diffraction pattern, covering both grains and the corresponding Kikuchi-pattern for each grain are shown in insets.

12.2. Grain Boundaries in $YBa_2Cu_3O_{7-\delta}$

interior, while boundary II is oxygen/hole deficient.

According to the CCSL model, an ideal $\Sigma 3$ boundary or a boundary close to the $\Sigma 3$ orientation is expected to be fully oxygenated [Zhu et al. 1993h]. We note that the observed grain boundary has a large deviation (8.91°) from the ideal misorientation. Calculation using Bollmann's O2-lattice theory [Bollmann 1970], showed that to accommodate the 8.91° deviation about the $[-0.949, 0.111, 0.099]$ axis from the ideal $\Sigma 3$ boundary, the required Burgers vectors of the GBDs are [010] and 1/3[001] (Displacement-Shift-Complete-lattice vectors, or DSC vectors). For the [010] set, the density of GBDs is the same for both boundary planes. In contrast, for the 1/3[001] set, the dislocation spacing for the [209]A/[230]B boundary plane is 27Å, but is about 100 times wider for the [001]A/[010]B boundary plane (2860Å). Thus, the hole depletion at the [209]A/[230]B boundary can be attributed to the closely spaced 1/3[001] dislocation array at the boundary. It was suggested that to generate one hole in $YBa_2Cu_3O_{7-\delta}$, five undistorted oxygen-chains are required [Cai and Welch 1995b]. Consequently, even for a boundary with a high oxygen concentration, GBDs may cause the disappearance of the oxygen pre-peak at the boundary due to severe distortion of the lattice.

Although the CCSL model is based on crystallography of grain boundaries and does not give detailed atomistic structure of a grain boundary, it can describe the gross features of the grain boundaries, such as local lattice constraint, oxygen contents and coincidence orientations of the large-angle grain boundaries in $YBa_2Cu_3O_{7-\delta}$. This finding demonstrates the significance of the geometrical approach in studying grain boundaries even in low-symmetry crystal systems, although other factors may also contribute to the observations. For example, the increase of the lattice parameters near the $\Sigma 5$ and $\Sigma 9$ boundaries may not be explained by the CCSL model alone. Grain-boundary volume expansion may be a general phenomena because a grain boundary between two close-packed crystals cannot be as closely packed as the crystal themselves, especially after the structural relaxation has taken place at the boundary. In addition, such expansion can be also due to the space charge at the boundary regions (common for ionic crystals) in these low carrier density oxides. To further understand the grain boundary structure and to illustrate the continuation of the CuO planes at boundaries in $YBa_2Cu_3O_{7-\delta}$, we are undertaking calculations of grain boundary potentials consisting of a long-range Coulomb interaction and short-range Buckingham potential using a standard shell model for ionic-crystals. The final thermodynamically "relaxed" boundary structures will result from the initial geometrical CCSL configurations.

12.2.5 Cation Segregation at the Boundary

Energy Dispersive X-ray Spectroscopy (EDS) measurements with 20Å spatial resolution were made to detect local variations in cation concentration at arbitrary large-angle grain boundaries in bulk $YBa_2Cu_3O_{7-\delta}$ (nominally $\delta \approx 0.0$) samples. Overall, we did not find any remarkable difference in Y and Ba concentrations at the grain boundaries in comparison with the crystal matrix. However, we observed Cu-rich regions at some grain boundaries which is consistent with previous investigations [Babcock et al. 1988, Alexander et al. 1991]. In one case, we found a variation in Cu composition along the boundary (Fig.12.14(a))[Zhu et al. 1993c]. To compare the relative change in local cation composition, we normalized the average of the peak intensity of the matrix area to $Y = 7.7\%$, Ba=15.4%, and Cu=23.1% by using the K-ratio, i.e., the ratio of characteristic intensities measured on the specimen and the standard. Seven locations were measured, spanning about 80Å along the boundary. The Cu concentration varied periodically at the boundary (the variation in the Y and Ba concentrations at the boundary was insignificant, and was within the uncertainty of the measurements), while it remained constant in the matrix. The average concentration of Y, Ba, and Cu from two traces parallel to the grain boundary from neighboring grains is shown in Fig.12.14(b).

The misorientation of the grain boundary was characterized as a rotation angle $39.81°$ about the $[0.9999, 0.0004, 0.0005]$ axis. It deviates from the ideal $\Sigma 5$ boundary (rotation angle: $36.87°$, rotation axis: [100], see Fig.12.15(a)) by $2.94°$ (O2-angle) rotated about [0.9998,- 0.0014, 0.0070] (O2-axis). The boundary plane normal is [010], which is perpendicular to the rotation axis, suggesting that the boundary can be approximated as a tilt boundary. For a pure tilt boundary, only one set of edge dislocations is need to accommodate the deviation from a

12.2. Grain Boundaries in $YBa_2Cu_3O_{7-\delta}$

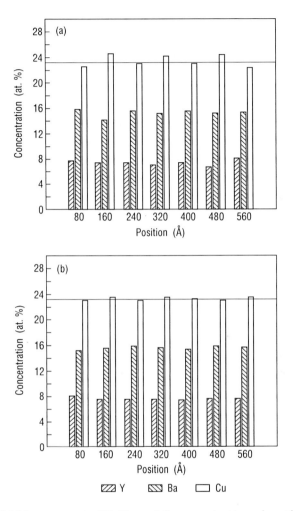

Figure 12.14 (a) Measurements of Y, Ba, and Cu concentrations along the boundary. Position denotes the distance from a reference starting point at the boundary. The Cu concentration varies periodically at the boundary (variation in the Y and Ba concentrations at the boundary was insignificant and was within the uncertainty of the measurement). The average concentration of Cu in the matrix is indicated by the horizontal line. (b) The average cation concentration from two traces parallel to the boundary from neighboring grains.

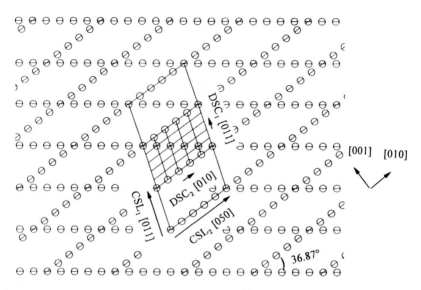

Figure 12.15 Projection of an ideal (unrelaxed) Σ5 CCSL grain boundary (rotation angle: $36.87°$, rotation axis: $[100]$, $c^2/a^2 = 135:15$) viewed along the common rotation axis. ⊘ and ⊖ represent lattice 1 and 2, respectively. The unit cell of the coincidence-site-lattice is shown in the middle of the drawing. The unit vectors of the CSL are $[011]$ and $[050]$, while the unit vectors of the DSC are $[011]$ and $[010]$.

coincidence orientation or from a perfect crystal. Based on the grain boundary dislocation theory as originally proposed by Frank [Frank 1950] we found that to accommodate the misorientation of $2.94°$, an array of edge dislocations with a Burgers vector of $[010]$ is required. The calculated spacing of the GBDs is about 75Å, which agrees the periodicity of the Cu concentration observed at the boundary. Although a weak contrast of the GBDs was observed when the boundary was tilted away from the edge-on position, it is too weak to be reproduced here (because two-beam diffraction condition for both grains could not be achieved in the allowable tilting range of the specimen). Thus, the observed Cu segregation at the boundary can be associated with the single array of edge dislocations there. The Cu solutes apparently can

12.2. Grain Boundaries in $YBa_2Cu_3O_{7-\delta}$

diffuse to the dilated region below the extra half plane and form clusters along the dislocation line, as schematically shown in Fig.12.15(b). The segregation of solute atoms associated with the GBDs was observed by Michael et al. [Michael et al. 1988], in a small-angle [001] twist boundary in a Fe-Au bicrystal using EDX, and Cu rich dislocation cores were observed in $YBa_2Cu_3O_{7-\delta}$ film by Gao et al. [Gao et al. 1991] using high-resolution electron microscopy. Segregation of solute atoms at defects is an intrinsic phenomena. However, such segregation at the GBDs may not always be observed depending on the spacing of the dislocations and the line direction of the dislocations as well as the extent of the segregation. For an arbitrary grain boundary, the segregated solutes may be delocalized due to the complication of the strain field associated with the three independent sets of dislocation arrays required at the grain boundary. Grain boundary segregation is known to have strong effects on the mechanical and superconducting properties of the boundary, such as embrittlement and flux pinning. Our observation suggests that grain boundaries which are crystallographically "special", i.e., being near a coincidence orientation, may not be special with respect to grain boundary segregation.

12.2.6 Strain Energy of the Grain Boundaries

Since grain boundaries are plannar defects, and the unit-cells of CSLs with low index rotation axes are characteristically anisotropic, the connection between Σ and energy is elusive, although it is noted that the elastic strain fields of such periodic boundaries are necessarily short-ranged. Nevertheless, the experimental data in support of a correction between low-energy and short-period coincidence site lattice are compelling.

GBD is an intrinsic feature of grain boundary structure and describes the topology of the boundary. It is worth noting that the observed Cu segregation and oxygen depletion at the boundary are well associated with the structure of GBDs. Furthermore, GBDs determine the elastic strain energy of the boundary, which is important in its structural transition. Grain boundary energy $E(\theta)$ can be considered as the sum of boundary core energy $E_c(\theta)$ and boundary elastic strain energy $E_{el}(\theta)$ (which is equivalent to the energy of GBDs), i.e.,

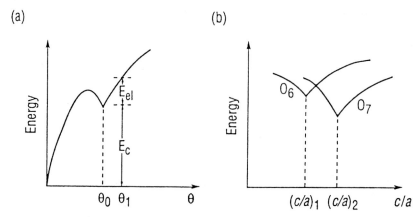

Figure 12.16 (a) Schematic plot of grain boundary cusps in the energy vs. misorientation angle curve. The grain boundary energy is the sum of core energy $E_c(\theta)$ and elastic strain energy $E_{el}(\theta)$. The deviation from a coincidence orientation θ_0 is accommodated by grain boundary dislocations that contribute to $E_{el}(\theta)$. (b) Schematic plot of grain boundary cusps in the energy vs. c/a ratio curve. O_7 and O_6 correspond to fully oxygenated and oxygen-deficient boundaries, respectively. If the O_7 boundary has lower energy than the O_6 boundary, then the latter is likely to transform into the former upon relaxation or during an oxygen annealing process.

$E(\theta) = E_c(\theta) + E_{el}(\theta)$. As shown by the early work of Read and Shockley [Read and Shockley 1950] the long-range elastic energy of GBDs should tend to produce cusps at a certain coincidence orientation, say θ_0, on curves of $E(\theta)$ vs θ, as schematically demonstrated in Fig.12.16(a). To demonstrate the equilibrium of the grain boundary structure, Brokman and Balluffi [Brokman and Balluffi 1981] constructed a boundary by first joining two crystals rigidly, and then allowing relaxation to a final structure. Using a pairwise energy model with four different types of pair potentials, they found that for a series of tilt or twist boundaries in cubic system, certain boundaries corresponding to low θ misorientations have relatively low energies. Upon relaxation, certain boundaries with a small deviation from low θ misorientations then become favored boundaries and all the other boundaries become nonfavored boundaries possessing dense arrays of GBDs. Their conclusion may hold true for all types of crystalline interfaces. However, for non-cubic crystals, there is an additional requirement, i.e., the

12.2. Grain Boundaries in $YBa_2Cu_3O_{7-\delta}$

constraint of the irrational ratio of the lattice parameters to the closest rational value to form a coincidence boundary. The basic idea is that any small deviation from an ideal (or rational) c^2/a^2 can be accommodated by an array of dislocations at the boundary, the same as the deviation of a misorientation from a coincidence orientation. For $YBa_2Cu_3O_{7-\delta}$, because the c/a ratio is very sensitive to the oxygen content, the rational ratio may be easily achieved by simply adjusting the oxygen content at the boundary [Zhu et al. 1991e, Zhu et al. 1993h, Zhu et al. 1993c].

An effort we try to make here is to understand how a grain boundary changes from one state to another during a relaxation process. Let us consider two ideal coincidence boundaries with very close misorientations, but with different c/a ratios, i.e., one corresponding to a fully oxygenated boundary (hereafter denoted as the O7 boundary), the other corresponding to an oxygen-deficient boundary (denoted as the O6 boundary). If the O7 boundary has lower energy than the O6 boundary, then the latter is likely to transform into the former upon relaxation or during oxygen annealing, as demonstrated in Fig.12.16(b). The tendency for such structural transition to occur may be determined by the elastic strain energy of grain boundary (resulting from accommodating the deviation from a favorable misorientation and a favorable c/a ratio), E_{el}: [Hirth and Lothe 1968]

$$E_{el} = \frac{\mu b^2}{4\pi(1-v)D}[\eta_0 \coth\eta_0 - \ln(2\sinh\eta_0)] \qquad (12.1)$$

where $\eta = \pi b/\alpha D$, b is the Burger vector of the GBDs (DSC vector for a large-angle grain boundary) and α is a factor describing the core of the dislocations. The spacing of GBDs, D, for a large-angle grain boundary can be determined by

$$D = \frac{\left|X^{(O2)}\right|^3}{X^{(O2)}}[n \times (n \times X^{(O2)})] \qquad (12.2)$$

where n is the boundary plane normal, and $X^{(O2)}$ is the O2-lattice vector, defined by

$$X^{(O2)} = (E^l - R_c E^l R^l)^l b^{dsc} \qquad (12.3)$$

where b^{dsc} is the Burgers vector of GBDs, R is a rotation matrix of the boundary and R_c is the rotation matrix of the corresponding coincidence boundary, and E is the constraint matrix [Bollmann 1982, Shin and King 1989].

Assuming there are two ideal tilt boundaries: $\Sigma 29$ (46.40°/[100], $c^2/a^2 = 15 : 135$ corresponding to fully oxygenated grain boundaries) and $\Sigma 25$ (47.16°/[100], $c^2/a^2 = 15 : 140$ corresponding to oxygen-deficient grain boundaries) with a [100] boundary plane normal and a [100] Burgers vector (a DSC vector) of GBDs. The difference in the misorientation between the two boundaries is less than 1°. Geometrically, since they have similar densities of coincidence sites per unit cell, the difference of their core energy, E_c, may be ignored. Thus, we can estimate their boundary energy by comparing their strain energy. A plot of grain boundary strain energy as a function of misorientation and c/a ratio (or temperature, since the c/a ratio decreases with temperature) is shown in Fig.12.17 using Eqs.(12.1)–(12.3). In each case, the strain energy tends to minimum, i.e., their dislocation spacing tends to infinity, when the ideal c/a ratios and the exact coincidence misorientations (marked as A and B, corresponding to $\Sigma 29$ and $\Sigma 25$, respectively) are achieved. For a given misorientation, the deviation from a coincidence orientation can be accommodated by GBDs, as well as oxygen content (c/a ratio) at the boundary.

At sintering temperature (950°C), the crystal lattice has a larger c-lattice parameter ($c/a > 2c/(a+b) > 3.05$) than that at low temperature, and the boundary formed with that c/a ratio will be oxygen deficient ($a^2 : c^2 = 15 : 140$) [Zhu et al. 1993h]. During slow cooling in an oxygen atmosphere, the specimen takes up oxygen and the c-lattice shrinks. For one misorientation, say 46.5°, oxygen may easily diffuse to the boundary to lower the boundary energy, which may result in a fully oxygenated grain boundary ($c/a \approx 3.0$, $a^2 : c^2 = 15 : 135$, $\Sigma 29$); for another misorientation, say 47.5°, the boundary will remain oxygen deficient ($c/a \approx 3.05$, $a^2 : c^2 = 15 : 140$, $\Sigma 25$), because the change of the c/a ratio from 3.05 to 3.0 (through taking up oxygen) is energetically unfavorable. Thus, during the oxygenation process, some boundaries can reach a fully oxygenated state, while others remain oxygen deficient. We estimated that among all possible CCSL boundaries ($\Sigma < 50$), only about 20% of the O6 boundaries ($a^2 : c^2 = 15 : 140$) are energetically

12.2. Grain Boundaries in $YBa_2Cu_3O_{7-\delta}$

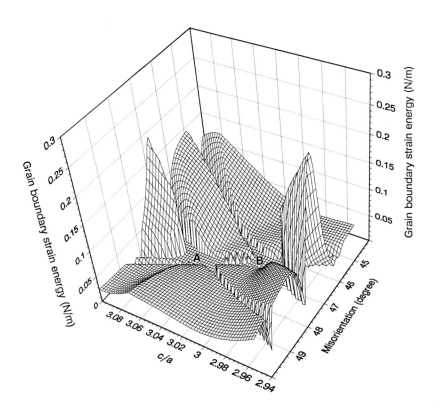

Figure 12.17 Grain boundary strain energy as a function of misorientation angle and c/a ratio and the exact coincidence misorientation (marked as A and B corresponding to $\Sigma 29$ and $\Sigma 25$, respectively), are achived. For given misorientation, the deviation from a coincidence orientation can be accommodated by GBDs, or equivalently by oxygen content (c/a ratio) at the boundary.

favorable for transforming into the O7 boundaries ($a^2 : c^2 = 15 : 135$) [Zhu et al. 1993h]. It has been observed that, statistically, prolonged oxygen annealing does increase the numbers of low Σ boundaries in bulk $YBa_2Cu_3O_{7-\delta}$ [Wang et al. 1992]. However, for standard sintered samples, especially when the grain size is small, the majority of large-angle grain boundaries may not be able to reach their equilibrium states. They usually show a large deviation from coincidence orientations and possess closely spaced GBDs. Such GBDs may be detrimental to the properties of the grain boundary. The strain field associated with a high density of GBDs can cause a depletion of hole density and suppress the superconducting order parameter, and hence, block the passage of superconducting current at the boundary [Chisholm and Pennycock 1991]. Thus, far fewer large-angle grain boundaries are expected to have superconducting characteristics in "real" bulk $YBa_2Cu_3O_{7-\delta}$ which has undergone a standard sintering.

12.3 Grain Boundaries in $Bi_2Sr_2CaCu_2O_{8+\delta}$

In recent years, significant improvements have been made in performance of $Bi_2Sr_2CaCu_2O_{8+\delta}$ (Bi2212) and $Bi_2Sr_2Ca_2Cu_3O_{10+\delta}$ (Bi2223) composite tapes for electric power and/or high-magnetic-field applications [Li et al. 1997a]; large-scale model devices, such as power-transmission lines and electric motors, are being constructed [IEEE Trans. 1997]. One of the most important issues that remains in further adapting of the polycrystalline conductors is to clearly identify the mechanisms that limit critical current density, J_c, of the materials. It has been known for several years that grain boundaries are detrimental to the transport of high currents in cuprate superconductors due to their tendency to act as weak links for misorientation angles greater than $\sim 10°$. Such weak-links were first demonstrated by Dimos et al. for $YBa_2Cu_3O_{7-\delta}$ (YBCO) bicrystal films [Dimos et al. 1988]. Later studies generally confirmed this early observation although a few exceptions, such as $90°$ boundaries, were also reported [Babcock et al. 1990, Eom et al. 1991]. These results corroborated the fact that it had been nearly impossible to fabricate polycrystalline YBCO wires that

12.3. Grain Boundaries in $Bi_2Sr_2CaCu_2O_{8+\delta}$

could carry meaningful currents. On the other hand the fact that the c-axis aligned Bi-cuprate platelets in the Ag-sheathed tapes, fabricated through the powder-in-tube process, can transport significant dissipation-less current promoted tremendous interest and success in developing Bi/2212 and Bi/2223 tapes which consist of highly c-axis aligned platelets.

Three models, the brick-wall [Malozemoff 1990, 1992, Bulaevskii et al. 1992], railway-switch [Hansel et al. 1993], and free way model [Cho et al. 1997, Malozemoff et al. 1997], have been proposed to describe current-transport mechanisms through the grain boundaries in these tapes. Their essential differences are the following: (1) In the brick-wall model, the highly c-axis aligned platelets are considered as a brick wall where the long face of each brick is the [001] twist boundary, and the short face is the [100] tilt boundary. The tilt boundaries are assumed to be non-current-carrying boundaries, and the current is transferred across the twist boundaries. Although the twist boundaries may eventually become the current-limiting factor at high currents [Bulaevskii et al. 1992], significant currents can move across the twist boundary because its area is significantly greater than that of the tilt boundary; (2) In proposing the railway-switch model, Hensel et al. [Hansel et al. 1993] pointed out that the brick-wall analogy was an oversimplification of the actual platelet structure in the tapes. They suggested that the [100] out-of-plane small-angle tilt boundaries, which are often observed in the tapes, are the means by which current is transferred from one grain to another, while the large-angle ones are not. They considered that the [001] twist boundaries are unlikely current-paths since the critical current along the c-axis is extremely low [Cho et al. 1997]; (3) More recently, the free-way model, a modified version of the brick-wall model, was elaborated [Malozemoff et al. 1997, Riley Jr. 1997]. It takes into consideration the fact that the platelets generally form into groups, or colonies [Feng et al. 1992]. Within a colony, the platelets are well aligned along the c-axis and each of them is separated by the [001] twist boundary. The interfaces between the colonies primarily consist of the out-of-plane tilt boundaries. Thus, the free-way model suggests that the transport of current on a macroscopic scale is accomplished through small-angle tilt boundaries at the colony interfaces. It was assumed that the current can flow across the [001] twist boundary within the colony since the ratio of the areas of the twist boundaries to these of the tilt boundaries is very large. In this section, we first discuss the misorientation distribution of the Bi-2212 and Bi-2223 tapes, and then discuss the structure and properties of the twist boundaries in Bi/2212 bicrystals and tapes [Zhu et al. 1997].

12.3.1 Misorientation Distribution in Bi-2212 and Bi-2223 Composite Tapes

Both tapes were fabricated by powder-in-tube processes. The Bi-2223/Ag tape was heated to $\sim 890^{\circ}C$ at which temperature the material in the tape was partially molten and was then slowly cooled to form the cuprate [Shibutani et al. 1994, Shibutani et al. 1993]. Multiple pressing with intermediate heating steps at $\sim 840^{\circ}C$ was used to form Bi-2223 through a pseudo-solid-state diffusion process [Haldar and Motowidlo 1992]. Significant differences were found in the surface morphologies between these tapes as illustrated in Figs.12.18(a) and (b), which were imaged using a scanning electron microscope (SEM) after the Ag sheaths were chemically removed. This suggests a difference in the mechanisms of formation for these cuprates. The Bi-2212 consists of large and well-aligned rectangular stacks of platelets, while Bi-2223 platelets were wavy, pancake-shaped and less well-aligned.

To analyze possible low-energy of the boundaries for Bi/2212 and Bi-2223, constrained CSLs were developed using a methodology we used was similar to that developed by Grimmer et al. and Singh et al. [Grimmer and Warrington 1987, Singh et al. 1990]. Table 12.1 lists the calculated coincidence site lattices for Bi-2212 and Bi-2223 for the [001] and [100] (or [010]) axes. Although both Bi-2212 [lattice parameters: $c = 30.89(2)$Å, $a = 5.411(2)$Å, and $b = 5.418(2)$Å] [Bordet et al. 1988a] and Bi-2223 [$c = 37.10$Å, and $b = 5.388$Å] [Hazen 1990] are pseudo-tetragonal, due to the incommensurate modulation along the b-axis the lattice has only a 2-fold symmetry rather than a 4-fold symmetry for [001] rotations, even though the basal plane can be considered as a square lattice. Thus, the lattice coincidence for Bi-based oxides about the [001] rotations differs from that of a cubic crystal. For [100] rotations, since c^2/a^2 is not a rational value for either Bi-2212 or Bi-2223, we need to use the constrained

12.3. Grain Boundaries in $Bi_2Sr_2CaCu_2O_{8+\delta}$

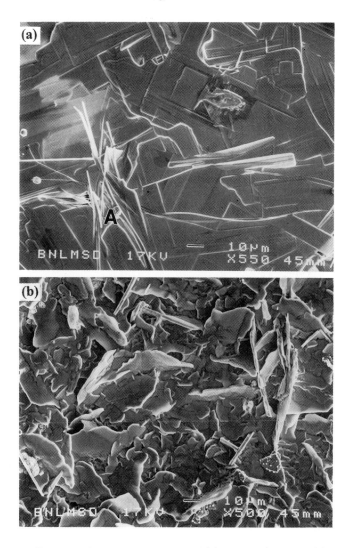

Figure 12.18 Scanning electron micrographs of (a) Bi-2212/Ag, and (b) Bi-2223/Ag tapes viewed along tape-normal incidence after Ag removal.

Table 12.1 Coincidence site lattices for $Bi_2Sr_2Ca_1Cu_2O_8$ and $Bi_2Sr_2Ca_2Cu_3O_{10}$ for rotations about the $[001]$ and $[100](or[010])$ axes with $\Sigma \leq 51$.

[001] axis Bi-2212 Bi-2223	[100] or [010] axis Bi-2212 $c^2/a^2 = 33:1$	Bi-2223 $c^2/a^2 = 47:1$
$\Sigma 1$, 90°	$\Sigma 7$, 55.15°	$\Sigma 24$, 16.60°
$\Sigma 5a$, 36.87°	$\Sigma 17$, 19.75°	$\Sigma 28$, 47.27°
$5b$, 53.13°	$\Sigma 19$, 65.10°	$\Sigma 36$, 72.21°
$\Sigma 13a$, 22.62°	$\Sigma 23$, 85.71°	$\Sigma 48$, 88.81°
$13b$, 67.38°	$\Sigma 29$, 82.07°	$\Sigma 51$, 32.53°
$\Sigma 17a$, 28.07°	$\Sigma 37$, 38.39°	
$17b$, 61.93°	$\Sigma 41$, 78.75°	
$\Sigma 25a$, 16.26°	$\Sigma 43$, 41.91°	
$25b$, 73.74°	$\Sigma 47$, 29.27°	
$\Sigma 29a$, 43.60°	$\Sigma 49$, 69.70°	
$29b$, 46.40°		
$\Sigma 37a$, 18.92°		
$37b$, 71.08°		
$\Sigma 41a$, 12.68°		
$41b$, 77.32°		

CSL model to define the coincidence site lattices [Zhu et al. 1993h, Zhu et al. 1991e]. We found that the closest rational value of c^2/a^2 is $33:1$ for Bi-2212, and $47:1$ for Bi-2223.

Experimentally, we noted that the edges of the Bi-2212 platelets are very straight and are parallel to the a- or b-axis. Thus, similar to Smith et al. [Smith et al. 1988], who examined the $YBa_2Cu_3O_7$ tablets, we measured the angles between the neighboring platelets, lying one on top each other, from SEM images such as Fig.12.18(a) using an image digitizing and analysis software [Rasband 1993]. Because the basal planes of these platelets are highly aligned, the rotations near the Bi-2212/Ag interfaces are essentially about the [001] axis. The measured rotations are plotted in Fig.12.19 as a histogram and compared with the calculated coincidence orientations. Clearly, the angular distribution

12.3. Grain Boundaries in $Bi_2Sr_2CaCu_2O_{8+\delta}$

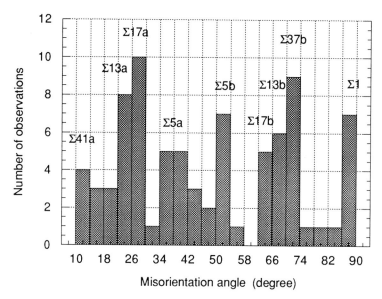

Figure 12.19 A histogram of the frequency of occurrence of Bi-2212 platelets (82 grain boundaries) measured by SEM images, as a function of the misorientation angle θ relative to their neighbors about the $[001]$ axis. The calculated CSL orientations are marked by Σ's.

of the boundary misorientations is not random, but coincides with the CSL orientations very well.

Since SEM observations cannot give the rotation component perpendicular to the tape surface normal (the $[001]$ axis), a more precise determination of boundary misorientations by transmission electron microscopy (TEM) using the Kikuchi-pattern-method [Zhu et al. 1991e] was carried out for boundaries with $[001]$ and $[100]$ rotations (twist boundaries and tilt boundaries, respectively); these are known to be the dominant factor in determining the intergranular critical current of the Bi-based superconductors. Considering the near coincidence orientations, we chose 4° as the permissible deviation from the rotation axes.

Figures 12.20(a)-(d) show the results of the TEM measurements.

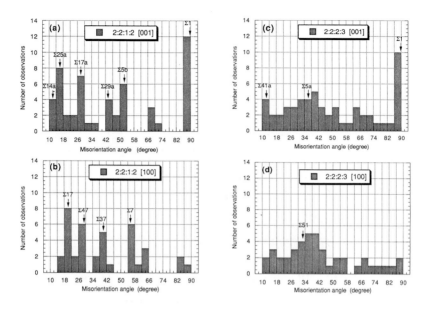

Figure 12.20 Angular distributions of grain boundaries. (a) Bi-2212 [001] twist boundaries (rotations about the [001] axis), 52 grain boundaries; (b) Bi-2212 [100] or [010] tilt boundaries (rotations about the [100] or [010] axis), 39 grain boundaries; (c) Bi-2223 [001] twist boundaries, 57 grain boundaries and (d) Bi-2223 [100] or [010] tilt boundaries, 43 grain boundaries.

The relative number of the types of the boundaries in each system does not reflect their relative abundance in the composites. The [001] twist boundaries are usually observed in the well-aligned regions, or colonies; while the [100] tilt boundaries are mostly found in the inter-colony regions, or in few areas that have clusters of unaligned platelets, such as the location marked as "A" in Fig.12.18(a).

The angular distributions of Bi-2212 (Figs.12.20(a) and 12.20(b)) show discrete peaks at the angles corresponding to the calculated coincidence orientations for the [001] twist and the [100] tilt boundaries. The small difference in the peak positions in Fig.12.19 and Fig.12.20(a) for the [001] twist boundaries are thought to be a consequence of the limitations of the measurements, rather than a significant feature of the data. If we convert 25b and 17b to 25a and 17a, respectively (be-

12.3. Grain Boundaries in $Bi_2Sr_2CaCu_2O_{8+\delta}$

cause the a- and b-axis cannot be distinguished in SEM images, we can consider that all rotations of angle greater than $45°$ are equivalent to rotations of $90°$), the observed preferred orientations in SEM and TEM measurements are essentially identical.

In contrast, the angular distribution for Bi-2223 does not exhibit clear peaks and is almost random, except for the $90°$ [001] boundaries which can be viewed as twin boundaries. This striking difference in the angular distributions between Bi-2212 and Bi-2223 can be traced to the way each of these Bi cuprates were formed in the composites. Bi-2212 platelets are formed out of a liquid and perhaps more importantly, as demonstrated clearly by Zhang and Hellstrom [Zhang and Hellstrom 1993], during slow cooling significant reorientation of the crystals takes place within the Ag sheath, promoting a high degree of alignment of the platelets along the tape surface. Such an adjustment in orientation allows boundaries to adopt a low-energy misorientation with high lattice coincidence. On the other hand, the Bi-2223 platelets are formed essentially in a solid-state reaction which does not allow the platelets to move from the nucleated orientations. In addition, unlike isotropic materials, the preferred selections of low-energy grain boundaries are very difficult during solid-state grain growth in these highly anisotropic materials. Thus, the grain boundaries in Bi-2223/Ag composite tapes are nearly random in their angular distributions (Figs.12.20(c) and (d)).

12.3.2 Superconducting Properties of Twist Boundaries of Bi-2212

As mentioned in the introduction of this section, the key to verifying the current-paths is to determine whether a significant current can be transported across the [001] large-angle twist boundaries which represent a large fraction of the boundaries in these tapes [Zhu et al. 1994b]. Although there have been extensive studies of grain boundaries in high-T_c superconductors [Hilgenkamp et al. 1996], investigations on the relation of structure and property have mainly focused on tilt boundaries in YBCO thin films (especially for structural analyses using high-resolution TEM). For the Bi- cuprate systems, such studies have been very limited. Misorientation-dependent critical currents were observed in artificially fabricated [001] twist boundaries in Bi-2212 [Tomita et al.

1996, Wang et al. 1994], and also in naturally grown [001] in-plane tilt boundaries [Wang et al. 1996a]. However, it was not clear, particularly for the former, whether the findings were statistically meaningful since only a handful samples were investigated. To overcome such deficiency, Li and coworkers systematically investigated the transport current as a function of misorientations for the [001] twist- and [001] tilt-boundaries. Surprisingly, they found that the critical current across the [001] twist boundaries in Bi-2212 bicrystals was independent of the boundary misorientations [Li et al. 1997b].

Using a similar method originated by Tomita et al. [Tomita et al. 1996] and Wang et al. [Wang et al. 1994], a high-quality, reproducible bicrystal was made by first cleaving a single crystal into two pieces along the ab plane, placing one piece atop the other rotated at a desired angle about the c-axis, and then sintering at 865°C in 7% O_2 [Li et al. 1997b]. Transport was measured using a six-probe configuration with voltage contacts placed on the inner surface of each crystal as shown in Fig.12.21 [Li et al. 1997c]. Such a configuration permitted simultaneous measurements of the resistance, R, and the voltage-current $(V - I)$ characteristics of the grain boundary, as well as those of constituent single crystals. The resolution of the voltage measurements was ~ 1 nV.

Figure 12.22 shows the ratio of the critical currents across a grain boundary, I_c^{GB}, to that within the constituent single crystals, I_{cg}, as a function of misorientation angle of the [001] twist boundary. This particular set of measurements was taken at temperature $T = 10$K, and at magnetic field $H = 9$ tesla ($H \| c$) with an I_c criterion of 10 nV ($\sim 2 \times 10^{-8}$V/mm). For simplicity, The crystal was considered to have tetragonal symmetry: thus, the smallest non-equivalent misorientation ranges from $0 - 45°$. In striking contrast to the results of Dimos et al. for YBCO thin films for various twist- and tilt- boundaries (measured at 0 tesla and 5K) [Dimos et al. 1988], the grain boundaries in these bicrystals carried practically the same critical currents as their constituent single crystals, regardless of misorientation angle. Similar independence of I_c on the misorientation was seen under different conditions including $H \perp c$. Moreover, all bicrystals exhibited similar $R-T$, $V - I$, and $I_c - T(H)$ characteristics [Li et al. 1997b]. This was an unexpected observation. Even for grain boundaries in simple cubic ma-

12.3. Grain Boundaries in $Bi_2Sr_2CaCu_2O_{8+\delta}$

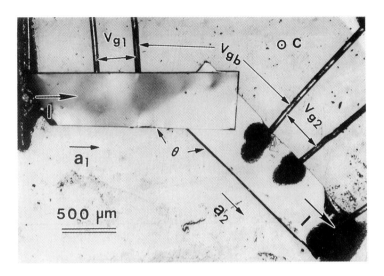

Figure 12.21 Experimental configuration for transport measurements of bicrystal twist boundary of Bi-2212. An [001] twist bicrystal and the six-probe configurations were shown. The common c-axis is perpendicular to the grain boundary [Li et al. 1997c].

terials, many studies have shown that the properties of boundaries vary significantly with their misorientation (for example [Watanabe 1993]).

In interpreting these observations, we noted that the voltage taps across the grain boundary encompassed a sizable fraction of the constituent single crystals. This inclusion caused a voltage rise, which was detected between the pairs of the wires by increasing current, because of the motion of vortices in the section of single crystals rather than in the boundary itself. This is, in part, due to the fact that the cross-sectional area of the boundary, A_{gb}, is much greater than that of the single crystal cross-section perpendicular to the ab plane, Ag. On the other hand, if the boundary was much weaker than the constituent single-crystals, the dissipation at the boundary would be easily detected [Tomita et al. 1996, Wang et al. 1994]. Indeed, these boundaries did not exhibit any poorly coupled region suggesting that the single crystals were very well

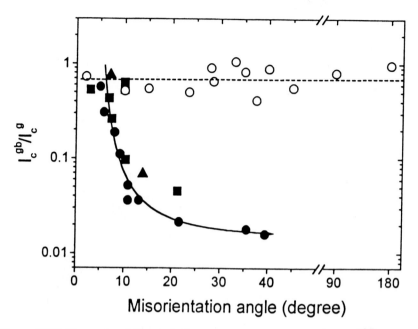

Figure 12.22 The ratio of the cirtical current across grain boundary I_c^{GB} vs. the critical currrent in the grain I_c^G as a function of misorientation angle of Bi-2212 [Li et al. 1997c] (open circle) and YBCO twist boundaries [Dimos et al. 1988] (filled circle [001]tilt, square [001]tilt and triangle [100]twist-boundaries).

coupled. To estimate the minimum critical current density, J_c, of these boundaries, we assumed that the value of A_{gb} being the entire overlapping area of the two crystals. Since $A_{gb}/A_g < 10$ in our experiments, it implies that the boundary could not be differentiated from the single crystals, i.e., the ratio of the critical currents shown in Fig.12.22 is independent of the misorientation, at least within a factor of 10 or less in critical current density. This conclusion is strikingly different from those reported previously where at least a drop of two orders of magnitude I_c^{GB}/I_c^G was observed at large misorientation angles [Hilgenkamp et al. 1996]. In this sense, our results are very surprising. More importantly, they suggest that the current can transport from one platelet to another along the c-axis within the colonies in the Bi-cuprate tapes since the area ratios of the twist boundary to the platelet cross-section is much greater (> 100) in these tapes than those in the synthetic bicrystals.

12.3.3 Structural Features of the Twist Boundaries

To find the origin of angle-independent superconductivity at the twist boundaries, the microstructure was carefully examined using advanced transmission electron microscopy including 0.17nm high-resolution imaging (HREM) and image simulation incorporating charge distribution, electron energy-loss spectroscopy (EELS), and energy dispersive x-ray spectroscopy (EDS) with a 2nm field-emission probe [Zhu et al. 1997].

Figure 12.23 is a medium-resolution image of a $37.45°$ twist boundary, one of the electromagnetically characterized bicrystal boundaries. The $37.45°$ misorientation was measured using the computer aided Kikuchi patterns technique [Zhu et al. 1991e]. As shown in insets of Fig.12.23 for crystal A and B, the beam direction is $3.66°$ off the $[\bar{1}10]$ direction of A, and $3.89°$ off the $[100]$ direction of B. The lattice spacing in A corresponds to d_{002}, while the dot-pattern in B represents the superlattice resulting from incommensurate modulation in Bi/2212. Fig.12.24(a)-(c) shows a set of high-resolution images of the same boundary. To examine the boundary structure at an atomic level, we first individually tilted the a-axis of crystal A, and of crystal B parallel to the incident beam, as shown in Fig.12.24(a) and (b), respectively. We then tilted both crystals off any major zone axes, Fig.12.24(c), to measure possible lattice expansion along the c-axis and the interplanar distance of BiO double layer, by avoiding the coupling of non-[001] Bragg beams while retaining the boundary edge-on. Although we cannot examine the atomic structure of both crystals simultaneously, a major difficulty in studying twist boundaries, by combining Fig.12.24(a) and (b), we retrieved the crystal structure in the vicinity of the boundary, and found that there is little lattice distortion there. The layered crystal structure is well ordered up to the boundary plane. The superlattice, seen as a body-centered black-cage-like pattern in Fig.12.24(a) and (b), originating from a pile-up of electrons in the double BiO layer [Zhu and Tafto 1996a], is little disturbed by the grain boundary. Such charge modulation, which is associated with oxy-

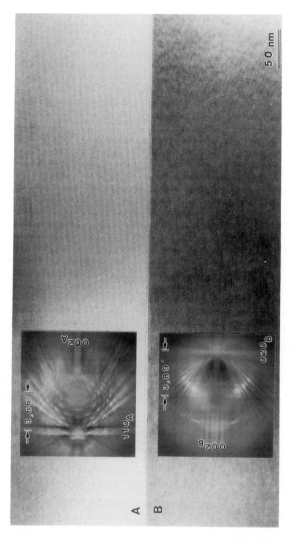

Figure 12.23 Medium-resolution image of the electromagnetically characterized $37.45°$ twist boundary in Bi-2212 bicrystals. Insets are Kikuchi-patterns for grain A and B. The beam direction is $3.66°$ off the $[\bar{1}10]$ direction of crystal A, and $37.45°$ off the [100] direction of crystal B. The lattice spacing in A corresponds to the (002) planes, and the dot-pattern in B represents the superlattice due to an incommensurate modulation.

12.3. Grain Boundaries in $Bi_2Sr_2CaCu_2O_{8+\delta}$

Figure 12.24 (a)-(c) High-resolution images of the same area from the 37.45^o twist boundary shown in Fig.12.23, with the boundary position marked by a pair of arrow heads. (a) The top crystal, and (b) the bottom crystal are viewed along the a-axis. (c) both crystals are viewed off any low-index zone. Note, the narrowly spaced double-dark-lines in (c) are the lattice image of the double BiO layers.

gen holes in the CuO_2 plane, is believed to be very sensitive to lattice displacement. To further understand the structure of the boundary at an atomic level, detailed image simulations were performed using an interfacial unit-cell involving 5204 atomic sites, including electrons and holes [Fig.12.25(a)]. The simulated high-resolution image is shown in Fig.12.25(b) (two unit-cell wide along the boundary). The good match between the calculation and the experimental observation verified our structural model of the boundary.

Because all the twist boundaries were examined with their rotation axis perpendicular to the direction of view, their projected HREM images were almost identical, regardless of their misorientation angles. Several notable structural features were observed. 1) All the boundary planes were microscopically flat (Fig.12.23), suggesting they are pure twist in character . 2) The boundaries were clean, structurally intact without any visible amorphous materials (Fig.12.23). 3) 2nm-probe EDS and EELS measurements showed that there was no detectable off-stochiometric composition including oxygen/hole concentration along and across the boundaries (Fig.12.26). 4) HREM image simulation revealed that the boundaries were located in the middle of the double BiO layers, without exception (Fig.12.25). 5) There was little boundary expansion, and the inter-planar distance of the double BiO layer ($d_{BiO} = 0.31 \pm 0.005$nm, measured with line-scan (Fig.12.27) at the boundary was the same as those far away from the boundary, within the measurement error. 6) Very often, there was an intercalation of a Ca/CuO_2 bi-layer near the boundary, either at one, or both sides, forming a local Bi-2223 structure (Fig.12.24 and Fig.12.25).

All these structural features of the twist boundaries were observed in our Bi-2212 tapes. Fig.12.28 is a HREM image of a $30.4°$ grain boundary observed in a Bi-2212 tape also with a Ca/CuO_2 bi- layer intercalated at the boundary. The atomic structure near the boundary is almost identical to that of the bicrystals. Similar HREM images of the [001] twist boundaries were also observed in sintered Bi-2212 bulk-polycrystals [Eibl 1990]. This suggests that the structure-property relationship observed in bicrystals can be applied to Bi-2212 tapes, and highlights the significance of our bicrystal study.

12.3. Grain Boundaries in $Bi_2Sr_2CaCu_2O_{8+\delta}$

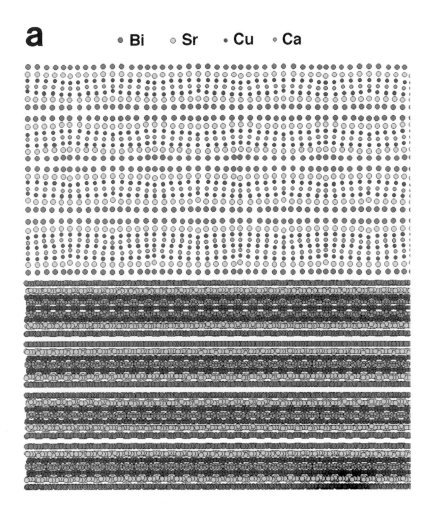

b

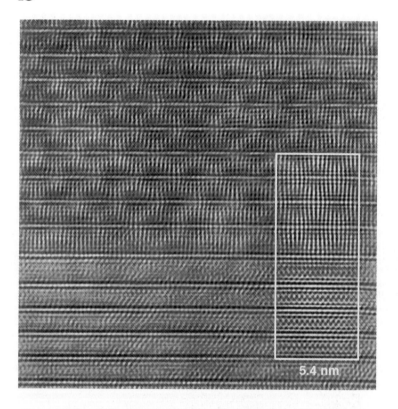

Figure 12.25 (a) A structural model of the $37.45°$ [001] twist boundary used in the calculation. The visible lattice displacement results from the structural modulation in Bi/2212. Charge transfer in crystals with holes evenly distributed in CuO_2 planes and electrons piled up in the BiO layers was also included in the calculations (for clarity, oxygen, electrons and holes are not shown). (b) Atomic image of the boundary. The embedded image is the calculated one. Note, the boundary is located in the middle of the BiO double layers. The good match with the experimental image suggests that there is little lattice expansion in the vicinity of the boundary.

12.3. Grain Boundaries in $Bi_2Sr_2CaCu_2O_{8+\delta}$ 393

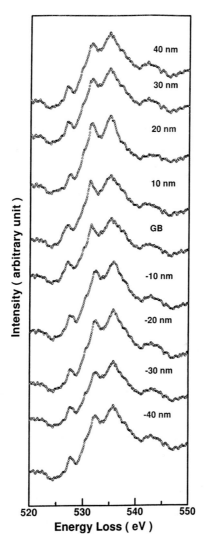

Figure 12.26 A series of nano-probe EELS spectra of the oxygen K-edge collected across a large-angle pure twist grain boundary. The distance from the boundary is given for each spectrum. The integrated intensity of the oxygen pre-peak ($\sim$528eV), which is a direct measure of the oxygen hole content, acquired from the boundary is very similar to that acquired away from the boundary, suggesting that there is no depletion in oxygen/hole concentration at the boundary.

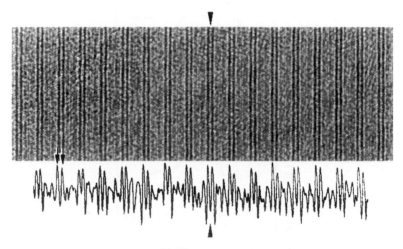

Figure 12.27 Lattice image of a $33.5°$ twist boundary (marked by a pair of arrow heads), similar to that in Fig.12.24(c) but without a local Bi-2223 structure, with a line scan across the boundary showing the inter-planar distance (marked by small arrows) of the double BiO layer at the boundary is the same as that in the interior of the constituent grains. Note, both grains are viewed off any low-index zone.

Strain Field of the Twist Boundaries

The structural integrity at twist boundary regions, singled by the absence of second phases and impurities, is a prerequisite to carrying the same current as their constituent crystals (a local Bi-2223 structure would not degrade the current-carrying capability, if not enhance it). However, it seems bizarre that a wide range of strain-fields associated with various misorientations of the boundary do not suppress the superconducting order-parameter, contradicting to our current understanding of grain boundaries (which mainly limited to cubic system), especially when the coincidence-site-lattice theory is concerned [Sutton and Baluffi 1995]. Thus, the strong-link behavior of large angle boundaries we observed must be due to the special characteristics of Bi-2212.

When two single crystals are brought together to form a bicrystal, the atoms at the boundary cannot be as closely packed as in their con-

12.3. Grain Boundaries in $Bi_2Sr_2CaCu_2O_{8+\delta}$

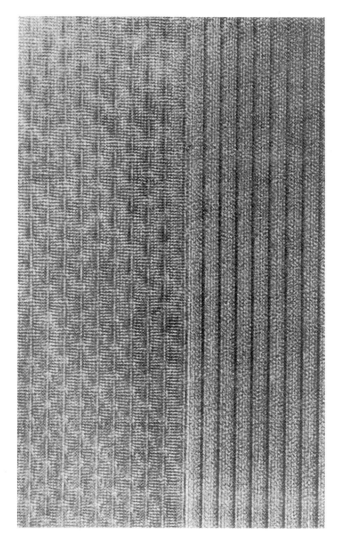

Figure 12.28 A 30.4^o pure twist boundary observed in a Bi-2212 tape also with an intercalation of a CuO_2/Ca layer near the boundary.

stituent crystals, resulting in lattice expansion in the vicinity of the boundary [Wolf and Merkie 1992]. The strain field generated by two crystals' lattice mismatch at the boundary varies with misorientation angles, and can be described in the framework of grain boundary dislocations (GBDs), which are often observed using TEM (Fig.12.29(a) is diffraction contrast of a 7.5° twist boundary with the view direction parallel to the boundary normal). For a pure twist boundary, two independent sets of parallel screw dislocations are required. Such a dislocation network does not generate normal stress ($\sigma_{xx} = \sigma_{yy} = \sigma_{zz} = 0$) which results in zero strain along the rotation axis ($\sigma_{zz} = 0$), consistent with our HREM observations. Using the linear, anisotropic elasticity theory [Ting 1996], and the treatment developed for dislocations in anisotropic hexagonal system with 5 independent elastic constants [Chou 1962], we derived the three non-zero shear- stresses, σ_{xy}, σ_{zx}, and σ_{zy}, for anisotropic crystals with tetragonal symmetry (6 independent elastic constants) generatred by two sets of screw dislocation arrays of a twist boundary, one perpendicular to the other:

$$\sigma_{xy} = \frac{C_{66}b}{2d} \frac{\sinh\gamma(\cos\alpha - \cos\beta)}{(\cosh\gamma - \cos\beta)(\cosh\gamma - \cos\alpha)}$$

$$\sigma_{zx} = \frac{\sqrt{C_{66}C_{44}}b}{2d} \frac{\sin\beta}{(\cosh\gamma - \cos\beta)}$$

$$\sigma_{zy} = -\frac{\sqrt{C_{66}C_{44}}b}{2d} \frac{\sin\alpha}{\cosh\gamma - \cos\alpha}$$

where $\alpha = 2\pi x/d$, $\beta = 2\pi y/d$, and $\gamma = (2\pi z/d)\sqrt{C_{66}/C_{44}}$ (C_{ij} denotes the elastic constants with $C_{44} = 6.18$ and $C_{66} = 15.1$ dyn/cm^2 for Bi-2212 [Baetzold 1997, Boekholt et al. 1991], and both x, y, z, and C_{ij} refer to the standard coordinate system for a tetragonal lattice), b is the amplitude of the Burgers vector, and d the spacing of the dislocations. Among the three shear stresses, only the basal-plane stress σ_{xy} is considered to be an important one which may alter atomic bonding in the basal planes, and thus, affect the hole concentration in CuO$_2$ layers. However, due to the high elastic anisotropy of the crystal, the shear stress σ_{xy}, or shear strain ϵ_{xy}, decays rapidly along the c-axis, and is reduced by more than three orders of magnitude from the BiO layer ($z = 0.150$ nm) at the boundary, to the nearest CuO$_2$ layer

12.3. Grain Boundaries in $Bi_2Sr_2CaCu_2O_{8+\delta}$

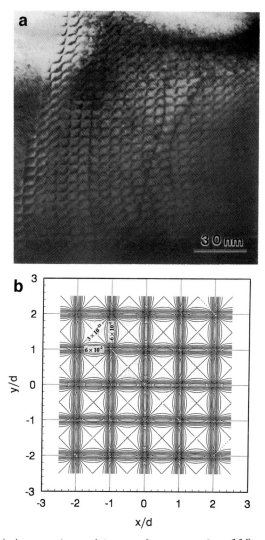

Figure 12.29 (a) An experimental image of two sets of $<110>$ type of screw dislocations from a $7.5°$ twist boundary in Bi-2212. (b) Calculated contour map of the in-plane strain field ϵ_{xy}, in units of $b/2d$, of the grain boundary at the nearest CuO_2 plane. The sign and amplitude of the strain at three locations are indicated.

($z = 0.618$ nm) from the boundary. Fig.12.29(b) is a two-dimensional contour map of the strain-field, in units of $b/2d$, at the first CuO_2 layer, showing maximum strain along the dislocation lines and zero strain at their nodes.

Fig.12.30(a) shows the maximum strain, ϵ_{xy}^{max}, as a function of the distance, z, from the boundary for different dislocation spacings, d. We note that larger strain generated by narrowly spaced dislocations decays faster than that caused by widely spaced ones, and the strain associated with an array of 2 nm-spaced dislocations is reduced to almost zero at a distance of a half unit-cell from the boundary. Fig.12.30(b) shows as a function of d at the first CuO_2 plane from the boundary. It is noteworthy that ϵ_{xy}^{max} saturates at a dislocation spacing of about 20 nm. Such geometric analysis is unambiguous for $\theta \leq 15°$, when the dislocation spacing d is larger than, or comparable to, interatomic spacing. For $\theta > 15°$, the periodic structure of the boundaries can be characterized using secondary GBDs with much smaller Burger vectors and larger spacing. Such secondary GBDs associated with large angle boundaries in Bi-2212 were frequently observed (for example, [Gao et al. 1993b]).

The calculated dislocation network for a twist boundary, based on the shear stress formulated above, using two-beam dynamic diffraction theory [Hirsch et al. 1965] is shown in Fig.12.31(a), and compared with the experimental observations (Fig.12.31(b)). The simulation confirmed several experimental parameters, such as the two sets of dislocations are screw in character with Burgers vector [110] and [1$\bar{1}$0], respectively, the foil normal was parallel to the boundary normal and both are parallel to [001], the beam direction was [5$\bar{3}$1], and the diffracting beam used to form the image was $g_A = g_B = [0\bar{2}0]$. For the calculation, the following were the other parameters we used: the dimensionless deviation parameters $w_A = 0.6$, $w_B = -0.2$ ($w = \xi_g s_g$, where ξ_g is the extinction distance and s_g is the deviation from the Bragg reflection), the absorption factor $\xi_g/\xi_g' = 0.1$, the foil thickness $t = 50$ nm, and the anisotropic parameter $\sqrt{C_{66}/C_{44}} = 1.56$. Although we present here a simulated dislocation network associated with a 7.5° small-angle twist boundary with a complete set of experimentally determined parameters, detailed image simulations in comparison with experimental observations show that the diffraction contrast from a large-angle grain

12.3. Grain Boundaries in $Bi_2Sr_2CaCu_2O_{8+\delta}$

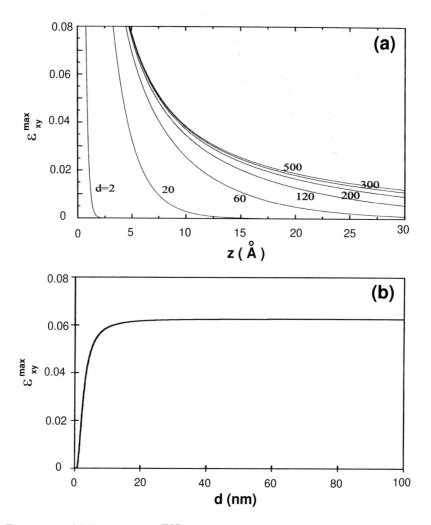

Figure 12.30 (a) Shear strain, ϵ_{xy}^{max}, as a function of the distance from the boundary for different dislocation spacings calculated using the linear, anisotropic elasticity theory; (b) ϵ_{xy}^{max}, as a function of dislocation spacing d at the same CuO_2 plane; note, ϵ_{xy}^{max}, saturates at $d = 20$nm.

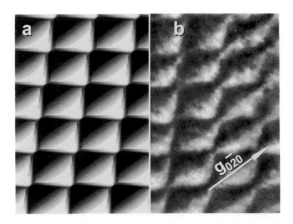

Figure 12.31 A simulated image (a) of a twist boundary using the two-beam dynamic electron-diffraction and anisotropic elasticity theory together with an experimental observation (b). The following parameters were used in the simulated image: Burgers vector of the two sets of screw dislocations $\mathbf{b}_1$=[110], $\mathbf{b}_2$=[1$\bar{1}$0], beam direction $\mathbf{B}$=[5$\bar{3}$1], boundary normal $\mathbf{n}$=[001], foil normal $\mathbf{m}$=[001], deviation parameter $w_A = 0.6$, $w_B = -0.2$, diffracting vector $\mathbf{g}_A = \mathbf{g}_B$=[0$\bar{2}$0], $\sqrt{C_{66}/C_{44}} = 1.56$, absorption factor $\xi_g/\xi_g' = 0.1$, and crystal thickness $t = 50$nm.

boundary is similar to that from a small-angle grain boundary, and there is no significant difference both in calculated and experimentally observed images between primary GBDs and secondary GBDs.

The Origin of the Robust Superconducting Behavior

The exact strain field at the boundary, unfortunately, cannot be predicted solely by using geometrical factors that describe boundary periodicity but depends also on its specific atomic configuration. The atoms at the boundary can be rearranged from partially ordered to totally disordered in relaxing to a low-strain configuration to reduce boundary energy. Thus, our calculation may provide only an upper limit for σ_{xy}. Understanding the detailed atomic configuration at the [001]

12.3. Grain Boundaries in $Bi_2Sr_2CaCu_2O_{8+\delta}$

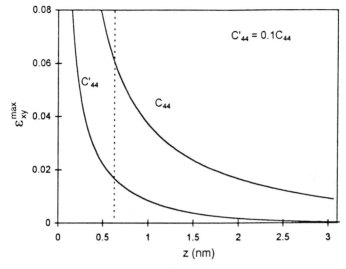

Figure 12.32 Basal plane shear strain, ϵ_{xy}^{max}, as a function of the distance, z, from the twist boundary with different shear moduli C_{44} and C'_{44} ($C'_{44} = 0.1C_{44}$). Note, ϵ_{xy}^{max} decays faster for a smaller shear modulus.

twist boundaries, or at the BiO layers, may not be essential, since they are electronically insulating layers. The most important issue here is the propagation of the boundary strain into the superconducting CuO_2 layers, which is mainly dominated by the shear modulus, C_{44}. The calculated value, ϵ_{xy}^{max}, at the first CuO_2 plane is about 6% (Fig.12.32), which appears large enough to cause hole depletion and is inconsistent with the hole measurements across the boundaries (Fig.12.26). The actual strain in the CuO_2 layers can be much less. We note that: 1) not all the basal-plane-layers in Bi-2212 are structurally identical; the bonding length between the double BiO layers is about twice as long as that of any other layers; 2) in as-hot-pressed Bi-2212 tapes, BiO layers were deformed the most and the sliding and breakage planes were always at double BiO layers; 3) Bi-2212 is known to be much more anisotropic than YBCO [Ledbetter and Lei 1991], but their anisotropy parameters C_{66}/C_{44} do not differ very much. Therefore, we believe that the elastic properties along the c-axis are very heterogeneous, and the physical origin of this surprisingly negligible effect of grain boundary strain on superconductivity results from the softness of the double BiO layers of this material. The shear modulus, C'_{44}, for BiO double layers, or BiO slabs, can be much smaller than the directional averaged C_{44}. Assuming $\epsilon = 2\%$ as a cut-off value (a displacement/hole-depletion criterion observed in YBCO [Chisholm and Pennycock 1991, Zhu et al. 1996], and that the boundary strain is mainly confined within the double BiO layers, we estimated C'_{44} is about one order of magnitude smaller than C_{44} (Fig.12.32). A smaller shear modulus results in faster decay of the grain boundary strain along the boundary plane normal. Our argument is supported by net-dipole-moment calculations showing that the cleaved BiO surface is energetically stable and does not require surface reconstruction [Cai 1997]. A grain boundary formed by such surfaces may be able to accommodate all interfacial lattice distortion. Thus, the origin of the unexpected robust superconductivity at the [001] twist boundaries can be attributed to the high anisotropy of the Bi-2212 crystals and to the softness of the double BiO layers at the boundaries.

From the advanced TEM measurements described in this section, we showed that the absence of weak-link behaviors of [001] twist boundaries in Bi-2212 results from the well confined grain-boundary mismatch in its double BiO layers. This in fact is fully consistent with the nature of intrinsic Josephson junctions in this layered material. Since Bi-2212 is highly anisotropic and the coherence length along the c-axis is less than 0.1nm, the crystal itself is virtually all Josephson junctions. The strength of Josephson coupling largely depends on the superconducting layers, as well as on the insulating barriers. Considering that the atomic configuration in basal-planes has a translation across each BiO double layer in the crystal itself, a similar in-plane rotation associated with a twist boundary with a very short range strain-field may not weaken the Josephson junctions. Consequently, the array of Josephson junctions can continue uninterruptedly across the boundaries, as evidenced by the transport measurements of our Bi-2212 bicrystals. This confirms the earlier suggestion by [Lay 1991] that a [001] twist boundary in the Bi-cuprates might not cause a significant perturbation in superconducting properties since the presence of the boundary resulted in no change in the atomic layer sequence along the c-axis.

12.4 Resistent-Shunted-Josephson-Junction Model for Grain Boundaries in Polycrystalline Superconductors

The superconducting tapes and wires that are important to large scale applications are generally polycrystalline samples with large number of grain boundaries of various misorientation angles. The critical current density of such sample, which is an important parameter of the quality of the superconducting tapes and wires, is determined by the collective properties of all the grain boundaries. It is therefore not enough to understand the properties of isolated grain boundaries. We also need to know the statistical distributions of grain boundaries in a polycrystalline sample and the overall superconducting properties of the polycrystalline sample.

In this section we illustrate a technique to calculate the critical current of a polycrystalline sample using the statistical data obtained from experiments [Cai and Welch 1992, Cai and Welch 1993].

The superconductivity is suppressed at most grain boundaries; thus such a grain boundary can be regarded as a very thin layer (usually in atomic scale) of normal phase (or equivalently, a superconducting phase with lower T_c). As pointed out by Likharev [Likharev 1979] a superconductor–normal conductor–superconductor (SNS) weak link exhibits properties of a Josephson-tunnelling-junction if the normal layer is sufficiently thin. Here we use a model which describes the I–V characteristics of $YBa_2Cu_3O_7$ ceramics as a 2-dimensional array of such Josephson junctions. Some aspects of this model were studied previously by other authors in the limit of Josephson-junction arrays with identical J_c [Teitel and Jayaprakash 1983, Li and Teitel 1990, Chung et al. 1989] and with simple defects [Xia and Leath 1989]. However, experimental studies have now yielded detailed information which permits improvement of this model to simulate the $YBa_2Cu_3O_7$ grain boundary more realistically [Dimos et al. 1990, Zhu et al. 1991e].

The two dimensional Josephson junction array model used here represents a particular class of $YBa_2Cu_3O_7$ ceramic compounds for which the c-axes of the grains are aligned in a common direction. Such kind of grain alignment has been used to obtain favorable orientations of the grains and/or obtain favorable grain boundaries. It can be obtained by a variety of techniques such as the application of a uniaxial pressure during processing or magnetic alignment of powder particles. In these samples the relative orientation of the two grains at grain boundaries (grain boundary angle) involves a rotation about the c-axis (i.e. a tilt-boundary). Conduction across these boundaries would be within the *ab*-plane, which has a larger coherence length than that in c-axis. In this section we study the effect of the misorientation of grains at grain boundaries on the critical current density of polycrystalline (ceramic) systems. The Josephson junction array model with parameters obtained from various experiments is used to calculate the transport properties of the system as limited by the grain boundaries.

The Ginzburg-Landau theory [Ginzburg and Landau 1950] is a good approximation if the amplitude of the wave function Ψ and its gradient $\nabla \Psi$ are small enough so that the free energy expansion is valid. Since local fluctuations dominate due to the very short coherence length in $YBa_2Cu_3O_7$ ceramics, we can safely ignore the effect of non-local superconductivity. Furthermore the magnetic penetration length λ is very long in these systems which reduces the amplitude of Ψ further, therefore we expect the G-L equations to be applicable for a wide range of temperature in this system.

For two identical superconductors in an SNS weak link, the supercurrent density across the grain boundary is given by [de Gennes 1966]

$$J_s = J_c \sin(\varphi_- - \varphi_+), \qquad (12.4)$$

where φ_+ and φ_- are the phase factors of the superconducting order parameter at the two surfaces in the junction. The maximum supercurrent density, *i.e.* critical current density J_c is given by

$$J_c = \frac{e^* h}{2\pi m^* M_{12}} \Psi_0^2, \qquad (12.5)$$

where the coefficient M_{12} depends on the transmission rate of the Cooper pairs from the superconductor to the normal layer, and the properties and thickness of the grain boundary. Ψ_0^2 is the superconducting order parameter in the bulk and m^* is the effective mass of the superconducting electrons. For the case of a tunneling junction at

12.4. RSJ Model

$T = 0$, Anderson [Anderson 1963] found $J_c = \pi\Delta/2eR$, where R is the tunneling resistance per unit area of the junction when both grains are in the normal state and Δ is the energy gap. This result was generalized to finite temperature by Ambegaokar and Bartoff [Ambegaokar and Baratoff 1963]. For a junction between two identical superconductors the result is

$$J_c = \frac{\pi\Delta(T)}{2eR}\tanh\left(\frac{\Delta}{2kT}\right). \tag{12.6}$$

We assume that the normal state resistance of the grain boundary satisfies the equation

$$R = \frac{b}{J_c} \tag{12.7}$$

where b is treated as a parameter independent of the grain boundary angle.

When the current density J across the grain boundary exceeds the critical current density J_c, there is a voltage across the boundary. The normal current is assumed to satisfy Ohm's law. We will study this model in the low temperature limit so that the effect of thermal noise on the current density can be ignored. Therefore the total current density across the grain boundary

$$J = J_s + \frac{V}{R}. \tag{12.8}$$

In our model we assume that the superconducting grains are uniformly distributed in two dimensions to form the sites of a square lattice. Each grain (site) is described by a complex superconducting order parameter $\Psi_i = |\Psi_i|e^{i\varphi}$. Each grain is coupled to its nearest-neighbor grains by a Josephson junctions (bonds) with critical current density $J_{cij}(\theta)$ and normal state resistance $R_{ij}(\theta)$ which depends on the tilt angle θ. Fig.12.33 shows a 10×10 array. The square lattice of bonds is connected to bus bars at the top and bottom and periodic boundary conditions are applied to the left and right sides of the lattice to eliminate edge effects. The external current I flows into the network through the bottom bus bar and out at the top.

Our calculation proceeds by directly solving the equations for a network of coupled Josephson junctions in the limit of zero capacitance: [Teitel and Jayaprakash 1983]

$$I_{ij} = V_{ij}/R_{ij} + I_{cij}\sin(\varphi_i - \varphi_j), \tag{12.9}$$

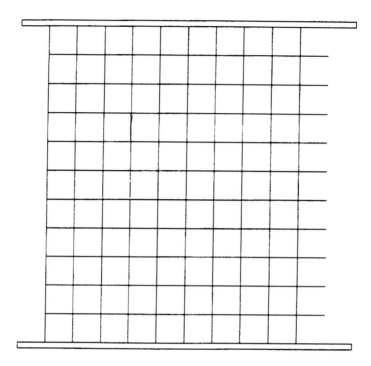

Figure 12.33 A 10 × 10 Josephson junction array. Each bond represents a grain boundary with tilt-angle θ_{ij}. The external current flows into the bottom bus bar and out of the top bus bar.

$$V_{ij} = \frac{h}{4\pi e}\frac{d}{dt}(\varphi_i - \varphi_j), \qquad (12.10)$$

$$\sum_j I_{ij} = I_{i;ext}, \qquad (12.11)$$

$$R_{ij} = \frac{b}{J_{cij}}. \qquad (12.12)$$

Where Eqn.(12.9) describes the current from grain i to grain j as the sum of a normal contribution V_{ij}/R_{ij} and a Josephson current. Eqn.(12.10) is the Josephson relation connecting the voltage difference V_{ij} between grains i and j and the phase difference $\varphi_i - \varphi_j$ between the phases of the superconducting order parameters. Eqn.(12.11) is Kirchhoff's law, expressing current conservation at grain i. Finally,

12.4. RSJ Model

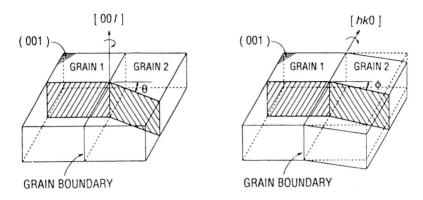

Figure 12.34 A tilt grain boundary is a grain boundary whose misorientation axis is in the plane of the boundary. The tilt angles θ and ϕ are defined as shown in the above figure.

Eqn.(12.12) is the relation between critical current of the grain boundary J_{cij} and the normal resistance R_{ij}.

The majority of the grain boundaries in c-axis aligned YBa$_2$Cu$_3$O$_7$ ceramics are low angle tilt boundaries as shown in Fig.12.34. The axis of misorientation is in the plane of the grain boundaries. For low angle tilt grain boundaries, the work of Dimos *et al* indicates that the critical current density of the grain boundary J_{cij} decreases rapidly as the angle θ_{ij} increases [Dimos et al. 1990]. However the critical current density of the boundary is almost independent of the tilt angle ϕ as shown in Fig.12.34. Using a least-squares fitting of their experimental data to the function

$$J_c(\theta) = J_c(0)\exp(-\alpha\theta) \qquad (12.13)$$

we get $\alpha = 0.2$(as shown in Fig.12.35). For simplicity, we set $J_c(0) = 1$.

We measured the tilt angles θ of over 200 grain boundaries of melt-textured YBa$_2$Cu$_3$O$_7$ ceramics [Zhu et al. 1991e]. It was shown that the majority of grain boundaries are low angle boundaries with $\theta < 20^0$. The tilt angle θ has a Gaussian distribution as shown in Fig.12.36:

$$P(\theta) = \frac{2}{\sqrt{2\pi}\sigma}\exp(-\frac{\theta^2}{2\sigma^2}) \qquad (12.14)$$

where $\sigma = 5^o$.

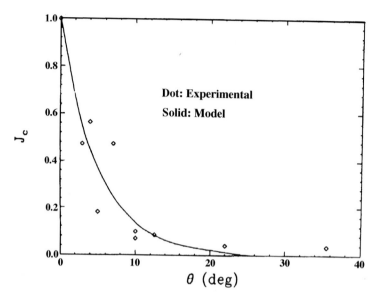

Figure 12.35 Plot of the ratio of the grain-boundary critical current density to the average value of the critical current density in the two grains vs the grain boundary tilt angle. The dots are data obtained on bicrystal samples of YBa$_2$Cu$_3$O$_7$ by Dimos et al. [Dimos et al. 1990] The solid line is the least square fit to the function $J_c = \exp(-\alpha\theta)$ for $\alpha = 0.2$.

Combining these two equations with Eqs.(12.9)–(12.12) yields coupled first-order nonlinear differential equations for the phases φ_i, which can be solved iteratively [Chung et al. 1989].

In this study periodic boundary conditions are used on the left and right edge of the lattice. The external current I_{ext} is injected into the bottom bus bar and extracted from the top bus bar. Therefore the external critical current density is $J_{ext} = I_{ext}/L$ for a $L \times L$ lattice. Eqs.(12.9)–(12.14) are integrated numerically using a fifth-order Gear algorithm [Gear 1971]. The phase factor φ of the superconducting order parameter at the top bus bar is kept fixed at zero, and the voltage drop of the system V is defined as $(h/4\pi e)(d\varphi/dt)$ at the bottom bus bar. The tilt angle θ for each grain boundary (i.e. each bond on the square network) is assumed to be randomly distributed between $0°$ and

12.4. RSJ Model

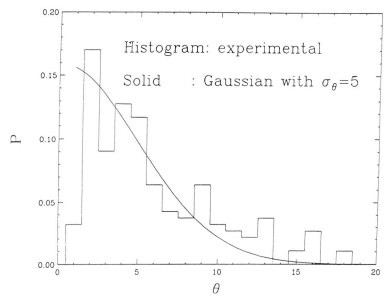

Figure 12.36 Plot of the tilt angle distribution of a melt-textured YBa$_2$Cu$_3$O$_7$ sample with over 200 grain boundaries. The histogram is the data obtained by Zhu *et al.* using Transmission electron microscopy [Zhu *et al.* 1991e]. The solid line is a least square fit to a Gaussian function with $\sigma = 5^o$.

20° with distribution function being a truncated Gaussian. Therefore $\sigma = 0^o$ represents the case of single crystal and $\sigma \to \infty$ represents the system with tilt angles uniformly distributed between 0° and 20°.

Figure 12.37 shows representative current-voltage curves of 10×10 arrays for several values of σ, the width of grain boundary angle distribution, measured in the units of J_{c0} and bL. The results are averaged over 100 sample distributions for each value of σ, and for each value of J over 1000 time steps. For small σ the array behaves very much like a single Josephson junction with J_c close to J_{c0}. As σ increases, more and more grain boundaries have large misorientation angles thus low critical current density J_{cij}. The critical current density of the whole system J_c, defined as the minimum external current density J_{ext} at which the voltage becomes finite, drops dramatically. For systems with $\sigma \to \infty$, J_c is less than 10% of J_{c0}, the critical current density

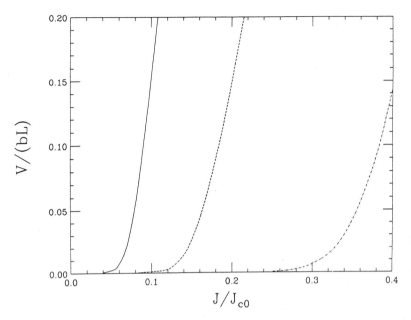

Figure 12.37 I–V characteristics of a 10×10 array for several different grain-boundary angle distributions. The solid line is the $I - V$ curve for $\sigma \to \infty$; the dashed line is the $I - V$ curve for $\sigma = 10^o$, and the dash-dot line is the $I - V$ curve for $\sigma = 5^o$.

of the single crystal. The characteristics of the $I - V$ curve above J_c for systems with large σ is also quite different from that of the single crystal because of the random distribution of normal state resistance. No sudden jump of voltage is found at J_c, instead the voltage shows a power-law like increase near and above J_c, as is found in polycrystalline $YBa_2Cu_3O_7$ samples.

At the critical current density J_c, a normal (resistive) region has grown (percolated) transversely across the sample, cutting the last vertical superconducting path between the bus bars. At this point a voltage appears across the sample for the first time and is supported by the percolating normal region. Fig.12.38 shows a typical path of normal junctions for a 10×10 array with $\sigma = 5^o$ at the critical current density $J_{ext} = J_c = 0.355$. We can see that the size of the clusters of normal domains in the Josephson junction array at its critical current density

12.4. RSJ Model

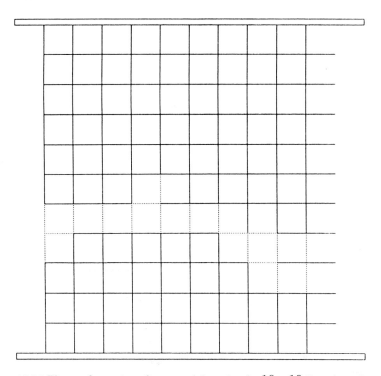

Figure 12.38 The configuration of a normal domain of a 10×10 Josephson junction array with $\sigma = 5^o$ at the critical current density (i.e. $J_{ext} = J_c = 0.355$). The dashed lines represent normal junctions and the solid lines represent superconducting junctions.

is quite different from that of clusters in a site or bond percolation system at its percolation threshold. The normal domain in the Josephson junction system is much smaller than the percolating clusters in the percolation system at the threshold and the shape of the cluster is very anisotropic. In this sense, the superconducting to normal transition is very similar to a mechanical or dielectric breakdown process [Duxbury et al. 1987]. Like the breakdown process, we can define a failure probability distribution $F(J_c, L)$ as the probability that a sample of size L becomes normal (i.e. the superconducting path will be broken, or the critical supercurrent exceeded) at a given value of current density J_c for 10. The numerical results for $F(J_c, L)$ for the case of $L = 10$

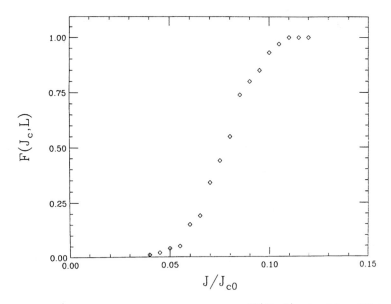

Figure 12.39 The failure probability distribution $F(J_c, L)$ of a 10×10 lattice vs J_c for an array with a grain boundary angle distribution width $\sigma = 5°$. This distribution was made from an ensemble of 100 random configurations. J_{c0} is the single-crystal critical density.

and $\sigma = 5°$ are shown in Fig.12.39. Duxbury, Beale and Leath (DBL) [Duxbury et al. 1987] have shown, from the statistics of extremes, that for cases where the probability of large critical clusters decays exponentially with cluster size, that the failure probability distribution is given in d dimensions by the formula

$$F_{DBL}(J_c, L) = 1 - \exp[-cL^d \exp(-k/J_c)], \qquad (12.15)$$

where c and k are constants. They argue that this distribution F_{DBL} provides a better fit to the data in general than the Weibull form (which should apply for the cases of site or bond percolations where the probability of large critical clusters decays algebraically with cluster size). The Weibull form is

$$F_W(J_c, L) = 1 - \exp(-cL^d J_c^m). \qquad (12.16)$$

12.4. RSJ Model

In order to test the applicability of these two formulae to the behavior of the Josephson junction array we plot the simulation data for $F(J_c, L)$ by forming the axillary variable

$$A(J_c) = -\ln\{-\ln[1 - F(J_c, L)]/L^d\}, \qquad (12.17)$$

which according to Eqn.(12.15) should behave as

$$A_{DBL}(J_c) = -\ln C + k/J_c, \qquad (12.18)$$

and according to Eqn.(12.16) should behave as

$$A_W(J_c) = -\ln C + m\ln(1/J_c). \qquad (12.19)$$

The same data shown in Fig.12.39 are plotted versus $1/J_c$ in Fig.12.40(a) to test the DBL formula and versus $\ln(1/J_c)$ in Fig.12.40(b) to test the Weibull formula. The results suggest that the DBL form is better fit to the simulation data, which indicates that the transition to the normal state at J_c is a breakdown process for a distribution of junction properties, where only a few bonds are turned normal at the transition point.

For $J > J_c$, V is the time average of a periodically oscillating voltage $V(t)$, which is show in Fig.12.41. For J close to J_c, $V(t)$ is almost harmonic, as show in Fig.12.41(a). This again suggests that when $J > J_c$, only a small number of grain boundaries has turned normal. The frequency of $V(t)$ is related to the J_{cij} of those particular boundaries ij that are turned normal. As J increases, more and more junctions become normal and contribute to $V(t)$. Since each junction has its distinct J_{cij}, therefore a distinct "natural frequency" of $V_{ij}(t)$, the total voltage drop across the sample shows apparently chaotic behavior as shown in Fig.12.41(b). Fig.12.42 shows the corresponding power spectra $S(\omega)$ for the system with $\sigma = 10^o$. Fig.12.42(a) clearly shows that for J above but close to J_c, the power spectrum is discrete with sharp peaks. For J well above J_c, however, those sharp peaks are nearly all wiped out by the noise and the power spectrum is almost continuous, as shown in Fig.12.42(b). It will be interesting to quantify the chaotic behavior of $V(t)$ for the system above the critical current density by calculating the Lyapunov exponent.

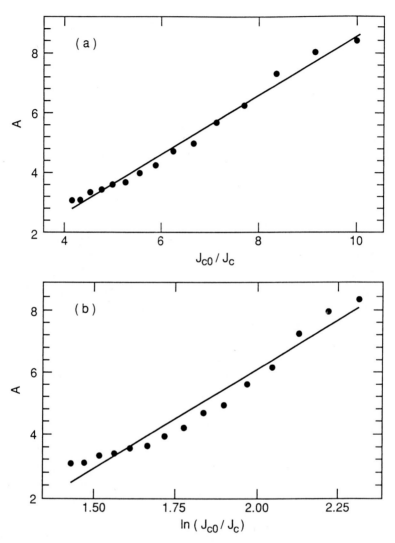

Figure 12.40 (a) A replot of the data in Fig.12.39 in terms of $A(J_c)$ plotted vs J_{c0}/J_c. (b) A replot of the data in Fig.12.39 in terms of $A(J_c)$ plotted vs $\ln(J_{c0}/J_c)$.

12.4. RSJ Model

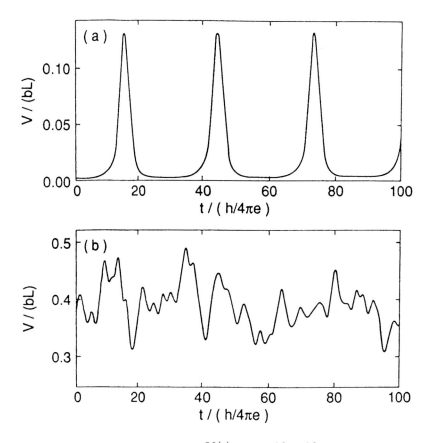

Figure 12.41 Time-dependent voltages $V(t)$ for the 10×10 array with the width of the grain boundary misorientation angle distribution $\sigma = 10^o$ for two levels of current density (a) $J = 0.36$; (b) $J = 0.48$.

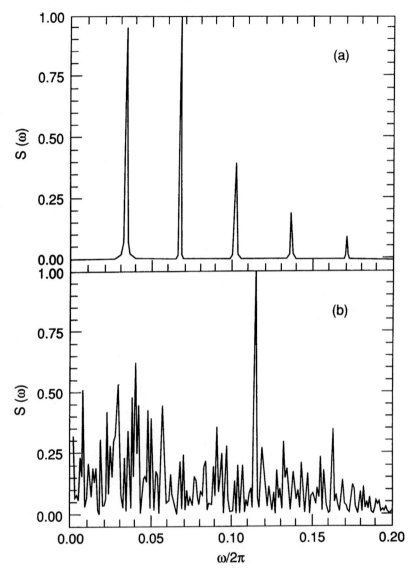

Figure 12.42 Power spectra for the 10×10 array with the width of the grain boundary misorientation angle distribution $\sigma = 10^o$ and (a) $J = 0.36$; (b) $J = 0.48$.

In this section we have studied the effect of a grain-boundary tilt-angle distribution on the critical current density of $YBa_2Cu_3O_7$ ceramics using a Josephson junction array model. The results are consistent with the view that there is a breakdown process at J_c. We think the method presented in this section can give some insight into the characteristics of superconducting ceramics made under various processing conditions, especially on the "reliability" of long superconducting wires or film. It will be interesting to apply this model to the case of $YBa_2Cu_3O_7$ ceramics obtained by various melt-texture techniques, though very little information about the systematic variation with processing condition of the distribution of grain boundary angle and other relevant properties is available. It will be also of interest to extend this model to finite temperature and magnetic field as in the case of ideal Josephson junction array model [Teitel and Jayaprakash 1983, Li and Teitel 1990, Chung *et al.* 1989] to study the effect of grain boundary misorientation angle dependence on the flux pinning.

12.5 Summary

As we can see from the previous section, the superconductivity of a polycrystalline sample is only as good as its weakest link. It does not take many bad grain boundary to reduce the critical current to a fraction of the J_c in single crystal. It is thus very important to understand the detailed structure and properties of grain boundaries in these materials.

In this chapter we decribed extensive theoretical as well as experimental studies of grain boundaris in $YBa_2Cu_3O_{7-\delta}$ and Bi-2212 systems. Much work remains to be done, however. Due to the lack of suitable theories of the electronic structure of the high-temperature superconductors, we still have very little theoretical understanding of the grain boundary chemistry of these compounds. While we understand the properties of low-angle grain boundaries relatively well, the reason that some large angle grain boundaries appear to be good conductors still eludes us. This is very much a work in progress. Processing techniques developed recently makes it possible now to produce very high-quality bicrystal samples. This opens up opportunities for the systematic studies of the properties of these grain boundaries.

Chapter 13

Artificially Created Defects

13.1 Introduction

Obtaining a high critical current density, J_c, requires strong pinning of flux lines in superconducting materials. To provide such pinning sites for the magnetic flux, defects have been deliberately and selectively created by irradiation with energetic electrons, γ rays, nucleon (protons and neutrons) and α particles since the early days of type-II superconductors (For detailed review, see [Sweedler et al. 1979]). Recently several groups have investigated the influence of radiation-induced point defects and small defect-aggregates on the critical current density of the bulk ceramics, single crystals, and thin films of high-T_c cuprates [Weber and Crabtree 1992, Civale et al. 1990, Dover et al. 1990, Schindler et al. 1990]. However, the effectiveness of such defects in pinning the flux lines has been disappointing in a temperature range where thermal fluctuation is important. Intuitively, one might think that the short coherence length ξ and large fluctuation effects in high-T_c cuprates suggest that linear defects along the entire length of the flux line might be more effective. Indeed, the linear tracks of amorphized material created by heavy ions of several hundred MeV energy have been found to provide strong flux pinning in the temperature and field regime where the effects of point defects are inconsequential [Civale et al. 1991, Budhani et al. 1992b, Budhani et al. 1992a, Budhani and Suenaga 1992, Zhu et al. 1993b]. In order to illustrate the relationships between superconducting properties and the defects produced by heavy-ion irradiation, we will present a systematic study of the defects in thin films and bulk samples of $YBa_2Cu_3O_{7-\delta}$ and $Bi_2Sr_2Ca_2Cu_3O_{10+\delta}$ created by irradiation with Au^{24+} (300-MeV), Ag^{21+} (276-MeV), Cu^{18+} (236-MeV), and Si^{13+} (182-MeV) ions [Zhu et al. 1993a, Zhu et al. 1993d]. As described in following sections the structure of the defects was characterized by transmission-electron microscopy (TEM) and high-resolution electron microscopy (HREM), combined with nanoprobe chemical analysis using energy-dispersive x-ray spectroscopy (EDS) and electron-energy-loss spectroscopy (EELS). These studies clearly show that the radiation-induced defects consists of severe chemical and structural local disorder. The degree of damage caused by the ions varies with the ion energy-deposition rate and species of ion, varying from Au, Ag, Cu, to Si in the order of decreasing severity. Furthermore, for all these ions, the density and size of the defects depend strongly on oxygenation of the sample, orientation of the crystal with respect to the ion beam, as well as on the presence of existing imperfections in the lattice. These changes can be attributed largely to differences of the thermal conductivity of the materials and a thermal spike model was developed to provide a theoretical framework for the discussion of the variation of size and shape of the amorphous ion tracks under various conditions.

13.2 Experimental Methods

13.2.1 Sample Preparation

Both thin films and bulk samples were used for this study. Thin films of $YBa_2Cu_3O_7$ were deposited *in situ* on (100)-cut $SrTiO_3$ plate by an off-axis magnetron sputtering technique. Standard sintering method were employed to produce polycrystalline $YBa_2Cu_3O_{7-\delta}$ [Ikeda et al. 1990] while a high-temperature pressing technique was used to produce c-axis aligned textured $Bi_2Sr_2Ca_2Cu_3O_{10+\delta}$ bulk samples [Zhu et al. 1993a]. Oxygen-deficient samples ($YBa_2Cu_3O_{6.3}$) were prepared by quenching the sintered samples from 950^oC into liquid nitrogen.

TEM specimens were made by slicing the sintered pellets, ultrasonically cutting out 3-mm disks, mechanically dimpling them to $\sim 10\mu m$ thick, and finally ion-milling them at 3-4 keV at a $8^o - 10^o$ tilt angle

13.3. Structure of the Defects

with a Gatan low-energy gun. Most of the $YBa_2Cu_3O_{7-\delta}$ TEM samples were stored in a desiccator for more than six months before they were used. However, in all cases, the samples were cleaned and examined before irradiation. Their oxygen substoichiometry δ was estimated as less than 0.1. Some samples were subjected to ozone oxygenation at 200^oC for 2–5 hours [Zhu *et al.* 1993e]. Flowing ozone gas was produced by an ozone generator, consisting of ultraviolet radiation from a low-pressure mercury lamp in the presence of 1 atm pressure of oxygen, These prethinned bulk samples and thin films were subsequently irradiated with heavy ions.

13.2.2 Heavy-ion Irradiation

Thin crystal, either thin films and ion-thinned TEM samples, were irradiated by various heavy ions at room temperature. Ions of Au^{24+}, Ag^{21+}, Cu^{18+}, and Si^{13+} were produced in a Tandem Van de Graff accelerator at Brookhaven National Laboratory. The ion beam was incident at 5° off the sample normal to avoid possible channeling. In Table 13.1 we list the flux (as measured with an annular Faraday cup), the energy, the charge state, and the range (average penetration depth) of these ions for $YBa_2Cu_3O_{7-\delta}$ and $Bi_2Sr_2Ca_2Cu_3O_{10+\delta}$. Since the thickness of the sample used there is very small with respect to the range of these ions, the linear energy transfer (LET) of the incident ions in the stopping medium can be regarded as constant. This ensure uniform damage tracks along the thickness of the samples.

13.3 Structure of the Defects

13.3.1 General Morphology

Figure 13.1 shows typical morphology of a $YBa_2Cu_3O_{7-\delta}$ bulk sample irradiated with 300 MeV gold ions. Diffraction-contrast imaging reveals the presence of massive disks induced by irradiation when the crystals were viewed along the direction of incident ion beam [Fig.13.1(a)]. Small variations in the sharpness at edges of the disk images suggest that some of the ion projectiles were weakly scattered after they entered the crystal. When the crystals were viewed away from the incident ion

Table 13.1 The energy, charge state, and fluence of various ions used in the present study and the estimated linear energy transfer and mean penetration depth, calculated using the tables of Ziegler for $YBa_2Cu_3O_{7-\delta}$ and $Bi_2Sr_2Ca_2Cu_3O_{10+\delta}$ (denoted by the asterisk).

Ion	Au	Ag	Cu	Si
Energy(MeV)	300	276	236	182
Charge	24	21	18	13
Fluence (ions/cm^2/s)	5.6×10^{10}	5.0×10^{10}	4.0×10^{11}	6.6×10^{10}
Energy loss (keV/nm)	34.8	23.8	13.4	3.8
	31.5*	21.8*	12.4*	2.5*
Range (μm)	14.1	15.7	19.2	37.5
	15.9*	17.7*	21.6*	42.0*

beam, the disks appear as rods, a clear indication that the defects are ion trajectories or ion tracks [Fig.13.1(b)]. The rods, running from the top to the bottom of the sample, give rise to thickness fringes as well as to contrast due to its intersection with the specimen surface and crystal matrix. Nanodiffraction from the damaged regions with a probe size less than 2 nm (smaller than the diameter of the defect) shows that they are amorphous. By tilting the specimen through 60° and observing the same defects, we conclude that the ion-induced defects are continuous columns with severe lattice distortion throughout the thickness of the specimen.

Careful observations show that, under two-beam conditions, these defects also generate dark images consisting of lobes of contrast over an area two to three times larger than the diameter of the ion track. We characterized the contrast surrounding the defects using the weak-beam technique under different diffraction conditions. Fig.13.2 shows the black contrast associated with the columnar defects imaged by using diffraction vectors **g** = 020 [Fig.13.2(a)], **g** = 110 [Fig.13.2(b)], **g** = 200 [Fig.13.2(c)], **g** = $\bar{1}$10 [Fig.13.2(d)] near the [001] axis (parallel to the incident ions). We found that the lobe contrast disappears only along the line perpendicular to the diffraction vector, **g**, and running across the center of the damaged area. The main features of the contrast are similar to those observed for spherical precipitates in alloys [Degischer 1972], but it has a different shape and is much smaller. We attribute

13.3. Structure of the Defects

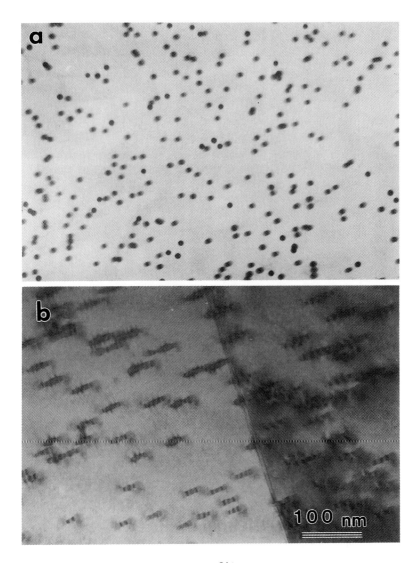

Figure 13.1 Typical morphology of the Au^{24+}-radiation-induced defects viewed (a) along the ion track, (b) about $18°$ away from the ion track.

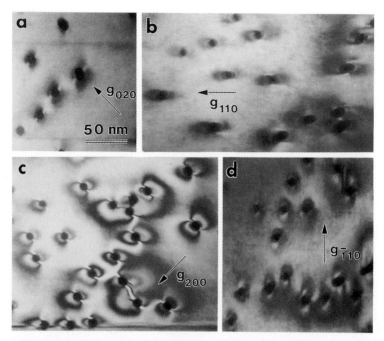

Figure 13.2 Strain contrast surrounding the columnar defects imaged using diffraction vectors (a) **g**=020, (b) **g**=110, (c) **g**=200, and (d) **g** = $\bar{1}10$. Note the contrast disappears in the direction perpendicular to the **g**.

the lobe contrast to radial strain/displacement fields surrounding the amorphous columnar defect (see section 13.5). the strain and structural disorder of the amorphous region propagates into the crystal lattice in a direction perpendicular to the ion track.

13.3.2 Radiation Damage Induced by Au, Ag, Cu, and Si Ions

In order to investigate the influence of the atomic number of the irradiating ion on the radiation damage, the structural defects induced in bulk $YBa_2Cu_3O_{7-\delta}$ by Au, Ag, Cu and Si ions were studied. The energies and energy deposition rates of these ions are listed in Table

13.3. Structure of the Defects

13.1. For Au and Ag ions, the irradiation yields a large number of columnar defects that are distributed randomly throughout the sample (see Fig.13.1). Nevertheless, each ion does not necessarily give rise to a single amorphous column. We found that the density of the defects strongly depends on the ion type and its energy, and the chemical state of the target material.

Si has the smallest atomic number of the four ions used. For Si-ion irradiation at 182 MeV, no columnar defect was observed in any samples we examined, at least for a fluence of 6.6×10^{10} ions/cm^2. Figs.13.3(a) and 13.3(b) show HREM images of the [001] and [010] projections, respectively, from Si-irradiated YBa$_2$Cu$_3$O$_{7-\delta}$ (with $\delta < 0.1$). No distinct structural damage running through the sample thickness was seen. Occasionally, we observed weak strain contrast at and near the specimen surface with [Fig.13.3(c)], and a high density of dislocation loops, with diameters ranging from 10 to 20 nm [Fig.13.3(d)]. Similar observations were made in samples irradiated by neutrons [Weber 1992] and protons [Kramer et al. 1992].

For Cu irradiation at 236 MeV, amorphous tracks were only occasionally observed for YBa$_2$Cu$_3$O$_{7-\delta}$ (with $\delta < 0.1$) when the ion beam was directed along a (or b) direction. The concentration of the damaged regions varied between samples and also between areas of the same sample. Figs.13.4(a) and 13.4(b) show two areas from the same sample viewed near the [001] axis of YBa$_2$Cu$_3$O$_{7-\delta}$ (with $\delta < 0.1$). The structural damage was more severe in Fig.13.4(b), where the black dots seen are strain contrast. However no amorphous columns were seen by HREM in either area. Columnar defects were occasionally observed only when the ion beam was directed along a (or b) axis.

Since the observed columnar defect density due to the copper and silicon ions was small and varies with samples, the relation between the ion dose and the number of defects per unit area was assessed only for Au and Ag ions (see Fig.13.5). The density of defects per unit area was measured over 100 areas of a fixed size (~ 0.2 μm^2) from TEM micrographs. For Au-ion radiation, the dose was 3.86×10^{10} ions/cm^2, a value that is very close to the average density we measured (3.7×10^{10} defects/cm^2). Thus each Au ion produces a single columnar defect. However, for the Ag-ion irradiation, the measured density of the defects was about 2.5×10^{10} defects/cm^2, only half of the ion dose

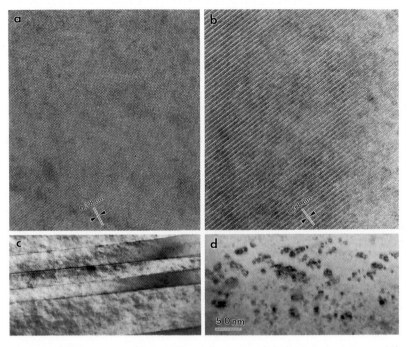

Figure 13.3 TEM micrograph on Si-irradiated samples: (a) lattice images in a [001] projection, (b) [100] projection, (c) low-magnification two-beam image, and (d) dislocation loops.

13.3. Structure of the Defects

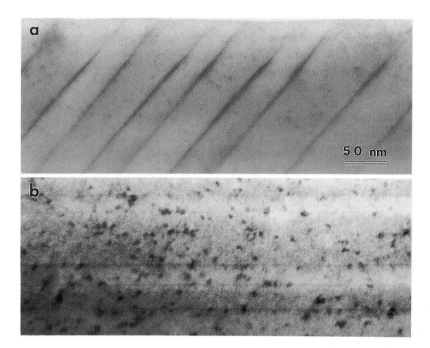

Figure 13.4 The degree of the radiation damage seen in TEM caused by Cu ions varies with the electron beam near the [001] axis of two different regions of a sample of $YBa_2Cu_3O_{7-\delta}$ ($0 < \delta < 0.1$). Note the defects do not exhibit the same type of strain contrast shown by "regular" columns, as in Fig.13.2. In some areas, even the weak strain contrast seen in (b) cannot be observed. [The approximately linear features seen in (a) and (b) are twin boundaries.]

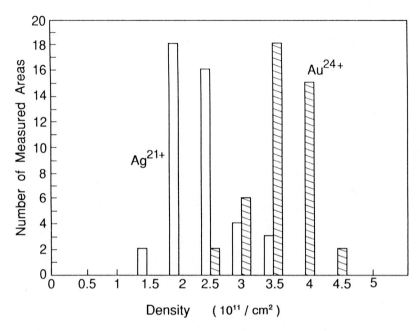

Figure 13.5 A comparison of the columnar defect densities observed by TEM in ∼ 100 areas of fixed size (∼ 0.2 μm^2) for Au- and Ag-irradiated samples of $YBa_2Cu_3O_{7-\delta}$ (with $\delta < 0.1$).

of 5×10^{10} ions/cm^2. It appears that many Ag ions do not produce columnar defects during the irradiation process. The probability of forming columnar defects with Ag is thus smaller than that for Au.

The size of the columnar defects also depends on the atomic number and energy of ions used for irradiation. We examined over one thousand [100] (or [010]) ion tracks produced by Au, Ag, and Cu irradiation. Their size was measured on HREM micrograph of lattice images surrounding the amorphous columns from regions less than 15 nm thick. Since the columnar defects have a well-defined facet along the $a - b$ plane, counting the number of $a - b$ planes lost in the amorphous region gives measurements of vary high accuracy (less than a c-lattice constant, i.e. 1.179 nm). Fig.13.6 shows the size (diameter of the amorphous column in units of the c-lattice parameter) distribution of the defects produced by Au, Ag, and Cu ions in $YBa_2Cu_3O_{7-\delta}$ (with

13.3. Structure of the Defects

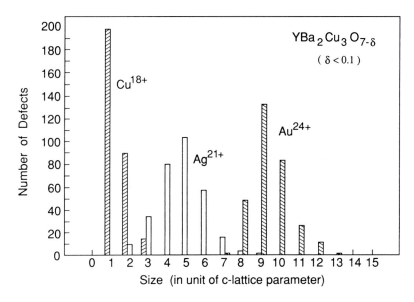

Figure 13.6 Size distribution of the defects induced by Au, Ag, and Cu in nominal YBa$_2$Cu$_3$O$_{7-\delta}$ ($0 < \delta < 0.1$). The dominant size of the defects was 10.6 nm (nine units of c) in diameter for Au-radiated samples, and 5.9 nm (five units of c) for Ag-radiated samples, and 2.36 nm (two units of c) for Cu-radiated samples.

$\delta < 0.1$). The dominant diameter of the defects was 10.6 nm (nine units of c) for Au-irradiated samples, while 5.9 nm (five units of c) for Ag-irradiated samples, and 2.36 nm (two units of c) for Cu-irradiated samples. This finding clearly shows that irradiation with 300-MeV Au ions causes more severe lattice damage compared with 276-MeV Ag and 236-MeV Cu ions. It is interesting to note that while the size distribution of Au- and Ag-ions produced defects show a more-or-less Gaussian distribution, the Cu-induced defects follow an approximate half-Gaussian distribution since those defects with a size less than a c-lattice parameter were not counted.

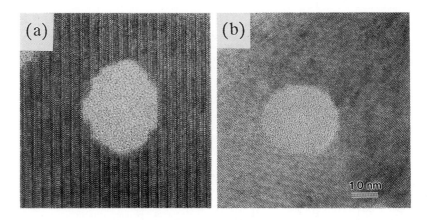

Figure 13.7 High-resolution images of Au^{24+} irradiation-induced defects in bulk $Bi_2Sr_2Ca_2Cu_3O_{10+\delta}$ recorded in (a) [100], and (b) [001] projections, respectively.

13.3.3 The Effect of Crystallographic Orientation

At low magnification, as shown in Fig.13.1, the defects appear to have a circular symmetry along the ion trajectory in all orientations. However, HREM shows that the morphology of the defects depends on the direction of the incident ions with respect to the crystallographic axes of the material. Fig.13.7(a) and 13.7(b) show high-resolution images of the Au^{24+} irradiation-induced defects in bulk $Bi_2Sr_2Ca_2Cu_3O_{10+\delta}$ viewed along the [100] and the [001] direction, respectively. The amorphism of the damaged area is clearly visible. When the ion track is parallel to the c axis [Fig.13.7(b)], the defects appear as circular disks ($\sim$ 16 nm in diameter). In contrast, when the ion track is parallel to the a or b axis [Fig.13.7(a)], the defects appear somewhat larger and elliptical, with a size about 18 nm × 21 nm. In the latter case, the edge of the amorphous region shows well-defined facets in the $a-b$ planes, or more precisely, between the BiO double layers. At the periphery near the long axis of the ellipse, the interface between the defect and the crystal matrix is diffuse (extending about two unit cells) and is similar to the interface observed [Fig.13.7(a)] when the defect is formed by an ion beam parallel to the c axis.

The variation of the high-resolution images of the defects in $YBa_2Cu_3O_{7-\delta}$ with respect to the crystallographic orientation was essentially identical to that seen in $Bi_2Sr_2Ca_2Cu_3O_{10+\delta}$, i.e., circular cross-sections for ions incident along [001] direction, and elliptical for those along the [100] or [010] direction. However, the size of the defects was smaller by a factor of 0.5 – 0.7 than those in $Bi_2Sr_2Ca_2Cu_3O_{10+\delta}$. The orientational dependence of the shape of the defect was very clearly illustrated for an irradiated $YBa_2Cu_3O_{7-\delta}$ (with $\delta \approx 0$) thin film containing a mixture of grains with a and c orientations (Fig.13.8), where the areas denoted A and C have the a (or b) and c axes parallel to the film normal, respectively. The difference in morphology of the defects between [001] and [100] (or [010]) orientations is clear. We conclude that an ion beam directed along the a or b axis causes more severe structural damage than a beam along the c axis. The interpretation of this observation is addressed in the following section.

13.3.4 The Effects of Oxygen Concentration in $YBa_2Cu_3O_{7-\delta}$

The extent of radiation damage depends on the oxygen stoichiometry of the sample. Differences between Ag-irradiation-induced defects in $YBa_2Cu_3O_{7-\delta}$ are shown in Fig.13.9 for samples with (a) $\delta \approx 0.7$, (b) $\delta < 0.1$, and (c) $\delta \approx 0$ (ozone oxygenated). All the micrographs show lattice images viewed along the [100] axis. In Figs.13.9(b) and 13.9(c), the direction of the incident beam was slightly off from the [100] projection and, therefore, only near the edge of the specimen (bottom part of the micrographs) the defects are end-on and appear as with contrast. The differences in the size and density of the damage among the various samples are remarkable. For $YBa_2Cu_3O_{6.3}$, the average size of the defects is close to that observed in the Au-irradiated $YBa_2Cu_3O_{7-\delta}$ ($\delta < 0.1$), while for ozone-treated $YBa_2Cu_3O_7$ the size is close to that observed in the Cu-irradiated material. The size distribution of the defects for these three samples is plotted in Fig.13.10 The sizes were measured by counting the number of $a - b$ planes across the area. The average diameter of the amorphous regions for $YBa_2Cu_3O_{6.3}$ was 7.8 nm, for $YBa_2Cu_3O_{7-\delta}$ ($\delta < 0.1$) was 5.3 nm, and for ozone-treated $YBa_2Cu_3O_7$ was 3.3 nm. It is important to note that a small decrease of the oxygen substoichiometry (δ from 0.1 to 0) significantly reduced the radiation damage.

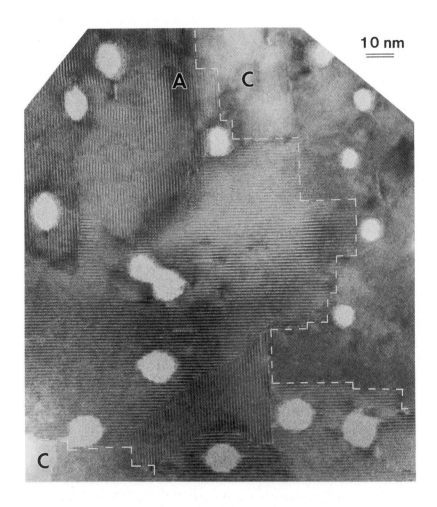

Figure 13.8 High-resolution image of Au^{24+}-irradiated $YBa_2Cu_3O_7$ thin film. The defects appear elliptical in shape when the incident ion is parallel to the [100] or [010] direction, but smaller and circular in shape when parallel to the [001] direction. The areas denoted A and C have the a (or b) and c axes parallel to the film normal, respectively.

13.3. Structure of the Defects

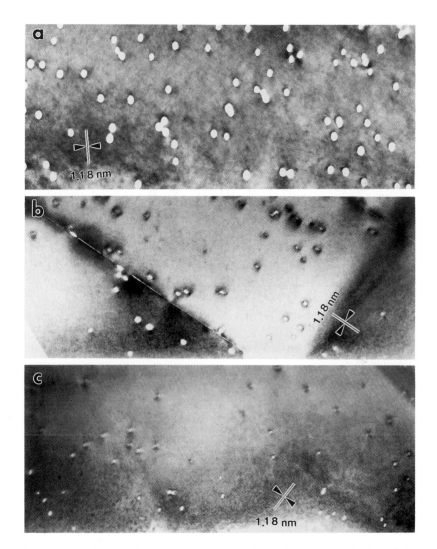

Figure 13.9 Columnar defects induced by Ag-ion radiation viewed along [100] direction. YBa$_2$Cu$_3$O$_{7-\delta}$ with (a) $\delta \approx 0.7$, (b) $\delta < 0.1$, and (c) ozone-treated ($\delta \approx 0$).

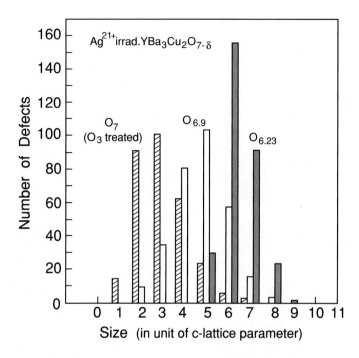

Figure 13.10 Size distribution of Ag-radiation-induced defects for the three samples shown in Fig.13.9.

As show in Fig.13.4, the probability of formation of the columnar defects varies from area to area for Cu-irradiated $YBa_2Cu_3O_{7-\delta}$. This variation appears to be due to oxygen inhomogeneity in the as-prepared TEM specimen from a nominally fully oxygenated sample. In general, no columnar defects were observed except when the ion beam was directed along the $a-b$ plane. Defects which appear to consist of columns continuously amorphized along the ion track are marked by arrows in Fig.13.11(a). However, the majority of the defects exhibit a lattice-image across the damaged region, presumably due to the supposition of an amorphous region and a crystal lattice. This may indicate an intermittent amorphization along the ion track. In contrast, in samples oxygenated with ozone, the damage induced by Cu irradiation was much less than that in oxygen-deficient samples. Even when the ion-

13.3. Structure of the Defects

beam is parallel to the a or b axis, only a low-density strain contrast (appearing as black dots) was visible [Fig.13.11(b)]. This is consistent with the observations of the effects of Ag-ion irradiation and suggests that the radiation damage of $YBa_2Cu_3O_{7-\delta}$ depends strongly on the oxygen concentration.

13.3.5 The Effect of the Pre-existing Crystal Imperfections

The pre-existing imperfections in the crystal also play an important role in forming the columnar defects. Such an effect is not clearly visible for Au irradiation because each Au ion always produce a single amorphous column, regardless of the target crystal. However, for a lighter and less energetic ion such as Cu, we demonstrated that the formation of the amorphous zone is very sensitive to the characteristics of the crystal, depending not only on the orientation of the crystal and oxygen concentration but also on the presence of imperfections in the crystal. Shown in Fig.13.12 is an example of a (001) lattice image from an area with pre-existing planar defects. In contrast with our observations of areas free of pre-existing defects such as ozone-treated sample, which show no amorphous column regardless of the orientation when irradiated with Cu, here we observed a high density of columnar defects with an average size ~ 2.2 nm. Furthermore, the distribution of these defects is not random. The radiation-induced defects seen in Fig.13.12 are formed only at the location (marked as B) of stacking faults (appearing as white lines, viewed here edge-on). In the area marked A, there were no stacking faults, and no columnar defects were formed there. Such stacking faults, characterized as having a plane normal of [001] and a displacement vector of $\frac{1}{6}[301]$, were sometimes observed in as grown samples and ozone-annealed samples. From these results, we conclude that structural imperfections in the as-grown crystal enhance radiation damage, especially in the case of planer defects such as stacking faults or grain boundaries when their plane normal is perpendicular to the incident ion beam.

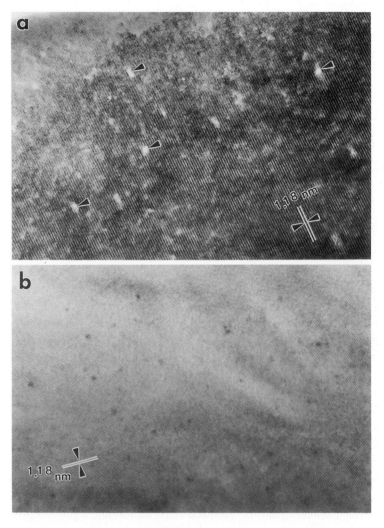

Figure 13.11 Lattice image of [001] projection observed in $YBa_2Cu_3O_{7-\delta}$. (a) Cu ion irradiated without ozone treatment. The arrows denote defects in which the amorphous column extends completely through the sample. Other defects can be seen which show a supposition of amorphous and crystalline regions along the ion track. (b) Cu ion irradiated after ozone treatment.

13.3. Structure of the Defects

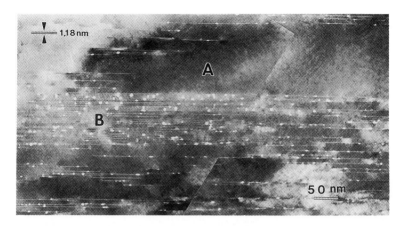

Figure 13.12 (100) lattice image of an ozone-treated sample of $YBa_2Cu_3O_{7-\delta}$ ($\delta \approx 0$) irradiated with Cu ions. Note the columnar defects were only observed at the location of the pre-existing stacking faults.

13.3.6 Creation of Stacking Faults

Planar defects associated with the amorphous columns are frequently observed in Au^{24+} and Ag^{21+} irradiated $YBa_2Cu_3O_{7-\delta}$ ($0.05 < \delta < 0.6$). These planar defects were created by the heavy ion irradiation, unlike the preexisting defects discussed in the previous section. They are clearly visible when the electron beam is parallel to the $a - b$ plane (Fig.13.13). The extension of the fault is about $3 - 5$ times as much as the radius of the amorphous column. Similar defects, characterized as intrinsic and extrinsic stacking faults with a [001] plane normal, were observed in quenched $YBa_2Cu_3O_{7-\delta}$.

Detailed analysis of high-resolution images of the faults yielded a determination of its character. The displacement vector of the faults is either $\mathbf{R} = \frac{1}{6}[031]$ or $\mathbf{R} = \frac{1}{6}[032]$. Thus the observed planar defect consists of two stacking faults bounded by a partial dislocation at each end. Since the motion of such a dislocation does not give rise to the observed fault, formation of the fault must be chemical in nature. The above structure models are also consistent with the observation of the Cu enrichment in such planar defects induced by heavy-ion irradiation by a high-resolution EDS measurement using a 2 nm probe.

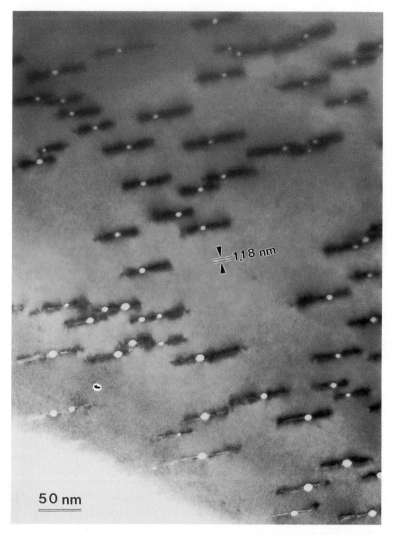

Figure 13.13 [001] lattice image of $YBa_2Cu_3O_{7-\delta}$ ($\delta < 0.1$) irradiated with Ag^{+21} ions showing stacking faults, view edge on, created by the radiation damage. The stacking faults are seen as dark wide lines in thick region and white narrow lines in thin region.

13.3.7 EELS Measurements

To characterize the chemical composition within and near the irradiation-induced defects, the amorphous region of Au-irradiated samples was analyzed using a nanoprobe TEM (2 nm) combined with EDS. There were no noticeable differences (< 3%) in the cation composition in both the amorphous regions and the surrounding crystalline matrix, either for an ion beam parallel to the $a-b$ plane or to the c direction. However, for anion oxygen, we observed with EELS a remarkable change across the damaged area in fine structure of the oxygen K-edge absorption spectrum, more precisely, the prepeak of the K-edge. The prepeak represents transition of electrons from oxygen $1s$ states into unoccupied states. The integrated intensity of the prepeak is a direct measure of the hole density near the Fermi level, and is usually proportional to the oxygen content. The integrated intensity can also be diminished by disorder in the CuO chains.

Figure 13.14 shows a series of EELS spectra acquired over a span of 60 nm across the amorphous region. The acquisition step is about 5 nm and the acquisition probe is 2 nm in diameter. To compare the strength of the prepeak at 528 eV, we have normalized the data with respect to the main-peak at 537 eV. The spectra can be divided into three groups, denoted as spectra A, B, and C. They are symmetrically acquired at the positions with respect to the center of the defect as shown in Fig.13.14. Spectra A, taken from positions more than 20 nm away from the center of the amorphous region, show a prepeak at about 528 eV [the oxygen prepeak would be much pronounced if a larger probe and a longer acquisition time were used [Zhu et al. 1993h]], similar to that observed in defect-free crystalline matrix. Spectra B, acquired $10-15$ nm away from the center of the amorphous region ($5-10$ nm from its edge), show the oxygen K-edge rising straight up toward the main peak without a prepeak. In the amorphous region (Spectra C), not only was the prepeak absent, but also the height of the oxygen main peak was drastically reduced (only 70% of that observed in the spectra A and B before normalization of the intensity). This suggests that there was a marked change in electronic structure and possibly a loss of oxygen atoms from the amorphous region. An important observation here is the existence of spectra B, where the drastic change in the oxygen pre-edge peak suggests a decrease in the number of holes in the

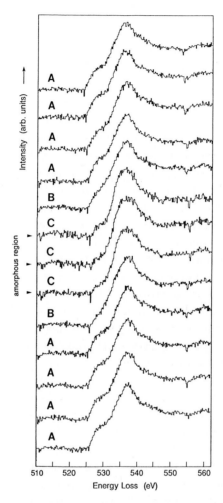

Figure 13.14 A series of EELS spectra acquired over a span of 60 nm at and near the amorphous region. The acquisition step is about 5 nm, and the acquisition probe diameter is 2 nm. To compare the strength of the prepeak each spectrum was normalized by the oxygen main peak at 537 eV. Spectra labeled *A*, *B* were acquired from within the crystalline matrix around the defects while those labeled *C* were acquired within the amorphous area.

crystal near the amorphous region. The reduction of the hole content outside the defect can be due to a reduction in oxygen concentration, or disorder in the crystal lattice, particularly the oxygen in the Cu-O chains or perhaps to elastic strain. A comparison of the prepeak of the spectra B with the observation of Nücker et al. [Nücker et al. 1988] and Takahashi et al. [Takahashi et al. 1990] suggest that the oxygen deficiency required to match the data is about $\delta \approx 0.7$. However, the depression of the hole concentration surrounding the amorphous region might also be attributed to the lattice distortion around the defects, as seen in Fig.13.2. Furthermore, theoretical estimates suggest that to generate one oxygen hole, an undistorted oxygen-chain should be at least five unit cells long, thus radiation-induced oxygen disorder may also be playing a role. Regardless of its origin, it is clear that the reduction of the hole concentration near the columnar defects enlarges the weak- or nonsuperconducting area to almost twice as large as that of the amorphous region.

13.4 Thermal Spike Model

An understanding of the formation and the structure of the defects due to heavy-ion irradiation requires some knowledge of the mechanism by which the projectile deposits energy and how this energy is dissipated in the target crystal.

The structure of the defects produced by heavy ion irradiation depends sensitively on the energy, charge state and mass of the incident ion and on the structural and physical properties of target materials. In this section we calculate the size and the shape of these defects under various experimental conditions using a simple model. The theoretical insights can help experimentalists to design and create ideal defect structure for flux line pinning in various superconducting materials.

Different models have been proposed to interpret the formation of irradiation-induced defects. One of the popular model is "Coulomb explosion" [Fischer and Spohr 1983], i.e. the violent disruption of a local region of the lattice by unbalanced electrostatic forces during the period before electrical neutrality is restored to a region around the ion track in which charge separation is induced by the passage of the ions. Recently, however, a defense has been made of the old notion of "thermal spikes" model [Chadderton 1988, Dienes and Vineyard 1957]. Though it uses continuum and linear heat-transfer approximations in discussion of the damage process caused by ion radiation, the thermal spike model does provide us a reasonable framework to quantitatively study the defect creation process. Furthermore, the thermal spike approximation has been proven useful in rationalizing phase transformations and the formation of metastable phases caused by fusion damage in metallic alloys [Dienes and Vineyard 1957]. Therefore, in this section we will study the formation of irradiation-induced defects in the framework of the thermal spike model.

13.4.1 Stopping Power of Heavy Ions

The stopping power of the heavy ions is defined as energy loss of the incident ions per unit length in the target materials. Heavy charged particle moving through a medium composed of atomic nuclei and electrons can lose energy by three main process. For very high energy ions the principal mode of energy loss is that due to interaction with the atomic *electrons* by Coulomb excitation or ionization. The ion may also lose energy by direct collisions with *nuclei* of the medium. The third energy loss process is emission of "bremstrahlung" and Cerenkov *radiation*. Therefore the total rate of energy loss is the combination of these three processes.

$$\left(-\frac{dE}{dx}\right)_{total} = \left(-\frac{dE}{dx}\right)_{electron} + \left(-\frac{dE}{dx}\right)_{nucl} + \left(-\frac{dE}{dx}\right)_{rad} \quad (13.1)$$

The intensity of the radiation is inversely proportional to the square of the mass of the incident particles. Therefore in the case of heavy ion (such as Au), the energy loss due to radiation is negligible. The energy loss due to nuclear collision can be estimated by Rutherford scattering law. Due to the small scattering cross-section of high energy ions, the energy loss due to nuclear collision is less than two orders of magnitude less than the energy loss due to electron excitation. In the following analysis, we will ignore the contribution of the energy loss due to nuclear collision.

The sample we are interested in here is very thin ($< 100nm$) compared with the estimated range of the incident ions. Therefore essentially all of the energy deposited into the sample by the incident ion

13.4. Thermal Spike Model

Table 13.2 Calculated stopping power and calculated and measured diameter D of the damaged area for $YBa_2Cu_3O_6$ and $YBa_2Cu_3O_7$ by various incident ions.

Ion	Au	Ag	Cu	Si
Energy (MeV)	300	276	236	182
Charge	24	21	18	13
Stopping power(KeV/Å)	3.48	2.38	1.34	0.38
D for O_6 cal.(Å)	200	150	50	0
D for $O_{6.3}$ meas.(Å)	140	70	–	–
D for O_7 cal.(Å)	160	80	0	0
D for $O_{6.9}$ meas.(Å)	110	55	22	0
D for O_7 meas.(Å)	–	30	0	0

can be regarded as electronic excitations [Chadderton 1988]. Using a classical model as described by Chadderton, The energy loss due to electronic excitation assumes the following form:

$$\left(-\frac{dE}{dx}\right)_{electron} = \frac{2Z_1^{2/3}Z_2^{1/3}h^3n_0}{\pi^2 e^2 m_e}\left(\frac{2E}{M_1}\right)^{1/2} \quad (13.2)$$

where Z_1 and Z_2 are the atomic numbers of the incident and target atoms, respectively. n_0 is the number of target atoms per unit volume. E is the energy of the incident ion. M_1 and m_e are the mass of the incident ion and electron, respectively. More accurate values of energy loss due to electron excitation can be obtained from Ziegler's table [Ziegler 1980]. As shown in Table 13.2, the stopping power depends sensitively on the atomic number and energy of the incident ions.

13.4.2 The Size of the Defect Area

Energy Transfer to the Lattice

As we have shown in the previous section, the energy transferred to the YBCO sample is predominately in the excited electrons. Here we discuss how the energy transfers from the gas of excited electrons to the lattice to create defects.

Because of the short time involved to excite electrons, we can assume that all the energy has been transferred to the electron gas at time $t = 0$. It is also reasonable to assume at the electron gas is in equilibrium at any time.

A simple model for the energy transfer was first described by Chadderton to evaluate the energy transfer from electron gas to the atoms [Chadderton 1988].

Assuming an electron gas of temperature T_1 is homogeneously mixed with a gas of heavy atoms of temperature T_2. The heat transport may be represented by the pair of differential equations:

$$\kappa_1 \frac{\partial^2 T_1}{\partial x^2} = C_1 \frac{\partial T_1}{\partial t} + b(T_1 - T_2), \tag{13.3}$$

$$\kappa_2 \frac{\partial^2 T_2}{\partial x^2} = C_2 \frac{\partial T_2}{\partial t} - b(T_1 - T_2), \tag{13.4}$$

where κ_1 and κ_2 are conductivity of electrons and atoms respectively, C_1 and C_2 are the heat capacities per unit volume, and b is a coefficient of heat transfer between electron and atoms. For a mixture of perfect gas at two temperatures $C_1 = C_2 = C$, so that we may let:

$$\kappa_1/C = D_1 \tag{13.5}$$

$$\kappa_2/C = D_2 \tag{13.6}$$

$$b/C = \mu. \tag{13.7}$$

Solutions to Eqn.(13.3) and (13.4) may be derived by substitutions of the form

$$T_1 = \alpha_1 \exp(ikx + \theta t) \tag{13.8}$$

$$T_2 = \alpha_2 \exp(ikx + \theta t) \tag{13.9}$$

into Eqn.(13.3) and (13.4). We then find that

$$\theta = -\mu - \frac{(D_1 + D_2)}{2} k^2 + \sqrt{\left(\frac{D_1 - D_2}{2}\right)^2 + \mu^2}. \tag{13.10}$$

For an instantaneous line source produced initially entirely in electron gas, the solution for the lattice temperature T_2 becomes

13.4. Thermal Spike Model

$$T_2 = \frac{Q}{2\pi C} \exp(-\mu t)$$
$$\times \int_0^\infty \exp[-(D_1 + D_2)k^2 t/2]$$
$$\times \frac{\sinh(\mu t\sqrt{1+\delta^2})}{\sqrt{1+\delta^2}} J_0(kr) k\, dk \qquad (13.11)$$

where

$$\delta = \frac{(D_1 - D_2)}{2\mu} k^2.$$

We will compare the lattice temperature (T_2) with the melting temperature of the YBCO compounds (T_m assumed to be $1000K$). The melting region has a radius of $R_m(t)$ where

$$T_2(R_m, t) = T_m. \qquad (13.12)$$

We also assume the following parameters

$$\mu = 10ps^{-1}, \qquad C = 1.84 \times 10^{-5} \frac{eV}{\text{Å}^2 \cdot K}$$

$$K_0 = Q/C = 1.68 \times 10^8 \text{Å}^2 \cdot K$$

$$\kappa_2 = 4 \frac{J}{s \cdot M \cdot K} = 0.0025 \frac{eV}{ps \cdot \text{Å} \cdot K}$$

$$D_2 = \kappa/C = 136 \text{Å}^2/ps.$$

The value of D_1 depends on the bounding of the compounds, as discussed below.

$YBa_2Cu_3O_6$: the Semiconductor

For a semiconductor, The thermal diffusivity of the electron gas is about the same order of magnitude of that of the lattice, i.e. $D_1 = D_2$. We therefore have

$$T_2 = \frac{Q}{2C} \cdot \frac{1}{4\pi D_1 t} \exp(-r^2/4D_2 t)[1 - \exp(-2\mu t)] \qquad (13.13)$$

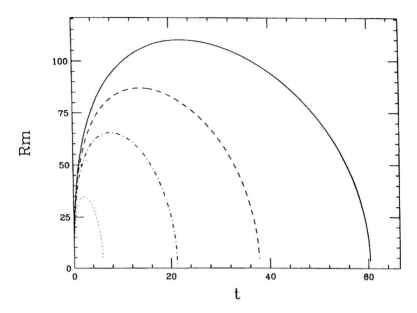

Figure 13.15 The $R_m(t)$ curve for YBa$_2$Cu$_3$O$_6$ irradiated by Au^{+24} (solid line), Ag^{+21} (dashed line), Cu^{+18} (dotdash line) and Si^{+13} (doted line).

When nearly all the energy were transferred to the lattice, the second term can be ignored. R_m can then be derived as

$$R_m(t) = \sqrt{4D_2 t \ln\left(\frac{Q}{8C\pi T_m t}\right)}. \qquad (13.14)$$

The time dependence of the radius of the melting region R_m for Au^{+24}, Ag^{+21}, Cu^{+18} and Si^{+13} ions irradiated YBa$_2$Cu$_3$O$_6$ are shown in Fig.13.15.

YBa$_2$Cu$_3$O$_7$: the Metal

For a metal, $D_1 \gg D_2$. We have,

$$T_2 = \frac{Q}{2\pi C} \exp(-\mu t) \int_0^\infty \exp\left(-\frac{D_1 k^2 t}{2}\right) \frac{\sinh(\mu t \sqrt{1 + \frac{D_1^2 k^4}{4}})}{\sqrt{1 + \frac{D_1^2 k^4}{4}}} J_0(kr) k dk, \qquad (13.15)$$

13.4. Thermal Spike Model

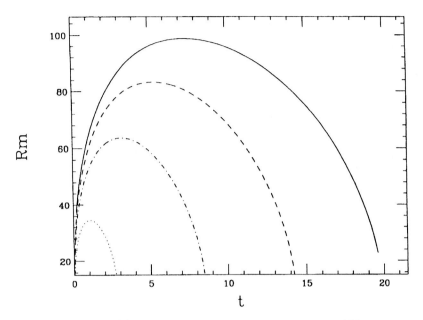

Figure 13.16 The $R_m(t)$ curve for YBa$_2$Cu$_3$O$_7$ irradiated by Au^{+24} (solid line), Ag^{+21} (dashed line), Cu^{+18} (dotdash line) and Si^{+13} (doted line).

where $D_1 = 5D_2$. The numerical results are shown in Fig.13.16. We also calculated the case for $D_1 = 10D_2$ and no significant change is found. This calculation shows that even though the energy loss and the electron-lattice couple in YBa$_2$Cu$_3$O$_7$ is almost the same as in YBa$_2$Cu$_3$O$_6$, higher diffusivity of electrons in the metallic YBa$_2$Cu$_3$O$_7$ cause the time for the $T_2 > T_m$ to be much lower than in the YBa$_2$Cu$_3$O$_6$ case. The disordered region in the YBa$_2$Cu$_3$O$_7$ sample is thus much smaller than in the YBa$_2$Cu$_3$O$_6$ sample. From the above equation we can see that the lattice temperature depends sensitively on the thermal diffusivity of both electron gas and lattice. For a metallic compound such as YBa$_2$Cu$_3$O$_7$, $D_1 \gg D_2$. For a semiconductor compound such as YBa$_2$Cu$_3$O$_6$, $D_1 = D_2$. For all the systems we considered, R_m increases rapidly at first, then decreases as the system cools down as shown in Fig.13.15 and Fig.13.16. It is clear that the lattice temperature of metallic YBa$_2$Cu$_3$O$_7$ is quite different compared to the semiconducting YBa$_2$Cu$_3$O$_6$ although they have the same energy loss Q and coupling constant μ.

The actual size of the damaged area depends on the following factor:

- The time during which the area has temperature $T > T_m$.
- The cooling rate at T_m.

It takes time for atoms to diffuse out of their lattice position and for the area to truly melt into liquid phase, and it take a fast cooling rate at T_m for the liquid to quench into an amorphous phase. Obviously the longer the area's temperature is above T_m, the more liquid like the area becomes, thus the slower the cooling rate can be allowed for the area to become amorphous. Therefore the two factors are interrelated. Since the typical time scale for atom diffusion is in the pico second range, we give an upper limit of the size of the damaged area by assuming that the area's temperature has to be above T_m for at least $10ps$, and the cooling rate at T_m, dR_m/dt, must be larger than $1\text{Å}/ps$ in order to become amorphous. Table.1 shows the calculated defect size for various incident ions and target materials. We can see that the model predicted the upper limit of the size of the damaged area correctly.

13.4.3 Anisotropy of the Size and Shape of the Defects

High-resolution electron microscopy shows a striking asymmetry in the morphology of these defects, which depends on the direction of the incident ions with respect to the crystallographic axes of the material. Fig.13.8 shows high-resolution images of the Au^{+24} irradiation-induced defects in bulk $YBa_2Cu_3O_7$ thin film containing a mixture of grains with a and c axis parallel to the film normal, as indicated by A and C, respectively.

This anisotropy of the size and shape of the defects is due to the anisotropy of the thermal conductivity of the layered superconducting compounds. It is well established that the thermal conductivity along c-axis is much smaller than that along the a or b axis [Ginsburg 1989]. When heat is transferred from an instantaneous line source, more energy will be transferred to the direction with higher thermal conductivity, in the expense of direction will lower thermal conductivity. Therefore we expect that, compared to the defect created by ions

13.4. Thermal Spike Model

parallel to the c-axis, the defect created by ions along a or b-axis will be smaller in the c direction but larger in a or b direction. The cross-section of such defect is predicted to be elliptical, with the diameter d_{a1} (along the c axis) and d_{a2} (along the b- (or a-) axis) satisfies the following:

$$d_{a1} < d_c, \qquad d_{a2} > d_c$$

Assuming the thermal diffusivity $D_a = D_b$ and D_c for heat conduction along the principal crystallographic axes, and ignoring the effects of the latent heat of melting, the thermal-spike model for semiconductors predicts that an ion track parallel to the c-axis will be a cylinder of circular cross-section with a diameter d_c:

$$d_c = \sqrt{\frac{4Q}{C\pi e \theta_{mp}}}$$

where θ_{mp} is the difference between the melting temperature and the room temperature ($\approx$ 300K), C is the heat capacity per unit volume, $Q = -dE/dx$, and $e = 2.78$. For an ion track along the a (or b) axis, the cross-section is predicted to be elliptical with d_{a1} and d_{a2} given by

$$d_{a1} = d_c (\frac{D_c}{D_a})^{1/4} \qquad d_{a2} = d_c (\frac{D_a}{D_c})^{1/4}.$$

For a 300 MeV Au ion, we estimate $d_c = $ 10nm and $d_{a2}/d_{a1} \approx 3-5$.

In summary, we have shown that the size and shape of the defect area in high T_c superconductors irradiated by high energy heavy ions can be evaluated using the thermal spike model. Though the results are only order of magnitude estimates, It provides an upper limit to the size of defect the heavy ion can produce. There are more questions to be answered in this interesting system. For example, the defect area is found to be much larger than the amorphous region. The stacking fault around amorphous region and its effect on flux pinning can only be characterized by atomic scale modeling.

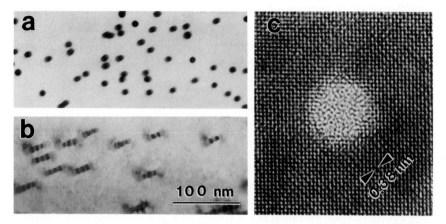

Figure 13.17 Columnar defects induced by Ag-ion irradiation, viewed along the ion-track (a) and 18° away from the ion-track (b). High-resolution image of the ion-track and the surrounding matrix viewed along the c-axis, suggesting the defect is amorphous (c).

13.5 Strain Field and Strain Contrast of Columnar Defects

In high-temperature superconducting cuprates, lattice defects, such as second-phase precipitates, dislocations, stacking faults, heavy-ion tracks, twin boundaries and grain boundaries engender elastic strains in the surrounding superconducting matrix. These strain fields alter the superconducting properties in the nearby matrix and thereby affect the flux-pinning strength of defects and the weak-link behavior of grain boundaries. Thus, understanding the strain fields associated with these defects is crucial to elucidating their behavior, and hence, to improving the superconducting properties of these materials. One good example is the study of amorphous columns induced by radiation with several-hundred-MeV heavy ions in $YBa_2Cu_3O_{7-\delta}$ ($\delta \approx 0$) (Fig.13.17) which, when introduced with the proper size and density, cause a radically-enhanced strength of flux-pinning and of critical current density [Zhu et al. 1993b]. An increase of normal-state resistivity and a decrease of the critical temperature in the samples were observed after sufficiently high radiation doses, and one factor in

13.5. Strain Field and Strain Contrast of Columnar Defects

this degradation of properties was suspected to be the strength of the strain field surrounding these columnar defects [Budhani et al. 1992b, Budhani et al. 1992a].

Examinations of samples by transmission electron microscopy (TEM) irradiated with 300 MeV Au^{+24} ions revealed a particular type of defect-associated diffraction contrast whose width is much greater than the defect size [Fig.13.18(a)]. When the columnar defect is nearly isolated (marked as A and B in Fig.13.18(a)), the contrast appears as a pair of lobes, approximately symmetrically-placed about a "notch-like" line of no contrast, which is perpendicular to the reciprocal lattice vector of the reflection used to form the image [Fig.13.18(c)]. Such contrast was interpreted as being due to the lattice distortion associated with the defects. In this section, we present a detailed analysis of the contrast produced by a strain field surrounding the columnar defects induced by heavy-ion irradiation [Zhu et al. 1996]. The analysis combines continuum elasticity theory and electron diffraction theory, including absorption, in the case of a cylindrically-symmetrical strainfield in an anisotropic tetragonal crystal to obtain an intensity map of its strain contrast in order to derive a parameter that describes the elastic state of the strained defect and the matrix. Although analyses of strain contrast for spherical precipitates and cuboidal inclusions were previously reported by Ashby and Brown [Ashby and Brown 1963] and [Sass et al. 1967], their analyses were limited to isotropic cubic systems and did not involve image simulations. Using state-of-the-art imaging simulation techniques developed at Brookhaven National Laboratory, we quantitatively describe the strain field associated with columnar defects in an anisotropic crystal. Although the method was exemplified with YBa$_2$Cu$_3$O$_{7-\delta}$ superconductors, with an appropriate modification it can be applied to columnar defects in any other material.

13.5.1 The Displacement Fields of Columnar Defects

For simplicity, we treat the superconducting cuprates as being crystals with tetragonal symmetry. Fig.13.19 shows the coordinate system used for the analysis. We assume that the surface of the specimen is unclamped so that there is no net force perpendicular to the surface.

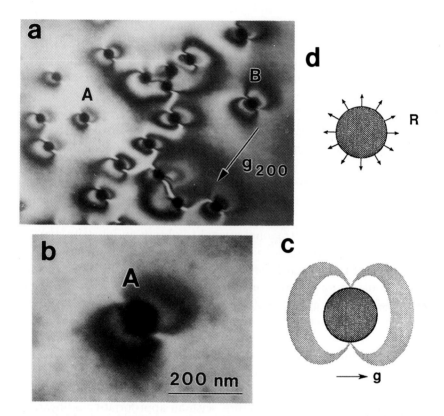

Figure 13.18 (a) Strain contrast surrounding the columnar defects imaged with g_{200}. (b) The enlarged strain contrast of the defect marked as A in (a). Note that the contrast disappears in the direction perpendicular to **g**, as sketched in (c). The "interstitial" type of radial displacement field associated with the amorphous column (d).

13.5. Strain Field and Strain Contrast of Columnar Defects

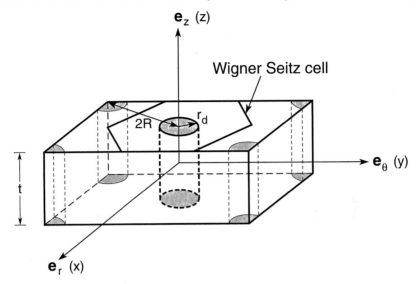

Figure 13.19 Schematic drawing of the columnar defects in the foil with the coordinate system used. $\mathbf{e}_z$ is parallel to the c-axis, and $x = r\cos\theta$ and $y = r\sin\theta$. We assume the defects form a triangular lattice and R is half of their average separation. The "Wigner Seitz cell" surrounding a columnar defect takes the form of a hexagonal-cell wall, made up of the planes vertically bisecting the line connecting the nearest neighbors of the defect.

We assume that the dimensions of a column of amorphous material, if it were free of stress, differ from the dimensions of the originally crystalline material from which it was formed and that the difference is anisotropic. If we are not too near the free surface, the strain in the matrix can be approximated as a state of plane strain with a uniform expansion or contraction superimposed. By treating the materials as elastically isotropic in the plane perpendicular to the c-axis (ab plane of the cuprates), then for cylindrically-symmetric defects the displacement field, $\mathbf{u}$, can be expressed by [Love 1944]

$$\mathbf{u} \approx u_r\mathbf{e_r} + u_z\mathbf{e_z} \approx (ar + \frac{b}{r})\mathbf{e_r} + ez\mathbf{e_z}. \tag{13.16}$$

For $r \leq r_d$, $b = 0$; for $r > r_d$, $b \neq 0$ (r_d is the radius of the columnar defects). The $\mathbf{e_r}$ and $\mathbf{e_z}$ are unit vectors in the direction of r and z, and e is a uniform dilation or compression. The coefficients a and b

must be obtained from the conditions of displacement compatibility and mechanical equilibrium.

To derive these coefficients, we imagine that we remove a column of material of radius r_d, amorphize it and re-insert it into the hole in the matrix from which it came. For simplicity, we assume that amorphization of the stress-free column causes a uniform dilation or compression of the originally crystalline material. The radial component of the amorphization strain we denote by ϵ_{ab}, while the longitudinal component is ϵ_c, in the case of a columnar ion track parallel to the crystal c-axis. Stress is required to change the shape of the column so that it fits the hole [Khachaturyan 1983]. We apply this stress, put the defect back and glue it to the matrix again. Now, there is a surface traction on the interface between matrix and amorphous material, and when these forces are canceled, displacements **u** are introduced into both the matrix and inclusion. Using superscripts d and m to refer to the defect and matrix, respectively, the compatibility condition for radial displacements at the interface at r_d can be written as

$$r_d + \delta r_d^d + u_r^d = r_d + u_r^m, \tag{13.17}$$

where $\delta r_d^d = \epsilon_{ab} r_d$ (ϵ_{ab} is the fractional change of dimension in the ab plane due to the amorphization). The compatibility condition for longitudinal displacements is

$$z + \delta z^d + u_z^d = z + u_z^m. \tag{13.18}$$

For $\delta z^d = \epsilon_c z$ mechanical equilibrium with normal stress, σ_r, at the defect/matrix interface, we have $\sigma_r^d = \sigma_r^m$ ($\sigma_r = \sigma_i = \sum_j c_{ij}\epsilon_j$, c_{ij} are the elastic coefficients), i.e.,

$$(c_{11}^d + c_{12}^d)a^d + c_{13}^d e^d = (c_{11}^m + c_{12}^m)a^m - (c_{11}^m - c_{12}^m)\frac{b^m}{r_d^2} + c_{13}^m e^m. \tag{13.19}$$

[In equation (13.19), the strains are derived from the displacement field in equation (13.16).] The condition of no net force normal to the free surface yields:

$$(2c_{13}^d a^d + c_{33}^d e^d)r_d^2 + (2c_{13}^m a^m + c_{33}^m e^m)R^2 = 0, \tag{13.20}$$

where R is half of the average value of the separation of the columnar defects. Assuming that the sample is not clamped at the edges, so that

13.5. Strain Field and Strain Contrast of Columnar Defects

the radial component of stress at the boundary of the "Wigner Seitz cell" (Fig.13.19) surrounding a columnar defect is zero, and approximates the hexagonal Wigner Seitz cell as a cylinder leads to:

$$(c_{11}^m + c_{12}^m)a^m - (c_{11}^m - c_{12}^m)\frac{b^m}{R^2} + c_{13}^m e^m = 0. \quad (13.21)$$

In principle, from the above five equations (Eqs.(13.17)–(13.21)), we can obtain the five unknowns a^m, b^m, e^m, a^d, and e^d ($b^d = b|_{r<r_d} = 0$). Since we do not know the elastic coefficients for amorphous cuprates, we will make the approximation that $c_{ij}^d \approx c_{ij}^m$. For low defect densities we can further simplify by treating the amorphous columns as isolated defects in an infinite matrix, i.e., $R/r_d \to \infty$, $a^m = e^m = 0$, from Eqs.(13.16)-(13.19) we have

$$\mathbf{u} = u_r \mathbf{e_r} = \frac{(c_{11} + c_{12})\epsilon_{ab} + c_{13}\epsilon_c}{2c_{11}} \frac{r_d^2}{r} \mathbf{e_r} \equiv \epsilon_{eff} \frac{r_d^2}{r} \mathbf{e_r}, \quad \text{for } r \geq r_d, \quad (13.22)$$

where ϵ_{eff} is a parameter describing the strength of the elastic strain field. Both the magnitude and sign of ϵ_{eff} can be determined through the image simulation of the strain contrast of the defect and by comparing with the result obtained using transmission electron microscopy. When $r < r_d$, from Eqn.(13.18) we can derive $e^d - \epsilon_c \neq 0$, and we obtain

$$\mathbf{u} = u_r \mathbf{e_r} + u_z \mathbf{e_z} = \epsilon_{eff} r \mathbf{e_r} - \epsilon_c z \mathbf{e_z} \qquad r < r_d. \quad (13.23)$$

In our study, the strain contrast of the columnar defect is the projected intensity along the c-axis so that the incident electrons used in imaging are not affected by the lattice displacement along the beam direction, and thus, we only need to consider the radial displacement u_r, i.e., $u_r = \epsilon_{eff} r$ $(r < r_d)$.

13.5.2 Intensity of the Diffraction Contrast of the Columnar Defects

The basis for computing the image profiles is the dynamical theory of diffraction contrast developed by [Howie and Whelan 1962] using the column approximation. We can rewrite the two-beam dynamical equations as:

$$\frac{dT}{dz} = \frac{\pi i}{\xi_0}T + \frac{\pi i}{\xi_g}S\exp(2\pi i s_g z + 2\pi i \mathbf{g}\cdot\mathbf{u})$$
$$\frac{dS}{dz} = \frac{\pi i}{\xi_0}S + \frac{\pi i}{\xi_g}T\exp(-2\pi i s_g z - 2\pi i \mathbf{g}\cdot\mathbf{u}), \quad (13.24)$$

where T and S are the amplitude of the transmitted and diffracted beam as a function of depth z in the crystal (z is parallel to the incident beam, see Fig.13.19). ξ_g is the extinction distance of the corresponding reflection $\mathbf{g}$ used to form the image. The $\mathbf{u}$ is the lattice displacement and s_g is the deviation from the Bragg position and ξ_0 is the mean refractive index of the crystal. Using complex crystal potentials, including a mean absorption coefficient ξ_0' and anomalous absorption coefficient ξ_g', we can substitute $(1/\xi_0 + i/\xi_0')$ for $1/\xi_0$ and $(1/\xi_g + i/\xi_g')$ for $1/\xi_g$, in which case Eqn.(13.24) becomes

$$\frac{dT}{dZ} = -\left(\frac{\xi_g}{\xi_0'}\right)T + \left(i - \frac{\xi_g}{\xi_g'}\right)S,$$
$$\frac{dS}{dZ} = \left(i - \frac{\xi_g}{\xi_g'}\right)T + \left[-\frac{\xi_g}{\xi_0'} + 2iS\xi_g + 2\pi i\frac{d}{dZ}(\mathbf{g}\cdot\mathbf{u})\right]S, (13.25)$$

where $Z = z\pi/\xi_g$. Using Eqn.(13.22) for the displacement field of the columnar defects, we have

$$\frac{d}{dZ}(\mathbf{g}\cdot\mathbf{u}) = \frac{-|\mathbf{g}|\epsilon_{eff}r_d^2 x \cdot 2(\frac{\xi_g}{\pi}Z)}{(x^2 + y^2 + \frac{\xi_g^2}{\pi^2}Z^2)^2}, \quad (13.26)$$

where $x^2 + y^2 = r^2$ and $x = r\cos\theta$; $\mathbf{g}$ is in the direction parallel to the x. When absorption is neglected, i.e., $\xi_g/\xi_0' = \xi_g/\xi_g' = 0$ and $s_g\xi_g \gg 1$, T can be considered as a constant. This is the kinematical approximation that is valid only when the transmitted intensity is very small in comparison with the direct intensity. Hence, contrast effects relying on changes in the diffracted intensity also will be small under kinematical conditions.

The intensities of the transmitted and diffracted beam, $|T|^2$ and $|S|^2$, respectively, can be evaluated as a function of the x and y position of the atomic columns after integrating equation (13.25) down columns of the crystals. The value of ϵ_{eff} for a crystal thus can be obtained by matching the experimental image with the simulated images for a particular imaging condition.

13.5.3 Simulation of the Strain Contrast of the Columnar Defects

An example of the strain contrast of columnar defects imaged with $\mathbf{g}_{200}$ is shown in Fig.13.18; images of such strain contrast with different reflections can be found in paper by Zhu et al. [Zhu et al. 1993d]. Because we considered the anisotropy of the defects with a tetragonal symmetry, here we analyzed only the strain contrast from the amorphous columns aligned along the c-axis. To image the strain contrast with the columnar defects edge-on, we mainly used reflections of $\mathbf{g}_{110}$ and $\mathbf{g}_{200}$ types ($\mathbf{g}_{100}$ are weak reflections for $YBa_2Cu_3O_{7-\delta}$). The line of no-contrast was always found to be perpendicular to the reflections, obeying the $\mathbf{g} \cdot \mathbf{u} = 0$ extinction criterion. This is because when a reflection $\mathbf{g}$ is used to form an image, the electron cannot detect displacements lying in the Bragg plane, and therefore the image is characterized by a line of no contrast normal to $\mathbf{g}$.

To simulate an image, parameters associated with the imaging condition were obtained as follows. The extinction distance ξ_g was first calculated from the structure factor, corrected for relativistic effects for $YBa_2Cu_3O_{7-\delta}$. For an experiment where s_g is not zero, the extinction distance was then corrected by the factor $\xi_g/(1+\xi_g^2 s_g^2)^{1/2}$, and the value of the effective extinction distance was confirmed by examining the period of the thickness contour at the specimen edge. The value of s_g was determined by measuring the distance of the corresponding Kikuchi line from the Bragg spot [Zhu et al. 1993f]. The radius of the amorphous column, r_d, was measured by high-resolution electron microscopy. The total thickness of the area was determined by the convergent beam electron diffraction technique.

Figure 13.20 shows the calculated images for a columnar defect induced by Au ions ($r_d = 50\text{Å}$). To focus on the lobe contrast outside the defect, we set the intensity inside the defect to be a constant. Figure 13.20(a)-(e) are the strain contrasts calculated with different values of the strain parameter ϵ_{eff} with $\xi_g/\xi_0' = \xi_g/\xi_g' = 0$ (a kinematic situation neglecting absorption), using the parameters described in the figure caption; these figures show the projected intensity in the ab plane of the crystal, and the positive x-axis is parallel to $\mathbf{g}_{200}$. The images have a two-fold symmetry in both x and y directions. The intensity of the strain contrast increases with the increase in the value of $|\epsilon_{eff}|$. When $|\epsilon_{eff}|$ is small ($|\epsilon_{eff}| < 0.01$), the intensity monotonically decreases from the periphery of the defect [Fig.13.20(a) and (b)], but when $|\epsilon_{eff}|$ is greater than 0.015, oscillations in the intensity appear in the direction perpendicular to the line of no-contrast. By comparing the shape and the period of the oscillation of the lobe contrast with the size of the amorphous column between the calculated contrast and the experimental one, we found that $|\epsilon_{eff}| = 0.020$ [Fig.13.20(c)] gave the best match for YBa$_2$Cu$_3$O$_{7-\delta}$ ($\delta \approx 0$).

The sign of the strain can be determined from the asymmetry of dark-field images of the defects by taking the absorption into account. When the effect of absorption is considered, it is found that the strain contrast of the bright-field and dark-field are not complementary, and the pair of lobe images is no longer symmetric with respect to the no-contrast line. In our case, the value of ϵ_{eff} was determined to be $+0.02$ with the absorption parameter $\xi_g/\xi_0' = \xi_g/\xi_g' = 0.01$ [Fig.13.18(f)]. On positive prints of the dark field micrograph, the contrast shows wide images which have their dark side on the side of the positive $\mathbf{g}$, suggesting that the atomic volume of the columnar defect is greater than that of the matrix, a finding consistent with the analysis from spherical inclusions [Ashby and Brown 1963]. The condition $\epsilon_{eff} > 0$ suggests that the defect is an "interstitial" type [Fig.13.18(d)] that causes matrix atoms to be displaced radially outward. However, the "interstitial" type of strain field in our samples irradiated with 300MeV Au-ions differs from the "vacancy" type of strain field observed by Frischherz et al. [Frischherz et al. 1993] in YBa$_2$Cu$_3$O$_{7-\delta}$ irradiated with 50KeV Kr ions. Such a difference in the characteristics of the defect may be attributed to the large difference in energy of the incident ions. In the former case, the interaction between the high-energy ions and the target crystal is mainly an electronic excitation, and the formation of the ion-track involves a local melting and recrystallization process [Zhu et al. 1993a]; for the latter, the radiation damage may be mostly elastic, involving only the knock-on mechanism.

13.5. Strain Field and Strain Contrast of Columnar Defects

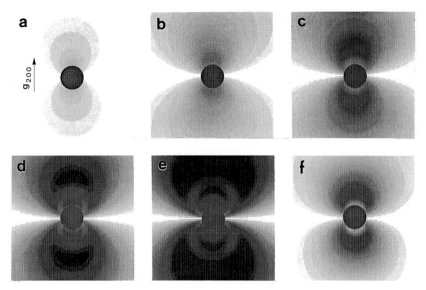

Figure 13.20 The simulated strain contrast of a columnar defect induced by irradiation with Au ions. $\mathbf{g}_{200}$, parallel to the positive x direction, $|\mathbf{g}| = 0.52 \text{Å}^{-1}$, $r_d = 50\text{Å}$, $\xi_g = 420\text{Å}$, and $s_g = 0.0031 \text{Å}^{-1}$ were used for the calculation. (a)-(e) Neglecting the absorption. $\xi_g/\xi_0' = \xi_g/\xi_g' = 0$. (a) $\epsilon_{eff} = 0.005$; (b) $\epsilon_{eff} = 0.01$; (c) $\epsilon_{eff} = 0.02$; (d) $\epsilon_{eff} = 0.03$; (e) $\epsilon_{eff} = 0.05$. (c) shows the best match with the experimental image. (f) Considering the absorption ($\xi_g/\xi_0' = \xi_g/\xi_g' = 0.01$). Other parameters used are the same as (c) ($\epsilon_{eff} = 0.02$). The contrast is asymmetric in the direction perpendicular to the line of no-contrast, while the contrast of (c) is symmetric.

Image matching can be made more quantitative by measuring the image widths (defined by a fractional intensity variation with respect to background, which varies with s_g) and shapes on a line parallel to $\mathbf{g}$ passing through the projected center of the defect [Ashby and Brown 1963, Chen and Carpenter 1991]. Fig.13.21 is a plot of the calculated intensity profiles and intensity contours of the columnar defect shown in Fig.13.20(f). A comparison between the digitized experimental images and the calculated intensity maps allows us to define, point by point, the local displacement and strain field surrounding the columnar defects. Such quantitative analysis was applied to different ion tracks

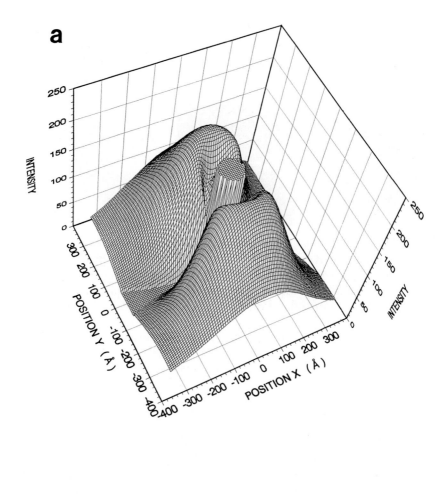

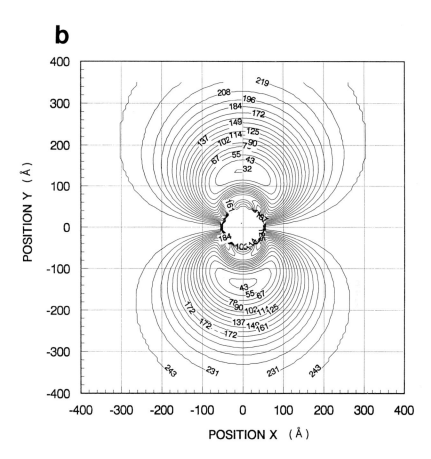

Figure 13.21 (a) The intensity profile of the strain contrast in Fig.13.20(f). (b) The intensity contours of the same strain contrast.

with various magnitudes of strain induced by radiation with Au, Ag, and Cu ions to derive the criterion for visibility of the strain contrast in superconducting cuprates. We found that to observe the strain contrast of a columnar defect, its 15% image width (15% of its maximum value) must be larger than the diameter of the amorphous column because, for faint images, the strongest contrast deviation from background always occurred near the periphery of the defect. In general, the strain contrast induced with Au ions was readily visible, while with Ag ions it was less clear. We could not see the strain contrast due to Cu ions. On the other hand, as we reported in [Zhu et al. 1993a], oxygen deficiency and imperfections in $YBa_2Cu_3O_{7-\delta}$ ($0 < \delta < 0.7$) crystals also promote the formation of the columnar defects after radiation with same types of ions; consequently, this yields larger values of ϵ_{eff}. For columnar defects with a large strain parameter ($|\epsilon_{eff}| > 0.01$), the contrasts can be imaged most clearly by using reflections with short extinction length (or low-order reflections), while defects with small strain parameters ($\epsilon_{eff} < 0.001$) can be imaged only by using long extinction distances (or high-order reflections) with a large $|s_g|$.

13.5.4 The Radial Displacement and Strain of the Columnar Defects

Figure 13.22 is a plot of the amplitude of the displacement and strain surrounding columnar defects versus the distance from the center of a typical ion track for $YBa_2Cu_3O_{7-\delta}$ ($\delta \approx 0$) irradiated with Au ions. In the crystal matrix outside the defect, the lattice displacement decreases with increasing distance as r^{-1}, while the absolute value of the lattice strain decreases as r^{-2}. We can define three different regions associated with the defect with radius A, B and C from its center (as marked in Fig.13.22). Radius A specifies an amorphous region, averaging about 50Å, as determined directly from high-resolution images. Radius B, covering a distance about $100 - 150$Å (with the corresponding strain $0.50\% - 0.22\%$, and displacement 0.50Å-0.33Å) away from the center of the defect, designates a region with severe lattice distortion, which also coincides with a hole-depletion zone. (The hole concentration was measured from the pre-peak of the oxygen K-edge using a 20Å probe across the defects with high-resolution electron energy-loss spectroscopy [Zhu

13.5. Strain Field and Strain Contrast of Columnar Defects

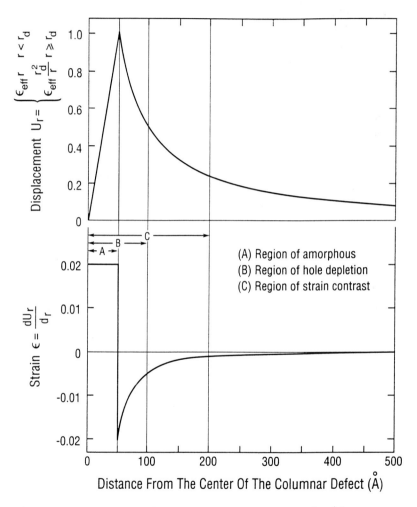

Figure 13.22 The magnitude of displacement, u_r, and strain, du_r/dr, as a function of the distance from the center of the columnar defect.

et al. 1993h]). Quantitative analysis of the integrated intensity of the oxygen pre-peak [Zhu *et al.* 1994c] shows that the hole concentration in the crystal matrix 100Å away from the defect is the same as that found in oxygen deficient samples of $YBa_2Cu_3O_{6.3}$. Radius C, about 200Å (0.13% of strain and 0.25Å of displacement), defines a region with visible strain contrast, as determined by contrast analysis and image simulation. The significance of our study in understanding the superconductivity of the high-temperature superconducting cuprates is that, for the first time, we have quantitatively and experimentally linked the hole deficiency to the lattice displacement/strain in these materials. The quantitative analyses enable us to demonstrate that the reduction of the hole density near the columnar defect enlarges the weak- or non-superconducting area to almost twice that of the amorphous region itself. Thus, our studies reconcile the previously-observed weak- or non-superconducting volume fraction in samples before and after heavy-ion irradiation. Without taking the regions of the distorted crystal lattice into account after the samples were irradiated with Au-and Ag-ions, the increase in resistivity and suppression of the critical temperature was not easily explained.

In summary, we present a formalism to describe the strain field and strain contrast associated with the columnar defects observed in heavy-ion-irradiated superconducting cuprates. Using the theories of continuum elasticity and dynamic diffraction, the lattice displacement and strain field surrounding the defects were quantitatively evaluated for an isolated column along the c-axis. Knowledge of the critical strength of the local displacement, which causes hole depletion and strain contrast in the high-temperature superconductors, helps us to understand both the TEM images and the change in the superconducting properties of these materials.

13.6. Planar Defects Induced by Heavy-ion Irradiation

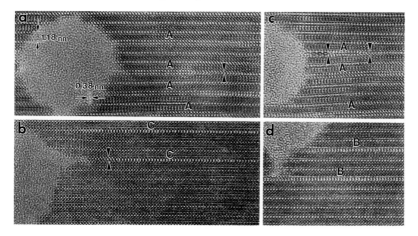

Figure 13.23 HREM images of the planar faults induced by gold-ion irradiation viewed along the [100] projection. A, B, and C denote various image features of the defects. The intercalation and the termination of the faults are marked by arrows.

13.6 Planar Defects Induced by Heavy-ion Irradiation

13.6.1 Morphology of the Planar Defects

Planar defects associated with the amorphous columns are often observed in heavy-ion irradiated $YBa_2Cu_3O_{7-\delta}$ (see Fig.13.13). These faults were not pre-existing defects, but were generated during the formation of the amorphous zone. The faults are best seen when the electron beam is parallel to the [100] or [010] axis. The extension of the faults usually ranged from 3 to 5 times the radius of the amorphous columns. The probability of the formation of such planar defects varied with the condition of the samples and the nature of the irradiating ion; they were frequently observed in $YBa_2Cu_3O_{7-\delta}$ ($0.05 < \delta < 0.6$) when it was irradiated with Au^{+24} (300MeV) and Ag^{+21} (276MeV) and the ion-beam was parallel to the a-b plane.

HREM revealed details of the faults. Examples of the images are shown in Fig.13.23 (induced by Au^{+24} irradiation, viewing along the [100] axis), and in Fig.13.24 (induced by Ag^{+21} irradiation, viewed along the [110] axis). Various image features are associated with the planar

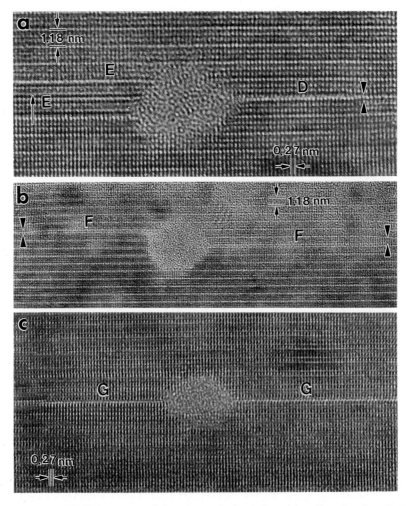

Figure 13.24 HREM images of the planar faults induced by silver-ion irradiation viewed along the [110] projection. E, D, F, and G denote various image features of the defects. The intercalation and the termination of the faults are marked by arrows.

faults, such as rows of double-dots, dumbbells, and elongated-dot contrast, as marked through A to G. Since different images were often observed in adjacent areas, and thus, were recorded under the same imaging conditions, i.e., with little difference in the values of the objective lens defocus and of specimen thickness, they probably represent different structures. However, by examining hundreds of lattice images of such defects [Zhu et al. 1993i], we found that all of them are, without exception, located in the a-b plane, and more than 80% involve intercalation of lattice planes in the direction parallel to the a-b plane, suggesting that they are extrinsic stacking faults. The intercalation can easily be confirmed by observing the termination of the faults, or the lattice bending at the end of the fault, as indicated by the arrow heads in Fig.13.23 and Fig.13.24.

13.6.2 Microchemical Analysis of the Planar Defects

With advances in chemical analysis techniques in TEM, it is possible to measure cation compositions of an area as small as ~ 20Å in diameter, such as the regions at the stacking faults. With the planar faults edge-on, we measured cation compositions at 60 locations including positions at, near, and far away from the faults. To observe the relative change in local cation compositions, we normalized the average of the peak intensity of the matrix area to Y=7.7%, Ba=15.4%, and Cu=23.1% by using the K-ratio, the ratio of characteristic intensities measured on the specimen and the standard [Cliff and Lorimer 1975]. Away from the fault, we observed a nominal composition of Y:Ba:Cu=1:2:3 [Fig.13.25(a)]. Near the fault (< 100Å away), we observed only a small variation ($< 2\%$) in Cu concentration relative to the concentration of Ba and Y, which is within the uncertainty of the measurements as seen in the matrix. In contrast, at the faults, more than 80% had Cu enrichment, and the rest were either Cu deficient or showed no change in Cu concentration. Measurements from 8 different locations (the data represents the average of 6 measurements for each location) in the matrix and at the faults are shown in Fig.13.25(a) and Fig.13.25(b), respectively. Fig.13.26(b) shows an example of the x-ray spectrum from a copper-rich planar fault, obtained with a ~ 20Å probe [corresponding

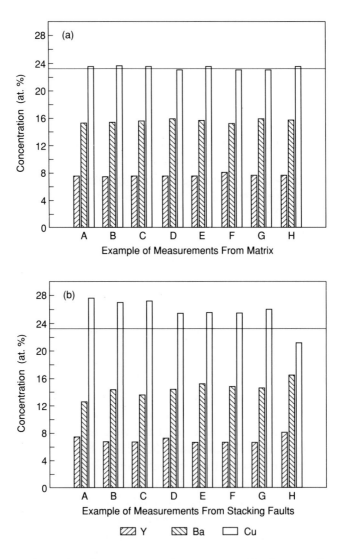

Figure 13.25 Examples of cation composition measured in the crystal matrix (a), and from planar defects (b). The horizontal line represents the stoichiometric Cu concentration (23.1% at.). Most planar defects are Cu-rich. A-H represent the measurements from 8 different locations.

13.6. Planar Defects Induced by Heavy-ion Irradiation

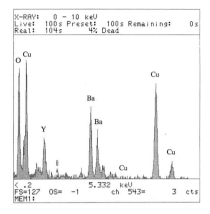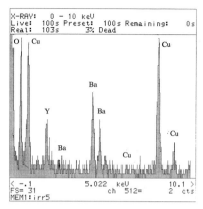

Figure 13.26 X-ray spectra from the planar defect with a ~ 20Å probe (a), and from the adjacent matrix using a ~ 40Å probe (b).

to the location A in Fig.13.25(a)]; one from the adjacent matrix, obtained with a ~ 40Å probe to increase the signal-to-noise ratio is shown in Fig.13.26(a) [corresponding to the location B in Fig.13.25(a)]. The maximum excess Cu concentration at the faults, compared to that of the matrix, was less than 6%, the drastically increased Cu concentration at the fault as reported previously by Dravid et al. [Dravid et al. 1992] was not observed in our study. Overall, the nano-probe EDS analysis and lattice images indicate that more than 80% of the faults are Cu-rich extrinsic faults, but also suggest the presence of intrinsic stacking faults, and faults due simply to the shear displacement from a perfect crystal.

13.6.3 Structure of the Planar Defects

To identify the detailed structures of the planar defects induced by heavy-ion radiation, we compared the observed HREM images with calculated ones. This comparison was necessary because HREM images with identical features observed in two-dimensional projections can result from different three-dimensional defects, while different images can result from the same structure with different orientations, or from structures with the same orientation but different defocus values and different specimen thickness. Thus, we first constructed various structural models for possible planar faults in $YBa_2Cu_3O_{7-\delta}$, some of which are depicted in Fig.13.27. Since the fault appears to be associated with the CuO planes, based on our chemical analysis, we rewrote the chemical formula for the oxide as a series of homologous structures represented by $YBa_2Cu_2O_6(CuO)_n$, where n is the number of CuO planes. For $n=0, 1, 2$, we have $YBa_2Cu_2O_6$ (122 phase), $YBa_2Cu_3O_7$ (123 phase), and $YBa_2Cu_4O_8$ (124 phase), respectively. Model A is a 123 phase with a displacement vector along the b-axis. Model B and C are intrinsic stacking faults. To avoid an unphysically short bonding length between two Ba atoms, a lattice displacement was added along [010] (model B), and along [110] (model C). Models D-H are extrinsic stacking faults with one extra Cu-O plane (124, Models D-F), and two extra Cu-O planes (125, Model G and H) combined with a different displacement component in the a-b plane. Among these models, A, D, and H may well have a high formation energy due to short interatomic spacings between oxygen and oxygen, or copper and copper atoms near the fault. However, this may be accommodated by local lattice relaxations, or by the depletion of the oxygen atoms.

The process of HREM image simulation and matching with the defects was started by matching the calculated image to the experimental image of the undisturbed area adjacent to the planar defects. This determines the appropriate values of defocus and the specimen thickness, and confirms the other parameters to be used to simulate the faulted area. The values which we used are spherical aberration Cs=0.9mm; defocus spread =15nm; beam divergence $\theta = 0.6$mrad; objective aperture size $d = 0.6$nm^{-1}; and, Debye-Waller factors taken from neutron diffraction studies [Capponi et al. 1987]. Fig.13.28(a) and (b) (corresponding to defect A and E in Fig.13.23 and Fig.13.24, respectively) show a high degree of matching between the calculated and experimental images of the faults. The results of image simulation confirmed that the images were taken at nearly optimum defocus ($f = -500$Å). Under these conditions (specimen thickness: 20Å), Ba and Y atoms appear as dark dots, while the Cu atoms are white dots. In Fig.13.28(a), viewing along the [100] direction, the stacking faults are shown as white dots marked by double arrow heads. The good match with the calculated image (inset of Fig.13.28(a)) suggests that the double dots correspond

13.6. Planar Defects Induced by Heavy-ion Irradiation

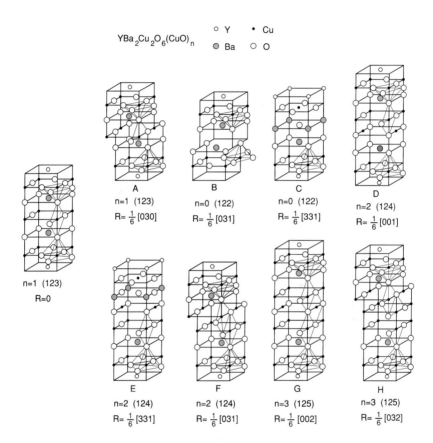

Figure 13.27 Various structural models used in simulating the images of planar defects. 123, 124, and 125 represent the local structure of $YBa_2Cu_3O_7$, $YBa_2Cu_4O_8$, and $YBa_2Cu_5O_9$.

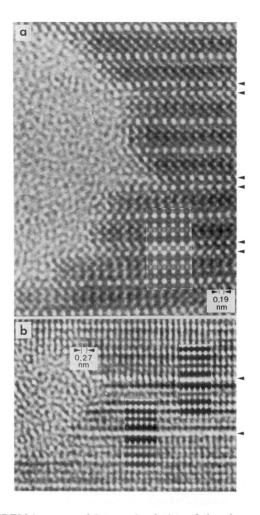

Figure 13.28 HREM images and image simulation of the planar defects. (a) 125 planar defects, marked as double-arrows, viewed along the [100] direction. The white dots represent Cu atoms. Inset shows the calculated image (Defocus: -500Å, specimen thickness: 20Å, model G, $\mathbf{R} = 1/6[002]$). (b) 124 planar defects, marked as single-arrow, viewed along the [110] direction. Inset shows the calculated image (Defocus: -500Å, specimen thickness: 20Å, model D, $\mathbf{R} = 1/6[001]$).

13.6. Planar Defects Induced by Heavy-ion Irradiation

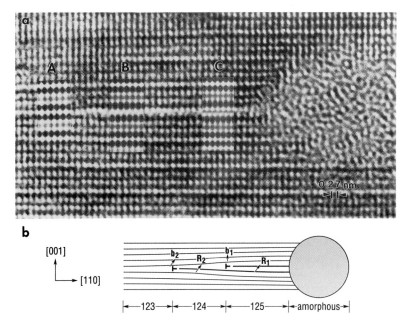

Figure 13.29 (a) A mixture of 124 and 125 planar defects viewed along the [110] direction. Inset A: 123 phase ($n = 1$, $R = 0$); inset B: 124 phase (model F, $\mathbf{R}_2 = 1/6[031]$); and inset C: 125 phase (model H, $\mathbf{R}_1 = 1/6[032]$). Defocus: -600Å, specimen thickness: 20Å. (b) Sketch of the observed stacking fault in (a). The stacking fault consists of two partial dislocations with a Burgers vector of $\mathbf{b}_1 = 1/6[001]$, and $\mathbf{b}_2 = 1/6[031]$, respectively.

to triple CuO planes forming a local 125 structure ($\mathbf{R} = 1/6[002]$, model G in Fig.13.27). A similar local 125 structure was also observed in YBa$_2$Cu$_4$O$_8$ [Ramesh et al. 1990], and in air-exposed YBa$_2$Cu$_3$O$_{7-\delta}$ [Zandbergen 1992]. The calculated image ([110] projection) of inset of Fig.13.28(b) shows a different planar fault with a local 124 structure ($\mathbf{R} = 1/6[001]$, model D of Fig.13.27), while Fig.13.29 (corresponding to defect D in Fig.13.24) is a planar fault that starts with a local 125 structure with a displacement $\mathbf{R}_1 = 1/6[032]$ (inset C, model H), then changes to a 124 structure (inset B, model F) with a fault displacement $\mathbf{R}_2 = 1/6[031]$, and finally terminates in the 123 matrix (inset A). Thus, the planar defect consists of two stacking faults bounded by two partial dislocations $\mathbf{b}_2 = 1/6[031]$ and $\mathbf{b}_1 = 1/6[001]$ at each end, respectively

[Fig.13.29(b)]. Since the motion of such dislocations cannot generate the observed fault, the formation of the fault must be non-conservative, i.e., due to a chemical change. The displacement vector of the fault suggests an edge-shear arrangement of the $Cu-O_5$ truncated octahedron. The observed 125 planar faults may correspond to locations A, B, and C of Fig.13.25(b), while the 124 faults correspond to that of locations D-G in our EDS data [Fig.13.25(b)]. Through image simulation, the structure of all the planar faults in Fig.13.23 and Fig.13.24 were determined: defects A and F are 125 structure; defects B, C, and E are 124; defect G is 122; and defect D is a mixture of 124 and 125. All the structures of the planar defects illustrated in Fig.13.27 were observed, although the majority was chemical faults comprising one or two extra CuO layers inserted between the double BaO layers of the original 123 unit cell.

13.7 Summary

The radiation damage processes accompanying the passage of an energetic heavy ion are likely to encompass energy deposition into the electron gas of the target, the transfer of the energy from the electron gas to heat the lattice ions, the transport of heat in the lattice, and the phase changes and defect formation which accompany the rapid heating and quenching of the lattice. In the context of our modified thermal-spike model, it is reasonable to assume that in the zone along the ion track almost the entire region of the thermal spike where the temperature rises to or above the melting point initially becomes molten. On the other hand, the molten zone does not all remain amorphous, but some epitaxial regrowth occurs during the cooling-down period, albeit with some lattice defects remaining. Thus, the diameter of the amorphous region is smaller than that of the original molten zone. Furthermore, the radial extension of the chemical faults are approximately equal to the size of the damaged area calculated without incorporating the concept of the regrowth, suggesting that the length of the faults is approximately equal to the molten region before regrowth take place. The formation of the chemical faults may start at the liquid and solid interfaces with off-stoichiometric Cu atoms and may involve nearest, or second-nearest neighbor diffusion. The observation that the major fraction of the faults

13.7. Summary

are extrinsic faults and are bounded by sessile dislocations provides supporting evidence for such a chemical process. The rapid lateral growth rate of a-b planes can account for all the faults being confined to the a-b plane. Also, it is likely that a minimum oxygen concentration also is required to form such planar defects because the intercalation takes place in the form of Cu-O chain plane and because we did not observe such defects when the oxygen sub-stoichiometry δ exceeded 0.6 in heavy-ion irradiated $YBa_2Cu_3O_{7-\delta}$. However, the existence of a large amount of local 124 and 125 structure associated with heavy-ion radiation points to the low formation energy of the extra Cu-O plane in a $YBa_2Cu_3O_{7-\delta}$ matrix. Similar chemical faults, extrinsic and intrinsic, were observed in quenched and sintered materials [Zhu et al. 1990f, Zandbergen et al. 1988].

The presence of a radial displacement field surrounding the amorphous columns, which was quantitatively characterized by image simulation of the lobe-contrast around the amorphous column using the two-beam dynamic diffraction theory, may also affect the formation of the planar defects. Such displacement, due to the volume expansion of the amorphous column in the crystal matrix could contribute to the formation of defects to release strain energy. We note that some extrinsic stacking faults have a radial-displacement-component perpendicular to the ion-trajectory, such as the faults shown in Fig.13.29. Very few stoichiometric planar faults, which were induced solely by displacement (model A in Fig.13.27), or by the motion of the partial dislocation which bounds the fault (disassociation of a perfect dislocation), were observed. The characteristics of the defects, determined as $\mathbf{R} = 1/6[030]$ and $\mathbf{b} = 1/2[100]$, agree with those observed in high-temperature (650-850°C) deformed $YBa_2Cu_3O_{7-\delta}$ samples [Yoshida et al. 1990]. However, the observation of the predominant extrinsic stacking faults in heavy-ion irradiated samples suggests less favorable conditions for forming displacement-induced faults, and strongly supports the likelihood of partial epitaxial regrowth of the molten zone after the passage of high-energy ions through the high T_c superconductors.

Chapter 14
Conclusions

In this book we have presented the results of extensive studies on the microstructure and defects of high-temperature superconductors using experimental techniques such as transmission electron microscopy as well as theoretical modeling methods. We also sought to link many structural properties of high-T_c materials to their superconducting properties. We have laid strong emphasis on the close coupling between experimental observations and computer simulations based on realistic phenomenological models. We believe such an approach combines the best features of the experimental techniques and theoretical modeling tools. In spite of the progress that has been made in formulating theories of the electronic structure of high-temperature superconductors, it would appear that any quantitative predictions based on the first-principle type calculations will remain unreliable for some considerable time in the future. The empirical models, as we discussed in this book, are quite powerful and can make sense of large amount of experimental data without using inordinately large amounts of computer time.

Because of their great potential for practical applications, high- temperature superconductors are the focus of great interest in the research community ever since their discovery. Unlike low-temperature superconductors, which are mostly intermetallic compounds, the high-T_c superconductors are ceramic materials. Therefore structure inhomogenieties such as grain boundaries, structural modulations and dislocations are facts of life in these materials. In addition, the superconducting coherence lengths of the high-T_c materials are quite short (typically in the order of tens of angstroms). The superconducting properties of these materials are very sensitive to the structural imperfections, even for defects in the atomic scale. On the other hand, the presence of strong flux pinning sites in the high-T_c material is essential to produce materials with high critical current density, which is prerequisite to any practical applications of this material. Therefore "designing" defects to act as strong pinning centers is another active area of research. Because of these, intensive research effort has been devoted to the study of microstructure and defects in the high temperature superconductors. Many aspects of the microstructure, such as twinning, grain boundary dislocations and charge modulations in ceramic materials have attracted the interest of materials scientist for many years. The enormous resources put into the high-T_c research have made possible the rapid progress in the last few years on these research topics. We have now a much better understanding of the microstructure and defects in ceramics materials. Many new experimental as well as theoretical techniques were developed that will benefit the whole materials research community for the years to come.

The research on the microstructure of high temperature superconductors is in rapid progress. It is impossible for us to cover all aspects of this exciting field. For example, high-T_c thin films, which are important for superconducting electronics applications, have quite different microstructural features, compared with the bulk materials as we discussed in this book, due to their interaction with substrates. We discussed only those aspects of the microstructure that we think are important for large scale applications. We also did not include the discussion on the recent progress in processing method, such as IBAD and RaBiT to make high-quality, long length YBCO thick films. All this topics are fascinating, however it is beyond the scopes of this book.

Looking forward, we still do not have a good understanding of the grain boundary structure and its relationship to superconducting properties of the high-T_c materials. The interaction between flux line lattice and various types of defects is of great interest lately because of its importance to high-current, large scale applications. We are still a long way from commercialize the great potential of the high-temperature superconductors, and understanding the microstructure and defects in this system is the essential step toward that goal.

Bibliography

Alexander, K.B., Kroeger, D.M., Bentley, J., and Brynestad, J., *Physica C* **180** (1991) 337.

Allen, M.P., and Tildesley, D.J., *Computer Simulation of Liquids*, Oxford University Press (1987).

Ambegaokar V., and Baratoff, A., *Phys. Rev. Lett.* **10** (1963) 486.

Amelinckx, S., and van Landuyt, J., in *Electron Microscopy in Mineralogy*, J.M. Christie, J.M. Colley, A.H. Hener, G. Thomas, and N.J. Tighe, eds., (1976) 68.

Amelinckx, S., and Dekeyser, W., *Solid State Phys.* **8** (1959) 325-499.

Amelinckx, S., and van Dyck, D., in *Electron Diffraction Techniques*, Cowley, J.M. ed., Oxford University Press, New York (1993).

Anderson, P.W., *Lectures at Ravello Spring School*, 1963.

Anstis, G.R., Lynch, D.F., Moodie, A.F., and O'Keefe, M.A., *Acta Cryst.* **A29** (1973) 138.

Ashby, M.F. and Brown, L.M., *Phil. Mag.*, **8** (1963) 1083.

Aukrust, T., Novotny, M.A., Rikvold P.A., and Landau, D.P., *Phys. Rev. B* **41** (1990) 8772.

Axe, J.D., Cox, D.E., Mohanty, K., Moudden, A.H., Moodenbaugh, A.R., Xu, Y., Thurston, T.R., *IBM J. Res. Dev.* **33** (1989a) 382.

Axe, J.D., Moudden, A.H., Hohlwein, D., Cox, D.E., Mohanty, K.M., Moodenbaugh, A.R., and Xu, Y., *Phys. Rev. Lett.* **62** (1989b) 2751.

Axe, J.D., *Proc. Lattice Effects in High-T_c Superconductors* Santa Fe, NM, Jan. 13-15, 1992, Y. Bar-Yam, T. Egami, J. Mustre-de Leon and A.R. Bishop Eds., World Scientific, Singapore (1992) p.517-530.

Babcock, S.E., Kelly, T.F., Lee, P.J., Seuntjens, J.M., Lavanier, L.A., and Larbalestier, D.C., *Physica C* **152** (1988) 25.

Babcock, S.E., and Balluffi, R.W., *Philo. Mag. A* **55** (1987) 643.

Babcock, S.E., and Larbalestier, D.C., *J. Mater. Res.* **5** (1990) 919.

Babcock, S.E., et al., *Nature* **347** (1990) 167-169; Larbalestier, D.C., Babcock, S.E., Cai, X.Y., Field, M.B., Gao, Y., Hienig, N.J., Kaiser, D.L., Merkle, K., Williams, L.K., and Zhang, N., *Physica C* **185-189** (1991) 315.

Baetzold, R.C., *Phys. Rev. B* **38** (11) (1988) 304.

Baetzold, R., *Phys. Rev. B* **42** (1990) 56.

Baetzold, R., unpublished.

Baetzold, R., private communication (1997).

Bakker, H., Welch D.O., and Lazarethe, O.W., *Solid State Comm.* **64** (1987) 237.

Balluffi, R.W., eds., *Grain Boundary Structure and Kinetics*, Metals Park, OH, American Society for Metals (1979).

Balluffi, R.W., and Bristowe, P.D., *Surf. Sci.* **144** (1984) 28.

Balluffi, R.W., Brokman, A., and King, A.H., *Acta Metall.* **30** (1982) 1453.

Banngärtel, G., and Bennemann, K.H., *Phys. Rev. B* **40** (1989) 6711.

Barry, J.C., *J. Electron Microscopy Tech.* **8** (1988) 325.

Batson, P.E., and Chisolm, M.F., *J. Electron Microsc. Technique* **8** (1988) 311.

Baumgart, P., Blumenröder, S., Erlr, A., Hillebrands, B., Splittgerber, P., Güntherodt, G., and Schmidt, H., *Physica C* **162-164** (1989) 1073.

Baumgärtel, G., and Bennemann, K.H., *Phys. Rev. B* **40** (1989) 6711.

Bednorz, J.G., and Müller, K.A., *Z. Phys.* **B64** (1986) 189.

Bednorz, J.G., Takashige, M., and Müller, K.A., *Euophys. Lett.* **3** (1987) 379.

Bell, J.M., *Phys. Rev. B* **37** (1988) 541.

Beyers, R., Ahn, B.T., Gorman, G., Lee, V.Y., Parkin, S.S.P., Ramirez, M.L., Roche, K.P., Vazquez, J.E., Gür, T.M., and Huggins, R.A., *Nature* **340** (1989) 619.

Beyers, R., Lim, G., Engler, E.M., Savoy, R., Shaw, T.M., Dinger, T.R., Gallagher, W.J., and Sandstrom, R.L., *Appl. Phys. Lett.* **50** (1987a) 1918.

Beyers R., et al., *Appl. Phys. Lett.* **51** (1987b) 614.

Bian, W., Zhu, Y., Wang, Y.L., and Suenaga, M., *Physica C* **248** (1995) 119.

Bianconi, A., Li, C., Longa, S.D., and Pompa, M., *Phys. Rev. B* **45** (1992) 4989.
Billinge, S.J.L., Kwei, G.H., Lawson, A.C., Thompson, J.D., and Takagi, H., *Phys. Rev. Lett.* **71** (1993) 1903.
Binder, K., (ed.), *Monte Carlo Methods in Statistical Physics*, Topics in Current Physics, Vol.**7** (Springer-Verlag, 1979).
Binder, K., (ed.), *Applications of the Monte Carlo Methods in Statistical Physics*, Topics in Current Physics, Vol.**36** (Springer-Verlag, 1984).
Binder, K., and Landau, D.P., *Phys. Rev. B* **21** (1980) 1941.
Boekholt, M., Harzer, J.V., Hillebrands, B., and Guntherodt, G., *Physica C* **179** (1991) 101-106.
Bollmann, W., *Crystal Defects and Crystalline Interfaces*, Springer, Berlin (1970).
Bollmann, W., *Crystal Lattice, Interfaces and Matrices*, published by author, (1982).
Bordet, P., Capponi, J.J., Chaillout, C., Chenavas, J., Hewat, A.W., Hewat, E.A., Hodeau, J.L., Marezio, M., Tholence, J.L., and Tranqui, D., *Physica C* **153-155** (1988a) 623.
Bordet, P., Chaillout, C., Chenavas, J., Hodeau, J.L., Marezio, M., Karpinski, J., and Kaldis, E., *Nature* **334** (1988b) 596.
Bordet, P., Hodeau, J.L., Strobel, P., Marezio, M., and Santoro, A., *Solid State Commun.* **66** (1988c) 435.
Bourgault, D., Groult, D., Bouffard, S., Provost, J., Studer, F., Nguyen, N., Raveau, B., and Toulemonde, M., *Phys. Rev. B* **39** (1989) 6549.
Brandon, D.G., *Acta Metall.* **14** (1966) 1479.
Brokman, A., and Balluffi, R.W., *Acta Metall.* **29** (1981) 1703.
Brown, I.D., *J. Solid State Chem.* **90** (1991) 155.
Bruggeman, G.A., Bishop, B.H., and Hartt, W.H., *The Nature and Behavior of Grain Boundaries*, edited by H.Hu, Plenum, New York (1972) pp.83-122.
Budhani, R.C., and Suenaga, M., *Solid State Commun.* **84** (1992) 8312.
Budhani, R.C., Suenaga, M., and Liou, S.H., *Phys. Rev. Lett.* **69** (1992a) 3816.
Budhani, R.C., Zhu, Y., and Suenaga, M., *Appl. Phys. Lett.* **61** (1992b) 985.
Budin, H., Eibl, O., Pongratz, P., and Skalicky, P., *Physica C* **207**

(1993) 208.
Bulaevskii, L.N., Clem, J.R., Glazman, L.I., and Malozemoff, A.P., *Phys. Rev. B* **45** (1992) 2545; Bulaevskii, L.N., Daemen, L.I., Maley, M.P., and Coulter, J.Y., *Phys. Rev. B* **46** (1993) 13798.
Buseck, P., Cowley, J., and Eyring, L., eds., *High-Resolution Transmission Electron Microscopy and Associated Techniques*, Oxford University Press, New York (1988).
Cahn, R.W., *Acta. Met.* **1** (1953) 49.
Cahn, R.W., *Advances in Physics* **3** (1954) 363.
Cai, Y., Chung, J.S., Thorpe, M.F., and Mahanti, S.D., *Phys. Rev. B* **42** (1990) 8827.
Cai, Z.-X., (1997), unpublished.
Cai, Z.-X., and Mahanti, S.D., *Solid State Commun.* **67** (1988) 287.
Cai, Z.-X., and Mahanti, S.D., *Phys. Rev. B* **40** (1989) 6558.
Cai, Z.-X., and Mahanti, S.D., *Computer Simulation Studies in Condensed Matter Physics* **III**, eds. D.P. Landau, K.K. Mon, and H.-B. Shüttler, Springer-Verlag,Berlin Heidelberg (1990) 210.
Cai, Z.-X., and Welch, D.O., *Phys. Rev. B* **45** (1992) 2385.
Cai, Z.-X., and Welch, D.O., *Computer Simulation Studies in Condensed Matter Physics* **IV**, eds. D.P. Landau, K.K. Mon, and H.-B. Shüttler, Spring-Verlag, Berlin, Heidelberg (1993) 743.
Cai, Z.-X., and Welch, D.O., *Physica C* **231** (1994a) 383.
Cai, Z.-X., and Welch, D.O., *Physica C* **234** (1994b) 373.
Cai, Z.-X., and Welch, D.O., *Structure and Properties of Interfaces in Ceramics*, eds. D.A. Bonnel, U. Chowdhry and M. Rühle, MRS Symposium Proceedings **357** (1995a) 453.
Cai, Z.-X., and Welch, D.O., *Bulletin of the American Physical Society* **40** (1995b) 743.
Cai, Z.-X., and Zhu, Y., *Mater. Sci. Eng. A* **238** (1997) 210.
Cai, Z.-X., Zhu, Y., and Welch, D.O., *Phys. Rev. B* **46** (1992a) 7841.
Cai, Z.-X., Zhu, Y., and Welch, D.O., *Philo. Mag. A* **65** (1992b) 931.
Cai, Z.-X., Zhu, Y., and Welch, D.O., *Materials Theory and Modelling*, J. Broughton, P. Bristowe and J. Newsam, eds. MRS Symposium Proceedings Vol.**291**, MRS (1993a) p.217.
Cai, Z.-X., Zhu, Y., and Welch, D.O., *Proceedings of the International Conference on Martensitic Transformations (1992)*, Monterey Institute for Advanced Studies (1993b) p.737.

Cai, Z.-X., Zhu, Y., and Welch, D.O., *Phys. Rev. B* **52** (1995) 13035.
Campo, R.A., Evetts, J.E., Glowacki, B.A., Newcomb, S.B., Somekh, R.E., and Stobbs, W.J., *Nature* (London) **329** (1987) 229.
Capponi, J.J., Chaillout C., Hewat, A.W., Lejay, P., Marezio, M., Nguyen, N., Raveau, B., Soubeyroux, J.L., Tholence, J.L., and Tournier, R., *Europhys. Lett.* **3** (1987) 1301.
Carter, C.B., *Acta Metall.* **36** (1988) 2753.
Cava, R.J., Batlogg, B., van Dover, R.B., Murphy, D.W., Sunshine, S., Siegrist, T., Remeika, J.P., Rietman, E.A., Zhurak, S., and Espinosa, G.P., *Phys. Rev. Lett* **58** (1987a) 1676.
Cava, R.J., Batlogg, B., Chen, C.H., Rietman, E.A., Zahurak, S.M., and Werder, D., *Phys. Rev. B* **36** (1987b) 5719.
Cava, R.J., Hewat, A.W., Hewat, E.A., Batlogg, B., Marezio, M., Rabe, K.M., Krajewski, J.J., Peck, W.F., and Rupp, Jr., L.W., *Physica C* **165** (1990) 419.
Chaillout, C., Alario-Franco, M.A., Capponi, J.J., Chenavas, J., Strobel, P., and Marezio, M., *Solid State Commun.* **65** (1988) 283.
Chaillout, C., Alario-Franco, M.A., Capponi, J.J., Chenavas, J., Hodeau, J.L., and Marezio, M., *Phys. Rev. B* bf 36 (1987) 7118.
Chadderton, L.T., *Nucl. Tracks. Radiat. Meas.* **15**, (1988) 11.
Chandrasekar, N., Suenaga, M., and Welch, D.O. unpublished.
Chaudhari, P., *Jpn. J. Appl. Phys.* **26-3** (1987) 2023.
Chaudhari, P., Koch, R.H., Laibowitz, R.B., McGuire, T.R., and Gambino, R.J., *Phys. Rev. Lett.* **58** (1987a) 2684.
Chaudhari, P., LeGoues, F., and Segumuller, A., *Science* **238** (1987b) 342.
Chaudhari, P., Mannhart, J., Dimos, D., Tsuei, C.C., Chi, J., Oprysko, M.M., and Scheuermann, M., *Phys. Rev. Lett.* **60** (1988) 1653.
Chen, F.-R., and King, A.H., *J. Electron Microsc. Technol.* **6** (1987) 55.
Chen, F.-R., and King, A.H., *Acta Crystallogr. B* **43** (1987) 416; *Philo. Mag. A* **57** (1988) 431.
Chen, C.H., Werder, D.J., Schneemeyer, L.F., Gallagher, P.K., and Wazczak, J.V., *Phys. Rev. B* **38** (1988) 2888.
Chen, Y.L., and Carpenter, R.W., *Proc. of the 49th Annual Meeting of the Electron Microscopy Society of America*, San Jose, 1991, Bailey, G.W. ed., p.872.

Chen, C.H., Cheong, S.W., Werder, D.J., Copper, A.S., and Rupp Jr., L.W., *Physica C* **175** (1991) 301.
Chen, C.H., Cheong, S.W., Werder, D.J., and Takagi, H., *Physica C* **206** (1993) 183.
Chen, C.T., Tjeng, L.H., Kwo, J., Kao, H.L., Rudolf, P., Sette, F., and Fleming, R.M., *Phys. Rev. Lett* **68** (1992) 2543.
Cheng, S.C., Sheinin, S.S., Jung, T., Yu, M.K., and Frank, J.P., *Physica C* **184** (1991) 385.
Chisholm, M.F., Pennycock, S.J., *Nature* **351** (1991) 47.
Chisholm, M.F., and Smith, D.A., *Philo. Mag. A* **59** (1989) 181.
Cho, J.H., Maley, M.P., Fleshler, S., Lacerda, A., and Bulaevskii, L.N., *Phys. Rev. Lett.*, (1997) in press.
Chou, Y.T., *J. Appl. Phys.* **33** (1962) 2747-2751.
Christian, J.W., *The Theory of Transformation in Metals and Alloys*, Pergamon, Oxford (1965).
Christian, J.W., *New Aspects of Martensitic Transformations* Japan Institute of Metals, Kobe (1976).
Chu, C.W., Gao, L., Chen, F., Huang, Z.J., Meng, R.L., Xue, Y.Y., *Nature* **365** (1993) 323.
Chumbley, L.S., Verhoeven, J.D., Kim, M.R., Cornelius, A.L., and Kramer, M.J., *IEEE Trans. Magn.* **25** (1989) 2337.
Chung, J.S., Lee, K.H., and Stroud D., *Phys. Rev. B* **40** (1989) 6570.
Civale, L., Marwick, A.D., McElfresh, M.W., Worthington, T.K., Malozemoff, A.P., Holtzberg, F., Thompson, J.R., and Kirk, M.A., *Phys. Rev. Lett.* **65** (1990) 1164.
Civale, L., Marwick, A.D., Worthington, T.K., Kirk, M.A., Thompson, J.R., Krusin-Elbaum, L., Sun, Y., Clem, J.R., and Holtzberg, F., *Phys. Rev. Lett.* **67** (1991) 648.
Cliff, G., and Lorimer, G.W., *J. Microsc.* **103** (1975) 203.
Cockayne, D.J.H., in *Diffraction and Imaging Techniques in Materials Science* **Vol. 1**, eds. S.A. Amerlinckx, R.Gevers, and J. van Landuyt, North-Holland, Amsterdam (1978) p. 153.
Cowley, J.M., *Acta Cryst.* **6** (1953) 516.
Cowley, J.M., *Acta Cryst.* **A32** (1976) 83.
Cowley, J.M., in *High-Resolution Transmission Electron Microscopy and Associated Techniques* eds. Buseck, P.R., Cowley, J.M., and Eyring L., Oxford University Press (1988).

Cowley, R.A., *Adv. Phys.* **29**, (1980) 1.
Cox, D.E., Zolliker, P., Axe, J.D., Moudden, A.H., Moodenbaugh A.R.,and Xu, Y., *MRS Symposium Proceedings* **156**, MRS, Pittsburgh (1989) 141.
Crawford, M.K., Harlow, R.L., McCarron E.M., and Farneth, W.E., Axe, J.D., Chou, H., and Huang, Q., *Phys. Rev. B* **44** (1991) 7749.
Daeumling, M., Seuntjens, and Larbalestier, D.C., *Nature* **346** (1990) 332.
d'Anterroches C., and Bourret, A., *Philo. Mag. A* **49** (1984) 738.
de Fontaine, D., Ceder, G., and Asta, M., *Nature* **343** (1990) 544.
de Fontaine, D., Mann, M.E., and Ceder, G., *Phys. Rev. Lett.* **63** (1989) 1300.
de Fontaine, D., Wille, L.T., and Moss, S.C., *Phys. Rev. B* **36** (1987) 5709.
de Gennes, P.G., *Superconductivity of Metal and Alloys*, Benjamin, New York (1966).
de Wolff, P.M., Janssen, T., and Janner, A., *Acta. Cryst. A* **37** (1981) 625.
Degischer, H.P., *Phil. Mag.* **26** (1972) 1137.
Deutscher G., and Müller, K.K., *Phys. Rev. Lett.* **59** (1987) 1745.
Dick, B.G., Overhauser, A.W., *Phys. Rev.* **112** (1958) 90.
Dienes G.J., and Vineyard, G.H., *Radiation Effects in Solids*, (Interscience Publishers, New York 1957) section 3.4.
Dimos, D., Chaudhari P., and Mannhart., J., *Phys. Rev. Lett.* **61** (1988) 219, *Phys. Rev. B* **41** (1990) 4038.
Dimos, D., Chaudhari, P., Mannhart, J., *Phys. Rev. B* **41** (1990) 4038.
Dinger, T.R., Worthington, T.K., Gallagher, W.J., and Sandstrom, R.L., *Phys. Rev. Lett.* **58** (1987) 2687.
Doyle, P.A., and Turner, P.S., *Acta Cryst.* **A24** (1968) 390.
Dou, S.X., Bourdillon, A.J., Sorrell, C.C., Ringer, S.P., Easterling, K.E., Savvides, N., Dounlop, J.B., and Roberts, R.B., *Appl. Phys. Lett.* **51** (1987) 535.
Duo, S.X., Liu, H.K., Zhang, Y.L., and W.-M., Bian, *Supercond. Sci. Technol.* **4** (1991) 203.
Dover, R.B., Gregory, E.M., White, A.E., Schneemeyer, L.F., Felder, R.J., and Waszczak, J.V., *Appl. Phys. Lett.* **56** (1990) 2681.

Dravid, V.P., Zhang, H., Marks, L.D., and Zhang, J.P., *Physica C* **192** (1992) 31.

Dunlap, B.D., Jorgensen, J.D., Segre, S., Dwight, A.E., Matykiewicz, J.L., Lee, H., Peng, W., and Kimball, C.W., *Physica C* **158** (1989) 397.

Duxbury, P.M., Beale, P.D., and Leath, P.L., *J. Phys. A* **20**, (1987) L441.

Eaglesham, D.J., Humphreys, C.J., Alford, N.McN., Clegg, W.J., Harmer, M.A., and Birchall, J.D., *Appl. Phys. Lett.* **51** (1988) 457.

Egerton, R.F., *Electron Energy-Loss Spectroscopy in Electron Microscopy* Plenum, New York (1986).

Eibl, O., *Physica C* **168** (1990) 239-248.

Eibl, *Physica C* **175** (1991) 419.

Eibl, O., van Aken, P., and Müller, W.P., *Phys. Stat. Sol.(a)* **128** (1991) 129.

Emery, V.J., *Phys. Rev. Lett.* **58** (1987) 2794.

Eom, C.B., Marshall, A.F., Suzuki, Y., Boyer, B., Pease, F.R.W., and Geballe, T.H., *Nature* **353** (1991) 544.

Faiz, M., Jennings, G., Campuzano, J.C., Alp, E.E., Yao, J.M., Saldin, D.K., and Yu, J., *Phys. Rev. B* **50** (1994) 6370.

Fang, M.M., Kogan, V.G., Finnemore, D.K., Clem, J.R., Chumbley, L.S., and Farrell, D.E., *Phys. Rev. B* **37** (1988) 2334.

Farneth, W.E., Borida, R.K., McCarron III, E.M., Crawford, M.K., and Flippen, R.B., *Solid State Comm.* **66** (1988) 953.

Feng, Y., Hautanen, K.E., High, Y.G., Larbaletier, D.C., Ray II, R., Hellstrom, E.E., and Babcock, S.E., *Physica C* **192** (1992) 293.

Feng, Y., High, Y.E., Larbalestier, D.C., Sung, Y.S., and Hellstrom, E., *Appl. Phys. Lett.* **62** (1993) 1553.

Fink, J., Nücker, N., Pellegrin, E., Romberg, H., Alexander, M., and Knupfer, M., *J. Electron Spectroscopy and Related Phenomena* **66**, (1994) 395.

Finnemore, D.K., Athreya, K., Sanders, K., Xu, Y., and Suenaga, M., (1988) unpublished.

Fischer, P., Karpinski, J., Kaldis, E., Jilek, E., and Rusiecki, S., *Solid State Comm.* **69** (1989)531.

Fischer, D.A., Moodenbaugh, A.R., and Xu, Y., *Physica C* **215** (1993)

279.
Fischer B.E., and Spohr, R., *Rev. Mod. Phys.* **55** (1983) 970.
Fleming, R.M., Schneemeyer, L.F., Gallagher, P.K., Batlogg, B., Rupp, L.W., and Wazczak, J.V., *Phys. Rev. B* **37** (1988) 7921.
Forwood, C.T., and Humble, P., *Phil. Mag. A* **31** (1975) 1025.
Forwood, C.T., and Clearebrough, L.M., *Electron Microscopy of Interfaces in Metals and Alloys*, IOP Publishing Ltd. (1991) Adam Hilger.
Alario-Franco, M.A., Chaillout, C., Capponi, C.C., and Chenavas, J., *Mat. Res. Bull.* **22** (1987) 1685.
Frank, F.C., 1950, *Proc. of symposium on Plastic Deformation of Crystalline Solids*, p.150, Mellon Institute, Pittsburgh, May, 1950.
Frank, F.C., *Philos. Mag.* **42** (1951) 809.
Frank, F.C., van der Merwe, J.M., *Pro. Roy. Soc.* **A198** (1949) 205.
Friedel, J., *Dislocations* Pergamon Press (1964) p.178.
Frischherz, M.C., Kirk, M.A., Zhang, J.P., and Weber, H.W., *Philo. Mag. A* **67** (1993) 1347.
Gallagher, P.K., O'Bryan, H.M., Sunshine, S.A., and Murphy, D.W., *Mat. Res. Bull.* **22** (1987a) 995.
Gallagher, W.J., Worthingtom, T.K., Dinger, T.R., Holtzberg, F., Kaiser, D.L., and Sandstrom, R.L., *Physica B* **148** (1987b) 221.
Gammel, P.L., et al., *Phys. Rev. Lett.* **59** (1987) 2592.
Gao, Y., Lee, P., Coppens, P., Subramanian, M.A., Sleight, A.W., *Science* **241** (1988) 954.
Gao, Y., Merkel, K.L., Bai, G., Chang, H.L.M., and Lam, D.J., *Physica C* **174** (1991) 1.
Gao, Y., Coppens, P., Cox, D.E., and Moodenbaugh, A.R., *Acta Cryst.* **A49** (1993a) 141.
Gao, Y., Wu, C.-T., Shi, Y., Merkle, K.L., and Goretta, K.C., *Appl. Superconductivity* **1** (1993b) 131-140.
Gao, L., *Phys. Rev. B* **50** (1994) 4260.
Reyes-Gasga, I., Krekels, T., van Tendeloo, G., van Landuyt, J., Amelinckx, S., Bruggink, W.H.M., and Verweij, H., *Physica C* **159** (1989) 831.
Gear, C.W., *Numerical Initial-Value Problems in Ordinary Differential Equations*, Prentice-Hall, Englewood Cliffs, NewJersey (1971).
Gerhauser, W., Ries, G., Neumuller, H.W., Schmidt, W., Eibl, O.,

Saemann-Ischnko, G., and Klaumunzer, S., *Phys. Rev. Lett.* **68** (1992) 879.

Gevers, R., van Landuyt, J., and Amelinckx, S., *Phys. Status Solidi* **11** (1965) 689.

Grimmer, H., and Warrington, D.H., *Acta Crystallogr. A* **43** (1987) 232.

Ginsberg, D.M., *Physical Properties of High Temperature Superconductors* **I** World Scientific, Singapore, New Jersey, London, Hong Kong (1989).

Ginzburg V.J., and Landau, L.D., *Zh. Eksperim. i Teor. Fiz.* **20** (1950) 1064.

Gjonnes, K., Boe, N., and Tafto, J., *Ultramicroscopy* **48** (1993) 37.

Gjonnes, J., and Hoier, R., *Acta Cryst.* **A27** (1971) 313.

Glauber, R.J., *J. Math. Phys.* **4** (1963) 294.

Gooding, R.J., *Script. metall. mater* **25** (1991) 105.

Grasso, G., Perin, A., and Flukiger, R., *Physica C* **250** (1995) 43; Grasso, G., Jeremie, A., and Flukiger, R., *Supercond. Sci. Technol.* **8** (1995) 827.

Grivel, J.-C., Jeremie, A., Hensel, B., and Flukiger, R., *Supercond. Sci. Technol.* **6** (1993a) 725.

Grivel, J.-C., Jeremie, A., Hensel, B., and Flukiger, R., *Supercond. Mater.*, eds. Etourneu, J., Torrance, J.B., and Yamauchi, H., I.I.I.T. International (1993b) p. 59.

Grivel, J.-C., and Flukiger, R., *Physica C* **229** (1994a) 177.

Grivel, J-C., and Flukiger, R., *Physica C* **235-240** (1994b) 505; Jeremie, A., Grivel, J.-C., and Flukiger, R., *Physica C* **235-240** (1994) 943.

Grivel, J.-C., and Flukiger, R., *Supercond. Sci. Technol.* **9** (1996) 555.

Guinea, F., *Europhys. Lett.* **7** (1988) 549.

Guo, Y.C., Liu, H.K., and Dou, S.X., *J. Mater. Res.* **8** (1993) 2187.

Guo, Y.C., Liu, H.K., and Dou, S.X., *Physics C* **235-240** (1994) 1231.

Gupta, R.P., and Gupta, M., *Phys. Rev. B* **49** (1994) 13154.

Gyorgy, E.M., van Dover, R.B., Jackson, K.A., Schneemeyer, L.F., and Waszczak, J.V., *Appl. Phys. Lett.* **55** (1989) 283.

Gyorgy, E.M., van Dover, R.B., Schneemeyer, L.F., White, A.E., O'Bryan, H.H., Felder, R.J., Waszczak, J.V., and Rhodes, W.W., *Appl. Phys. Lett.* **56** (1990) 2465.

Haldar, P., and Motowidlo, L., *J. of Metals* **44** (1992) 54.

Hammersley, J.M., and Handscomb, D.C., *Proc. Camb. Phil. Soc.* **60** (1967) 115.

Hansel, B., Grivel, J.-C., Jeremie, A., Perin, A., Pollin, A., and Flukiger, R., *Physica C* **205** (1993) 329; Hansel, B., Grasso, G., and Flukiger, R., **Phys. Rev. B 51** (1995) 15456.

Hardy, V., Grout, D., Hervieu, M., Provost, J., and Raveau, B., *Nucl. Instrum. Methods B* **54** (1991) 472; *Physica C* **191** (1992) 255.

Hawley, M., Raistrick, I.D., Beery, J.G., and Houlton, R.J., *Science* **251** (1991) 1587.

Hazen, R.M., *Crystal Structures of High-Temperature Superconductors*, in *Physical Properties of High Temperature Superconductors* II, ed. by D. M. Ginsberg, Chapter 3. ed., World Scientific, Teanek, NJ, (1990).

Heine, K., Tenbrink, N., Thoner, M., *Appl. Phys. Lett.* **55** (1989) 2441.

Hensel, B., Roas, B., Henke, S., Hopfengrtner, R., Lippert, M., Ströbel, J., Vildic, M., and Saemann-Ischenko, *Phys. Rev. B* **42** (1990) 4135.

Hewat, E.A., Dupuy, M., Bourret, A., Capponi, J.J., and Marezio, M., *Solid State Commun.* **64** (1987) 517.

Hlilgenkamp, H., Mannhart, J., and Mayer, C., *Phys. Rev. B* **53** (1996) 14586.

Hiroi, Z., Takano, M., Takeda, Y., Kanno, R., and Bando, Y., *Jpn. J. Appl. Phys. Lett.* **27** (1988) 580.

Hiroi,Z., Takano, M., and Bando, Y., *Physica C* **158** (1989) 269.

Hirsch, P.B., Howie, A., Nicholson, R.B., Pashley, D.W., and Whelan, M.J., *Electron Microscopy of Thin Crystals*, Butterworths, London, 1965.

Hirth, J.P., and Lothe, J., *Theory of Dislocations*, McGraw Hill, New York (1968).

Hirth, J.P., and Baluffi, R.W., *Acta Metall.* **21** (1973) 929.

Horiuchi, S., Meada, H., Tanaka, Y., and Matsui, Y., *Jpn. J. Appl. Phys.* **27** (1988) L1172.

Howie, A., and Whelan, M.J., *Proc. Roy. Soc.* **A263** (1962) 217.

Hu, Q.Y., Li, H.K., and Dou, S.X., *Physica C* **250** (1995) 7.

Huang, Q., Lynn, J.W., Meng, R.L., Chu, C.W., *Physica C* **218** (1993) 356.

Hulbert, S.F., *J. Br. Ceram. Soc.* **6** (1969) 11.
Humble, P. and Forwood, C.T., *Phil. Mag. A* **31** (1975) 1011.
Hwang, N.M., Roth, R.S., and Rawn, C.J., *J. Am. Cerm. Soc.* **73** (1990) 2531; Roth, R.S., Rawn, C.J., Ritter, J.J., and Burton, B.P., *J. Am. Cerm. Soc.* **72** (1989) 1545.
Ichimiya A., and Uyeda, R., *Z. Naturforsch* **32a** (1977) 750.
IEEE Trans. Appl. Supercond. **7** (1997).
Iijima, S., Ichihashi, T., Kubo, Y., and Tabuchi, J., *Jpn. J. Appl. Phys.* **26** (1987) L1790.
Iijima, S., and Ichihashi, T., Proc. of *the XII International Congress for Electron Microscopy* Vol.4, eds, Peachy, L.D., and Williams, D.B., San Francisco Press (1990) p.14.
Iijima, Y., Tanabe, N., Kohno, O., and Ikeno, Y., *Appl. Phys. Lett.* **60** (1992) 769.
Ikeda, H., Yoshizaki, R., and Yoshikawa, K., *Cryogenic Eng.* **25** (1990) 99.
Imai, K., Nakai, I., Kawashima, T., Sueno, S., and Ono, A., *Jpn. J. Appl. Phys.* **27** (1988) L1661.
Ishibashi, Y., *J. Phys. Soc. Jpn.* **59** (1990) 800; Ishibashi Y., and Suzuki, I., *J. Phys. Soc. Jpn.* **53** (1984) 903.
Iwazumi, T., Nakai, I., Izumi, M., Oyanagi, H., Sawada, H., Ikeda, H., Saito, Y., Abe, Y., Takita K., and Yoshizaki, R., *Solid State Commun.* **65** (1988) 213.
Jacobs, A.E., *Phys. Rev. B* **31** (1985) 5984.
Jiang, X., Wochner, P., Moss, S.C., and Zschack, P., *Phys. Rev. Lett.* **67** (1991) 2167.
Jin, S., Kammlott, G.W., Nakahara, S., Tiefel, T.H., and Graebner, J.E., *Science* **253** (1991) 427.
Jones, P.M., Rackham, G.M., and Steeds, J.W., *Proc. Roy. Soc. Lond. A* **354** (1977) 197.
Jorgensen, J.D., *Physics Today* **44** (1991) 34.
Jorgensen, J.D., Beno, M.A., Hinks, D.G., Soderholm, L., Volin, K.J., Hitterman, R.L., Grace, J.D., Schuller, I.K., Segre, C.U., Zhang, K., and Kleefisch, M.S., *Phys. Rev. B* **36** (1987a) 3608.
Jorgensen, J.D., Schuttler, H.-B., Hinks, D.G., Capone II, D.W., Zhang, K., Brodasky, M.B., and Scalapino, D.J., *Phys. Rev. Lett.* **58** (1987b) 1024.

Jorgensen, J.D., Veal, B.W., Kwok, W.K., Grabtree, G.W., Umezawa, A., Nowicki, L.J., and Paulikas, A.P., *Phys. Rev. B* **36**, (1987c) 5731.

Jorgensen, J.D., Veal, B.W., Paulikas, A.P., Nowicki, L.J., Crabtree, G.W., Claus, H., and Kwok, W.K., *Phys.Rev.B* **41** (1990) 1863.

Jou, C.J., and Washburn, J., *J. Mater. Res.* **4** (1989) 795.

Kaiser, D.L., Holtzberg, F., Chisholm, M.F., and Worthington, T.K., *J. Cryst. Growth* **85** (1987) 393.

Katsuyama, S., Ueda Y., and Kosuge, Y., *Physica C* **165** (1990) 404.

Kawasaki, K., In *Phase Transitions and Critical Phenomena*. (C. Domb and M. S. Green, eds.) Vol II. Academic Press (1972) p.443.

Keimer, B., Birgeneau, R.J., Cassanho, A., Endoh, Y., Greven, M., Kastner, M.A., and Shirane, G., *Z. Phys. B* **91** (1993) 373.

Kerr, W.C., Hawthorne, A.M., Gooding, R.J., Bishop A.R., and Krumhansl, J.A., *Phys. Rev. B* **45** (1992) 7036.

Kerr W.C., and Rave, M.J., *Phys. Rev. B* **48** (1993) 16234.

Kes, P.H., *Physica C* **153-5** (1988) 1121.

Khachaturyan, A.G., *Theory of Structural Transformations in Solids*, Wiley, New York (1983).

Khachaturyan, A.G., and Morris, J.W., *Phys. Rev. Lett.* **59** (1987) 2776.

Khachaturyan, A.G., and Morris, J.W., *Phys. Rev. Lett.* **61** (1988) 215.

Kikuchi, S., *Jap. J. Phys.* **5** (1928) 23.

King, A.H., and Zhu, Y., *Phil. Mag. A* **67** 1993 1037.

Kishio, K., Shimoyama, J., Hasegawa, T., Kitazawa, K., and Fueki, K., *Jpn. J. Appl. Phys.* **26** (1987) L1228.

Konczykowski, M., Rollier-Albenque, F., Yacoby, E.R., Shanlov, A., Yeshurun, Y., and Lejay, P., *Phys. Rev. B* **44** (1991) 7167.

Kosevich, A.M., and Boiko, V.S., *Soviet Phys. Uspekhi* **14** (1971) 286.

Krakauer, H., Pickett, W.E., and Cohen, R.E., *J. Supercond.* **1** (1988) 111.

Kramer, M.J., Chumbley, L.S., McCallum, R.W., Nellis, W.J., Weir, S., and Kvam, E.P., *Physica C* **166** (1990) 115.

Kramer, M.J., McCallum, R.W., Margulies, L.,Arrasmith, S.R., and Holesinger, T.G., *J. Electron. Mater.* **22** (1993) 1269.

Kramer, M.J., Qian, Q., Finnemore, D., and Snead, C.L., *Physica C* **203** (1992) 83.

Kreckels, T., van Tendeloo, G., Broddin, D., Amelinckx, S., Tanner, L., Menbod, M., Vanlathem, E., and Deltour, R., *Physica C* **173** (1991) 361.

Krivoglaz, M.A., *Theory of X-ray and Thermal Neutron Scattering by Real Crystals*, Plenum Press, New York (1969).

Kubo, Y., Ichihashi, T., Manako, T., Baba, K., Tabuch, J., and Igarashi, H., *Phys. Rev. B* **37** (1988) 7858.

Kulik, J., Huang, Z.J., Bechtolf, J., Xue, Y.Y., and Hor, P.H., and Bailey, G.W., *Proc. of 49th Annual Meeting of the Electron Microscopy Society of America*, San Francisco Press (1991) p. 1084.

Kwok, W.K., Crabtree, G.W., Umezawa, A., Veal, B.W., Jorgensen, J.D., Malik, S.K., Nowiciki, L.J., Paulikas, A.P., and Nunez, L., *Phys. Rev. B* **37** (1988) 106.

Kwok, W.K., Welp, U., Crabtree, G.W., Vandervoort, K.G., Hulscher, R., and Liu, J.Z., *Phys. Rev. Lett.* **64** (1990) 966.

Lairson, B.M., Streiffer, S.K., and Brarman, J.C., *Phys. Rev. B* **42** (1990) 10067.

Landau, L.D., and Lifshitz, E.M., *Statistical Physics*, Pergamon Press, Oxford, New York (1965), *Course of theoretical Physics*, Vol.5.

Larbalestier, D.C., Babcock, S.E., Cai, X.Y., Field, M.B., Gao, Y., Heinig, N.F., Kaiser, D.L., Merkle, K., Williams, L.K., and Zhang, N., *Physica C* **185-189** (1991) 315.

Lay, K.W., *AIP Conf. Proc.* **219**, ed. Y. K. Kao (American Institute of Physics) 1991, p.119.

Ledbetter, H., and Lei, M., *J. Mater. Res.* **6** (1991) 2253-2255.

Lee, D.F., Chaud, X., and Salama, K., *Jpn. J. Appl. Phys.* **31** (1992) 2411.

Levesque, D., and Verlet, L., *Phys. Rev. A* **2** (1970) 2514.

Levesque, D., Verlet, L., and Kürkijarvi, J., *Phys. Rev. A* **7** (1973) 1690.

Li, J.C.M., in *Electron Microscopy and Strength of Crystals* Thomas, G., and Washburn, J., eds., John Weily & Son, New York (1963) p.713.

Li, Y.-H., and Teitel, S., *Phys. Rev. Lett.* **65** (1990) 2595.

Li, Q., Riley Jr., G.N., Parrella, R.D., Fleshler, S., Rupich, M.W., Carter, W.L., Willis, J.O., Coulter, J.Y., Bingert, J.F., Skka, V.K., Parrell, J.A., and Larbalestier, D.C., *IEEE Trans. Appl. Super-*

cond. **7** (1997a) 2026.
Li, Q., Tsay, Y.N., Zhu, Y., Suenaga, M., Gu, B.D., and Koshizuka, N., *IEEE Trans. on Appl. Supercond.* **7** (1997b) 1584; and unpublished work.
Li, Q., Tsay, Y.N., Suenaga, M., Gu, G.D., and Koshizuka, N., *Physica C* **282** (1997c) 1495.
Lifshitz, I.M., *Zh. Eksper. Teoret. Fiz.* **18** (1948) 1134.
Likharev, K.K., *Rev. Mod. Phys.* **51**, 101 (1979).
London, F., London, H., *Physica* **2** (1935) 34.
Love, A.E.H., *The Mathematical Theory of Elasticity*, Wiley, New York (1944).
Luo, J.S., Dorris, S.E., Fischer, A.K., LeBoy, J.S., Wiesmann, H.J., and Suenaga, M., *J. Electr. Mater.* **24** (1995) 1817.
Luo, J.S., Merchant, N., Maroni, V.A., Gruen, D.M., Tani, B.S., Carter, W.L., and Riley Jr., G.N., *Applied Superconductivity* **1** (1993a) 101.
Luo, J.S., Merchant, N., Maroni, V.A., Riley Jr., G.N., and Carter, W.L., *Appl. Phys. Lett. 63* (1993b) 690.
Luo, J.S., Merchant, N., Escorcia-Aparicio, E.J., Maroni, V.A., and Tani, B.S., *J. Mater. Res.* **9** (1994) 3059.
Luo, J.S., Merchant, N., Haash, M., and Rupich, M., *High Temperature Superconductors: Synthesis, Processing, and Large-Scale Applications*, Proc. of the 1996 TMS Symposium, Anaheim, CA, eds., Balachandran, U., McGinn, P.J., and Abell, J.S., TMS, PA (1997) p.33.
Maeno, Y., Tomita, T., Kyogku, M., Awaji, S., Aoki, Y., Hoshino, K., Minami, A., and Fujita, T., *Nature* (London) **328** (1987) 512.
Malozemoff, A.P., in *High Temperature Superconducting Coupounds II*, ed. Whang, S.H., TMS Publications, Warrenddale PA (1990) p. 3; in *Superconductivity and its Applications*, ed. Kao, Y.H., American Institute of Physics, New York (1992) p.6.
Malozemoff, A.P., Riley Jr., G.N., Fleshler, S., and Li, Q., To be published in *Proc. SPA*, Xi'an, China, March 6-8, 1997.
Matsui, Y., Maeda, H., Tanaka, Y., and Horiuchi, S., *Jpn. J. Appl. Phys.* **27** (1988a) L372.
Matsui, Y., Maeda, H., Tanaka, Y., Takayama-Muromachi, E., Takekawa, S., and Horiuchi, S., *Jpn. J. Appl. Phys.* **27** (1988b)

L827.

Matsushita, T., Fumaki, K., Takeo, M., and Yamafuji, K., *Jpn. J. Appl. Phys.* **26**, (1987) L1524.

Maeda, H., Tanaka, Y., Fukutomi, M., and Asano, T., *Jpn. J. Appl. Phys.* **27** (1988) L209.

McCallum, R.W., Dessis, Margulies, L., and Kramer, M.J., *Proc. of Processing and properties of Long Lengths of Superconductors*, the 1993 Fall TMS Meeting, Pittsburgh, PA, Oct. 17-21, 1993, eds., Balachandran, U., Collings, E., and A Goyal (1993).

Majewski, P., Kaesche, S., Su, H.-L., and Aldinger, F., *Physica C* **221** (1994) 295.

Manaila, T., Malis, O., Manicu, M., and Devernyi, A., *Phys. Status, Solidi A* **147** (1995) 31; Malis, O., Manicu, M., Manaila, R., and Devernyi, A., *Phys. Solidi A* **147** (1995) 325.

Manolikas, C., van Tendeloo, G., and Amelinckx, S., *Sol. State Comm.* **58** (1986) 851.

Matsui *et al. Jpn. J. Appl. Phys.* **27** (1988) L372.

Merchant, N., Luo, J.S., Maroni, V.A., Riley Jr., G.N., and Carter, W.L., *Appl. Phys. Lett.* **65** (1994) 1039.

Metropolis, N., Rosenbluth, A.W., Rosenbluth, M.N., Teller, A.H., and Teller, E., *J. Chem. Phys.* **21** (1953) 1087.

Meyer, H.A., (ed.), *Symposium on Monte Carlo Methods*, Wiley (1953).

Micheli, P.F., Tarascon, J.M., Greene, L.H., Barboux, P., Rotella, F.J., and Jorgensen, J.D., *Phys. Rev. B* **37** (1988) 5932.

Miehe, G., Vogt, T., Fuess, H., and Wilhelm, M., *Physica C* **171** (1990) 339.

Michael, J.R., Lin., C.-H., Sass, S.L., *Scripta Metall.* **22** (1988) 1121.

Mironva, M., Lee, D.F., and Salama, K., *Physica C* **211** (1993) 188.

Mitchell, T.E., Roy, T., Schwarz, R.B., Smith, J.F., and Wohlleben, D., *J.Electron Microsc. Tech.* **8** (1988) 317.

Monod, P., Ribault, M., D'Yvoire, F., Jegoudez, J., Collin, G., and Revcolevschi, A., *J. de Physique* **48** (1987) 1369.

Moodenbaugh, A.R., and Fischer, D.A. *Physica C* **230** (1994) 177.

Moodenbaugh, A.R., Xu, Y., Suenaga, M., Folkerts, T.J., and Shelton, R.N., *Phys. Rev. B* **38** (1988) 4596.

Morgan, P.E.D., Doi, T., and Housley, R.M., *Adv. In Supercond.* **VI** (1994) 327, eds., Fujita, T., and Shioha, Y., Springer-Verlag, Tokyo

(1994).

Morgan, P.E.D., Housley, R.M., Porter, J.R., and Ratto, J.J., *Physica C* **176** (1991) 279; Morgan, P.E.D., Piche, J.D., and Housley, R.M., *Physica C* **191** (1992) 179.

Morris J.R., and Gooding, R.J., *Phys. Rev. Lett.* **65** (1990) 1769.

Mouritsen, O.G., *Computer Studies of Phase Transitions and Critical Phenomena*, Springer-Verlag (1984).

Murakami, M., Gotoh, S., Fujimoto, H., Yamaguchi, K., Kishizuka, N., and Tanaka, S., *Supercond. Sci. Technol.* **4** (1991) S49.

Nakahara, S., Jin, S., Sherwood, R.C., and Tiefel, T.H., *Appl. Phys. Lett.* **54** (1989) 1926.

Nakanishi, S., Kogachi, M., Sasakura, H., Fukuoka, N., Minamigawa, S., Nakahigashi, K., and Yanase, A., *Jpn. J. Appl. Phys.* **27** (1988) L329.

Nishiyama, Z.,*Martensitic Transformation*, Academic Press, New York (1978).

Nozaki, H., Ishizawa, I., Fukunaga, O., and Wada, H., *Jpn. J. Appl. Phys.* **26** (1987) L1180.

Nücker, N., Fink, J., Fuggle, T.C., Durham, P.J., and Temmerman, W.M., *Phys. Rev. B* **37** (1988) 5158.

Nücker, N., Romberg, H., Xi, X.X., Fink, J., Gegenheimer, B., Zhao, Z.X., *Phys. Rev B* **39** (1989) 6619.

Nücker, N., Pellegrin, E., Schweiss, P., Fink, J., Molodtsov, S.L., Simmons, C.T., Kaindl, G., Frentrup, W., Erb, A., and Muller-Vogt, G., *Phys. Rev. B* **51** (1995) 8529.

Oda, Y., Fujita, H., Toyoda, H., Kaneko, T., Kohara, T., Nakada, I., and Asayama, K., *Jpn. J. Appl. Phys.* **26** (1987) L1660.

Oh, S.S., and Osamura, K., *Supercond. Sci. Technol.* **4** (1991) 239.

Oles, A.M., and Grzelka, W., *Phys. Rev. B* **44** (1991) 9531.

Olsen, A., Goodman, P., and Whitfield, H.J., *J. Solid State Chem.* **60** (1985) 305.

Onnes, H.K., *Commun. Phys. Lab. Univ. Leiden* **120b** (1911) 3.

Onnes, H.K., *Commun. Phys. Lab. Univ. Leiden* **133a** (1913) 3.

Osamura, K., Katsumata, Y., Nonaka, S., and Okuda, H., *Adv. In Supercond.* **VIII** (1997), in press.

Ossipyan, Y., Shekhtman, V.Sh., and Shmyt'ko, I.M., *Physica C* **153-155** (1988) 970.

Pande, C.S., Singh, A.K., Toth, L.E., Gubser, D.U., and Wolf, S.A., *Phys.Rev. B* **36** (1987) 5669.

Pellegrin, E, Nücker, N., Fink, J., Simmons, C.T., Kaindl, G., Bernhard, J., Renk, K.F., Kumm, G., and Winzer, K., *Phys. Rev. B* **48** (1993) 10520.

Pham, A.Q., Studer, F., Merrien, N., Maignan, A., Michel, C., and Raveau, B., *Phys. Rev. B* **48** (1993) 1249.

Pickett, W.E., *Rev. Mod. Phys.* **6** (1989) 433.

Pickett, W.E., Cohen, R.E., and Krakauer, H., *Phys. Rev. Lett.* **67** (1991) 228.

Polonka, J., Xu, M., Li, Q., Goldman, A.I., and Finnemore, D.K., *Appl. Phys. Lett.* **59** (1991) 3640.

Putilin, S.N., Antipov, E.V., Chmaissen O., and Marezio, M., *Nature* **362** (1993) 226.

Rabier, J., and Denanot, M.F., *Revue Phys. Appl.* **25** (1990) 55.

Radaelli, P.G., Wagner, J.L., Hunter, B.A., Beno, M.A., Knapp, G.S., Jorgensen, J.D., and Hinks, D.G., *Physica C* **216** (1993) 29.

Rahman, A., *Phys. Rev.* **136** (1964) A405.

Rahman, A., *J. Chem. Phys.* **45** (1966) 258.

Ramesh, R., Jin, S., Nakahara, S., and Tiefel, T.H., *Appl. Phys. Lett.* **57** (1990) 1458.

Rasband, W., NIH Image program, ver.1.54g, (1993).

Raveau, B., *Physics Today* **45** (1992) 53.

Read Jr., W.T., *Dislocations in Crystals*, McGraw-Hill (1953).

Read, W.T., and Shockley, W., *Phys. Rev.* **78** (1950) 275.

Rechav, B., Yacoby, Y., Stern, E.A., Rehr J.J., and Newville, M., *Phys. Rev. Lett.* **72** (1994) 1352.

Reichardt, W., Pintschovius, L., Hennion, B., and Collin, F., *Supercond. Sci. Technol.* **1** (1988) 173.

Rez, D., Rez, P., and Grant, I., *Acta Cryst.* **A50** (1994) 481.

Riley Jr., G.N., submitted to *Jour. of Materials Research.* (1997).

Robertson, I.M., and Wayman, C.M., *Phil.Mag. A* **48** (1983) 421, 443, 629.

Roitburd, A.L., *Sov. Phys. Solid Sate* **10** (1969) 2870.

Roy, T., and Mitchell, T.E., *Philos. Mag. A* **63** (1991) 225.

Roy, T., Sickafus, K., Clinard, F.W., and Mitchell, T.E., Bailey, G.W., Eds., *Proc. 47th Annual Meeting of the Electron Microscopy Soci-*

ety of America (1989) p.184.

Rozeveld, S.J., Howe, J.M., and Schmauder, S., *Acta Metall. Mater.* **40** (1992) Suppl. pp. S173.

Saini, N.L., Law, D.S-L., Menovsky, A., Franse, J.J., and Garg, K.B., *Physica C* **235-240** (1994) 1017.

Salomons, E., Koeman, N., Brouwer, R., de Groot, D.G., and Griessen, R., *Solid State Comm.* **64** (1987) 1141.

Sarikaya, M., Kikuchi R., and Aksay, I.A., *Physica C* **152** (1988) 161.

Sasaki, K., Kuroda, K., Saka, H., and Imura, T., *J. Electron. Microsc.* **36** (1987) 232.

Sass, S.L., Mura, T., and Cohen, J.B., *Phil. Mag.* **16** (1967) 679.

Sayers, D.E., Stern, E.A., Lytle, F.W., *Phys. Rev. Lett.* **27** (1971) 1204.

Schmahl, W.W., Putnis, A., Salje, E., Freeman, P., Graeme-Baber, A., Jones, R., Singh, K.K., Blunt, J., Edwards, P.P., Loram, J., and Mirza, *Phil. Mag. Lett.* **60** (1990) 241.

Schilling, A., Cautoni, M., Gao, J.D., and Ott, H.R., *Nature* **363** (1993) 56.

Schindler, W., Raos, B., Saemann-Ischenko, G., Schultz, L., and Gerstenberg, H., *Physica C* **169** (1990) 117.

Schneider T., and Stoll, E., *Phys. Rev. B* **13** (1976) 1216.

Schuller, I.K., Hinks, D.G., Beno, M.A., Capone II, D.W., Soderholm, L., Locquet, J.P., Ruynserade, Y., Segre, C.U., and Zhang, K., *Solid State Commun.* **63** (1987) 385.

Semenovskaya, S., and Khachaturyan, A.G., *Phys. Rev. Lett.* **67** (1991) 2223; *ibid.* **64** (1991) 29.

Shapiro, S.M., Larese, J.Z., Noda, Y., Moss, S.C., and Tanner, L.E., *Phys. Rev. Lett.* **57** (1986) 3199.

Shaw, T.M., Shimde, S.L., Dimos, D., Cook, R.F., Duncombe, P.R., and Kroll, C., *J. Mater. Res.* **4** (1989) 248

Shelton, R.N., Peng, J.L., Xu, Y., Moodenbaugh, A.R., and Suenaga, M., unpublished.

Sheng, Z.Z., and Hermann, A.M., *Nature* **332** (1988a) 55.

Sheng, Z.Z., and Hermann, A.M., *Nature* **332** (1988b) 138.

Shin, K., and King, A.H., *Mater. Sci. & Eng.* **A113** (1989) 121.

Shibutani, K., Egi, T., Hayashi, S., Fukumoto, Y., Shigaki, I., Masuda, Y., Ogawa, R., and Kawate, Y., *IEEE Trans. Appl. Supercond.* **3** (1993) 935.

Shibutani, K., Wiesmann, H.J., Sabatini, R.J., Suenaga, M., Hayashi, S., Ogawa, R., Kawate, Y., Motowidlo, L., and Hadlar, P., *Appl. Phys. Lett.* **64** (1994) 924.

Siegrist, T., Schneemeyer, L.F., Waszczak, J.V., Singh, N.P., Opila, R.L., Batlogg, B., Rupp, L.W., and Murphy, D.W., *Phys. Rev. B* **36** (1988) 8365.

Singh, A., Chandrasekhar, N., and King, A.H., *Acta Crystallogr. B* **46** (1990) 117.

Skjerpe, P., Olsen, A., Tafto, J., Suenaga, M., Zhu, Y., and Moodenbaugh, A.R., *Phil. Mag. B* **65** (1992) 1067.

Smith, D.A., and Pond, R.C., *Int. Metals. Rev.* **205** (1976) 61.

Smith, J.F., and Wohlleben, D., *Z. Phys. B* **72** (1988) 323.

Smith, D. A., Chisholm, M. F., and Clabes, J., *Appl. Phys. Lett.* **53** (1988) 2344.

Spence, J.C.H., and Zuo, J.M., *Electron Microdiffraction*, Plenum Press, New York (1992).

Specht, E.D., Sparks, C.J., Dhere, A.G., Brynestad, J., Cavin, O.B., Kroeger, D.M., and Oye, H.A., *Phys. Rev. B* **36** (1987) 5723.

Specht, E.D., Sparks, C.J., Dhere, A.G., Brynestad, J., Vavin, O.B., Kroeger, D.M., and Oye, H.A., *Phys. Rev. B* **37** (1988) 7426.

Streiffer, S.K., Zielinski, E.M., Lairson, B.M., and Bravman, J.C., *Appl. Phys. Lett.* **58** (1991) 2171.

Ströbel, P., Toledano, J.C., Morin, D., Schneck, J., Vacquier, G., Monnerieau, O., Primot, J., and Fournier, T., *Physica C* **201** (1992) 27.

Sun, C.P., and Balluffi, R.W., *Philo. Mag. A* **46** (1982) 49, 63.

Sun, Y.S., and Hellstrom, E.E., *Physica C* **255** (1995) 266.

Sutton, A.P., and Balluffi, R.W., *Interfaces in Crystalline Materials*, Oxford University Press, Oxford (1995).

Sutton, A.P., and Vitek, V., *Phil. Trans. Roy. Soc. A* **309** (1983) 1, 37, 55.

Sweedler, A.T., Snead Jr., C.L., and Cox, D.E., Irradiation Effects in Superconducting Materials, in *Treatise on Materials Science and Technology*, Vol. **14**, T. Luhman and D. Dew-Hughes eds., (1979) p.349-426.

Tafto, J., Suenaga, M., and Sabatini, R.L., *Appl. Phys. Lett.* **52** (1988) 667.

Takahashi, T., Matsuyama, H., Watanable, T., Katayama-Yoshida, H., Sato, S., Kosugi, N., Yagishita, A., Shamoto, S., and Sato, M., *Proceedings of the 3rd International Symposium on Superconductors*, eds. K. Kajimura and H. Hayakawa Springer-Verlag, Berlin (1990) p.75.

Tanner, L.E., 1966, *Phil. Mag. A* **14** (1966) 111.

Tarascon, J.M., Barboux, P., Miceli, P.F., Greene, L.H., Hull, G.W., Eibschutz, M., and Sunshine, S.A., *Phys. Rev. B* **37** (1988a) 7458.

Tarascon, J.M., McKinnon, W.R., Barboux, P., Hwang, D.M., Bagley, B.G., Greene, L.H., Hull, G.W., LePage, Y., Stoffel, N., and Giroud, M., *Phys. Rev. B* **38** (1988b) 8885.

Teitel S., and Jayaprakash, C., *Phys. Rev. B* **27** (1983) 598.

Terasaki, O., and Watanabe, D., *Acta Cryst.* **A35** (1979) 895.

Thomas, G., and Goringe, M., *Transmission Electron Microscopy of Materials*, Wiley, New York (1979) p.37.

Thomsen, C., Cardona, M., Liu, R., Gegenheimer, B., and Simon, A., *Physica C* **153-155** (1988) 1756.

Thorpe, M.F., Jin, W., and Mahanti, S.D., *Phys. Rev. B* **40** (1989) 10294.

Thurston, T.R., Haldar, P., Wang, Y.-L., Suenaga, M., Jisrawi, N., and Wildgruber, U., *J. Mater. Res.* **12** (1997) 891.

Thurston, T.R., Wildgruber, U., Jisrawi, N., Haldar, P., Suenaga, M., and Wang, Y.-L., *J. Appl. Phys.* **79** (1996) 3122.

Tietz, L.A., Cater, C.B., Lathrop, D.K., Russek, S.E., and Buhrman, R.A., in *Proc. Of 46th Annual Meeting of Electron Microscopy Society of America*, edited by G.W. Bailey San Francisco (1988) p.870.

Ting, W., Fossheim K., and Lægreid, T., *Physica B* **165-166**, 1293 (1990).

Ting, T.C.T., *Anisotropic Elasticity*, Oxford University Press, New York (1996).

Tokumoto, M., Ihara, H., Matsubara, T., Hirabayashi, M., Terada, N., Oyanagi, H., Murata, K., and Kimura, Y., *Jpn. J. Appl. Phys.* **26** (1987) L1565.

Tomita, N., Takahashi, Y., Mori, M., and Ishida, Y., *Jpn. J. Appl. Phys.* **31** (1992) L942; Xu, B.S., Ichinose, H., Tanaka, S., Ishida, Y., *J. Japan Inst. Metals* **130** (1996) 121.

Torardi, C.C., Subramanian, M.A., Calabrese, J.C., Gopaakrishnan, J., McCarron, E.M., Morrissey, K.J., Askew, T.R., Flippen, R.B., Chowdhry, U., and Sleight, A.W., *Phys. Rev. B* **38** (1988) 225.

Torrance, J.B., Tokura, Y., Nazzal, A.I., Bezinge, A., Huang, T.C., and Parkin, S.S.P., *Phys. Rev. Lett.* **61** (1988) 1127.

Tranquada, J.M., Heald, S.M., Moodenbaugh, A.R., and Xu, Youwen, *Phys. Rev. B* **38** (1988) 8893.

Tsuda, K., and Tanaka, M., *Acta Cryst.* **A51** (1995) 7.

van Bakel, P.E.M, Hof, P.A., van Engelen, J.P.M, Bronsveld, P.M., and De Hosson, J.Th.M., *Phys. Rev. B* **41** (1990) 9502.

van Dyck, D., van Tendeloo, G., and Amelinckx, S., *Ultramicroscopy* **15** (1984) 357.

van Landuyt, J., Gevers, R., and Amelinckx, S., *Phys. Status Solidi* **13** (1966) 467.

van Tendeloo, G., Brodden, D., Zandbergen, H.W., and Amelinckx, S., *Physica C* **167** (1990) 627.

van Tendeloo, G., Zandbergen, H.W., and Amelinckx, S., *Solid State Commun.* **63** (1987a) 603.

van Tendeloo, G., Zandbergen, H.W., and Amelinckx, S., *Solid State Commun.* **63** (1987b) 389.

Veal, B.W., Paulikas, A.P., You, H., Shi, H., Fang, Y., and Downey, J.W., *Phys. Rev. B* **42** (1990) 6305; Veal, B.W., You, H., Paulikas, A.P., Shi, H., Fang, Y., and Downey, J.W., *Phys. Rev. B* **42** (1990) 4770.

Verlet, L., *Phys. Rev.* **159** (1967) 98; *ibid.* **163** (1968) 201.

Vincent, R., Bird, D.M., and Steeds, J.W., *Philo. Mag. A* **50** (1984) 765.

Vinnikov, L.Ya., Gurevich, L.A., Yemelchenko, G.A., and Ossipyan, Yu.A., *Solid State Commun.* **67** (1988) 421.

Wagner, J.L., Radaelli, P.G., Hinks, D.G., Jorgensen, J.D., Mitchell, J.F., Dabrowski, B., Knapp, G.S., and Beno, M.A., *Physica C* **210** (1993) 447.

Wang, J.-Y., King, A.H., Zhu, Y., and Suenaga, M., presented in MRS Fall Meeting, Dec. 1992, Boston, MA (1992).

Wang, M., Xiong, G., Tang, X., and Hong, Z., *Physica C* **210** (1993a) 413.

Wang, Z.L., Goyal, A., and Kroeger, D.M., *Phys. Rev. B* **47** (1993b)

5373.

Wang, J.L., Cai, X.Y., Kelley, R.J., Vaudin, M.D., Babcock, S.E., and Larbalestier, D.C., *Physica C* **230** (1994) 189.

Wang, Y.-L., Bian, W., Zhu, Y., Fukomoto, Y., Wiesmann, H.J., and Suenaga, M., *J. Electr. Mater.* **24** (1995) 1817.

Wang, J.L., Tsu, F.I., Cai, X.Y., Kelley, F.J., Vaudin, M.D., Babcock, S.E., and Larbalestier, D.C., *J. of Mater. Res.* **11** (1996a) 868.

Wang, Y.-L., Bian, W.-M., Zhu, Y., Cai, Z.-X., Welch, D.O., Sabatini, R.L., and Suenaga, M., *Appl. Phys. Lett.* **69** (1996b) 580.

Warlimont, H., and Delaey, L., *Martensitic Transformations in Copper-Silver and Gold-Based Alloy*, Pergamon, New York (1974).

Warren Jr., W.W., Walstedt, R.E., Brennert, G.F., Cava, R.J., Batlogg, B., and Rupp, L.W., *Phys. Rev. B* **39** (1989) 831.

Watanabe, T., *Materials Science Forum*, **126-128** (1993) 295-304.

Wayman, C.M., *Introduction to Crystallography of Martensitic Transformation*, Macmillan, New York (1964).

Weber, H.W., *Super. Sci. Technol.* **5** (1992) S19.

Weber, H.W., and Crabtree, G.W., *Studies of High-Temperature Superconductors*, Vol. **9** A.V. Narlikar eds., Nova Scientific, New York (1992) p. 37-39.

Werder, D.J., Chen, C.H., Cava, R.J., and Batlogg, B., *Phys. Rev. B* **37** (1988a) 2317.

Werder, D.J., Chen, C.H., Cava, R.J., and Batlogg, B., *Phys. Rev. B* **38** (1988b) 5130.

Whangbo, M.-H., *J. Sol. St. Chem.* **97** (1992) 490.

Whangbo, M.-H., and Torardi, C.C., *Acc. Chem. Res.* **24** (1991) 127.

Wille, L.T., Berera, A., and de Fontaine, D., *Phys. Rev. Lett.* **60** (1988) 1065.

Wolf, D., and Merkle, K.L., in *Materials Interfaces*, ed. Wolf, D., and Yip, S., Chapman and Hall, London (1992) p.87-p.150.

Wooster, W.A., *Diffuse X-ray Reflections From Crystals*, (Clarendon Press,1961).

Wördenweber, R., Sastry, S.v.S., Heinemann, K., and Freyhardt, H.C., *J. Apps. Phys.* **65** (1989) 1648.

Wu, M.K., Ashburn, J.R., Torng, C.J., Hor, P.H., Meng, R.L., Gao, L., Huang, Z.L., Wang, Y.Q., and Chu, C.W., *Phys. Rev. Lett.* **58** (1987) 908.

Wu, L., Wang, Y.L., Bian, W., Zhu, Y., Thurston, T.R., Sabatini, R.L., Haldar, P., and Suenaga, M., *J. Mater. Res.* **12** (1997a) 3055.

Wu, L., Zhu, Y., and Tafto, J., submitted to *Phys. Rev. Lett.* (1997b).

Wu, X.D., Foltyn, S.R., Arendt, P.N., Blumenthal, W.R., Campbell, I.H., Cotton, J.D., Coulter, J.Y., HUlts, W.L., Maley, M.P., Safar, H.F., and Smith, J.L., *Appl. Phys. Lett.* **67** (1995) 2397.

Xia W., and Leath, P., *Phys. Rev. Lett.* **63** (1989) 1428.

Xiao, Gang, Cieplak, M.Z., Musser, D., Gravrin, A., Streitz, F.H., and Chien, C.L., *Phys. Rev. Lett.* **60** (1988) 1446.

Xu, M., Finnemore, D.K., Balachandran, U., and Haldar, P., *Appl. Phys. Lett.* **66** (1994) 3359.

Xu, M., Finnemore, D.K., Balachandran, U., and Haldar, P., *J. Appl. Phys.* **78** (1995) 360.

Xu, Y., Suenaga, M., Tafto, J., Sabatini, R.L., Moodenbaugh A.R., and Zolliker, P., *Phys. Rev. B* **39** (1989) 6667.

Xu, Y., Subatini, R.L., Moodenbaugh, A.R., Zhu, Y., Shyu, S.G., Suenaga, M., Dennis, K.W., and McCallum, R.W., *Physica C* **169** (1990) 205.

Yamaguchi, S., Terabe, K., Saito, A., Yahagi, S., and Iguchi, *Jpn. J. Appl. Phys.* **27** (1988) L179.

Yang, C.Y., Heald, S.M., Tranquada, J.M., Welch, D.O., and Suenaga, M., *Phys. Rev. B* **39** (1989) 6681.

Yang, C.Y., Moodenbaugh., A.R., Wang, Y.L., Xu, Y., Heald, S., Welch, D.O., Suenaga, M., Fischer, D.A., and Penner-Haha, J.E., *Phys. Rev. B* **42** (1990) 2231.

Yoshida, T., Kuroda, K., and Saka, H., *Phil. Mag. A* **62** (1990) 573.

Young, C.T., Steele, J.H., and Lytton, J.L., *Metall. Trans.* **4** (1973) 2081.

Zaanen, J., Paxton, A.J., Jepsen, O.J., and Anderson, O.K., *Phys. Rev. Lett.* **60** (1988) 2865.

Zandbergen, H.W., Gronsky, R., Wang, K., and Thomas, G., *Nature* **331** (1988) 596.

Zandbergen, H.W., *Physica C* **193** (1992) 371.

Zhang, W., and Hellstrom, E.E., *Physica C* **218** (1993) 141.

Zhang, W., and Hellstrom, E.E., *Supercond. Sci. Technol.* **9** (1996) 1.

Zhao, Z., Chen, L., Yang, Q., Huang, Y., Chen, G., Tang, R., Liu, G., Cui, C., Chen L., Wang, L., Guo, S., Li, S., and Bi, J., *Kexue*

Tongbao **32** (1987) 1098.
Zhu, J., and Cowley, J.M., *Acta Cryst. A* **38** (1982) 718.
Zhu, W. and Nicholson, P.S., *J. Mater. Res.* **7**, (1992) 38.
Zhu, Y., *Philo. Mag. A* **69** (1994) 717.
Zhu, Y., and Cai, Z.-X., *Ultramicroscopy* **52** (1993) 539.
Zhu, Y., and Suenaga, M., *Phil. Mag. A* **66** (1992) 457.
Zhu, Y. and Suenaga, M., *Physica C* **252** (1995) 117.
Zhu, Y., and Tafto, J., *Phys. Rev. Lett.* **76** (1996a) 443.
Zhu, Y., and Tafto, J., *Philo. Mag. A* **74** (1996b) 307.
Zhu, Y., and Tafto, J., *Philo. Mag. B* **75** (1997) 785.
Zhu, Y., Suenaga, M, and Xu, Y., *Phil. Mag. Lett.* **60** (1989a) 51.
Zhu, Y., Suenaga, M., Xu, Y., Sabatini, R.L., and Moodenbaugh, A.R., *Appl. Phys. Lett.* **54** (1989b) 374.
Zhu, Y., Moodenbaugh, A.R., Suenaga, M., and Tafto, J., *Physica C* **167** (1990a) 363.
Zhu, Y., Pan, M., Li, Z.G., Suenaga, M., and Welch, D.O., *Proc. 12th International Congress for Electron Microscopy, Seattle, Aug. 1990*, L.D. Peachey and D.B. Williams, eds., Vol. **4** (1990b) p18.
Zhu, Y., Suenaga, M., and Xu, Y., *J. Mater. Res.* **5** (1990c) 1380.
Zhu, Y., Suenaga, M, Xu, Y., and Kawasaki, M., *Mat. Res. Soc. Symp. Proc.* **159** (1990d) 413.
Zhu, Y., Suenaga M., and Moodenbaugh, A.R., *Phil. Mag. Lett.* **62** (1990e) 51.
Zhu, Y., Zhang, H., Moodenbaugh, A.R., and Suenaga, M., *Proc. XIIth International Congress for Electron Microscopy*, August, 1990, Seatle, San Francisco Press (1990f) p.72.
Zhu, Y., Suenaga, M., Tafto, J., and Welch, D.O., *Phys. Rev. B* **44** (1991a) RC 2871.
Zhu, Y., Suenaga, M., and Moodenbaugh, A.R., *Ultramicroscopy* **37** (1991b) 341.
Zhu, Y., Suenaga, M., and Tafto, J., *Philo. Mag. Lett.* **64** (1991c) 29.
Zhu, Y., Tafto, J., and Suenaga, M., *MRS Bulletin* **11** (1991d) 54.
Zhu, Y., Zhang, H., Wang, H., and Suenaga, *J. Mater. Res.* **12** (1991e) 2507.
Zhu, Y., Suenaga, M., and Moodenbaugh, A.R., *Proc. of International Workshop on Superconductivity*, co-sponsored by ISTEC and MRS, June 23-26, 1992, Honolulu, Hawaii, (1992) p.165-168.

Zhu, Y., Budhani, R.C., Cai, Z.-X., Welch, D.O., Suenaga, M., Yoshizaki, R., and Ikeda, H., *Phil. Mag. Lett.* **67** (1993a) 125.

Zhu, Y., Cai, Z-X., Budhani, R.C., Suenaga, M., and Welch, D.O., *Phys. Rev. B* **48** (1993b) 6436.

Zhu, Y., Corcoran, Y., and Suenaga, M., *J. Interface Science,* **1** (1993c) 361.

Zhu, Y., and Cowley, J.M., *Phil. Mag. A* **69** (1993d) 397.

Zhu, Y., Sabatini, R.L., Wang, Y.L., and Suenaga, M., *J. Appl. Phys.* **73** (1993e) 3407.

Zhu, Y., Suenaga, M., and Tafto, J., *Phil. Mag. A* **67** (1993f) 573.

Zhu, Y., Suenaga, M., and Tafto, J., *Phil. Mag. A* **67** (1993g) 1057.

Zhu, Y., Wang, Z.L., and Suenaga, M., *Phil. Mag. A* **67** (1993h) 11.

Zhu, Y., Zhang, H., Suenaga, M., and Welch, D.O., *Phil. Mag. A* **68** (1993i) 1079.

Zhu, Y., Moodenbaugh, A.R., Cai, Z.-X., Tafto, J., Suenaga, M., and Welch, D.O., *Phys. Rev. Lett.* **73** (1994) 3026.

Zhu, Y., Suenaga, M., and Sabatini, R.L., *Appl. Phys. Lett.* **65** (1994b) 1832.

Zhu, Y., Zuo, J.M., Moodenbaugh, A.R., and Suenaga, M., *Phil. Mag. A* **70** (1994c) 969.

Zhu, Y., Cai, Z.X., and Welch, D.O., *Philo. Mag. A* **73** (1996) 1.

Zhu, Y., Wu, L., Wang, J.-Y., Li, Q., Tsay, Y.N., and Suenaga, M., *Microscopy & Microanalysis* **3** (1997) 423.

Ziegler, J.F., *Handbook of Stopping Cross-Sections For Energetic Ions in All Elements*, Pergamon (1980).

Zou, J., Cockayne, D.J.H., Auchterlonie, G.J., McKenzie, D.R., Dou, S.X., Bourdillon, A.J., Sorrell, C.C., Easterling, K.E., and Johnson, A.W.S., *Philos. Mag. Lett.* **57** (1988) 157.

Zuo, J.M., Spence, J.C.H., and Hoier, R., *Phys. Rev. Lett.* **62** (1989) 547.

Zuo, J.M., *Ultramicroscopy* **41** (1992) 211.

Index

Artificially created defects, 65, 421
    Planar defects, 439, 466, 471
    Strain field, 424, 452, 464
    Structural features, 423
    Thermal spike model, 443

Bi-2212
    Charge distribution, 294
    Grain boundary, 378
        Misorientation distribution, 380
        Strain field, 396
        Structural features, 389
        Superconducting properties, 385

Bi-2223
    Fabrication, 243
    Formation mechanism, 273
    grain alignment, 246
    Intercalation, 267, 276
    Layer rigidity model, 277

Burgers vector, 60, 72, 73, 326, 329

Charge distribution, 291
    Bi-2212, 294
    $YBa_2Cu_3O_{7-\delta}$, 303

Coincidence Site Lattice, 70

Computer simulation, 34
    Molecular dynamics method, 38
    Monte Carlo method, 35

Crystal structure of HTC superconductors, 41
    $Bi_2Sr_2Ca_{n-1}Cu_nO_{2n+6}$, 48
    $(La_{1-x}M_x)_2CuO_4$, 42
    Lattice mismatch, 54
    Layered structure, 53
    $YBa_2Cu_3O_{7-\delta}$, 44

Diffraction simulation, 118, 173
Dislocations, 58

EELS, 330, 331, 336, 392, 441

Flux-line pinning, 24
Frank formula, 73

Ginzburg-Landau theory, 14–18
    Coherence length, 17
    Free energy, 15
    Order parameter, 15

Grain boundary
    Basics, 69
    Bi-2212, 378
        Misorientation distribution, 380

Strain field, 396
Structural features, 389
Superconducting properties, 385
Cation segregation, 370
Coincidence Site Lattice, 70
Crystallography, 316, 323
Dislocation model, 73
Grain boundary dislocations, 322, 398
Misorientation distribution, 383
O-Lattice, 72, 320
$YBa_2Cu_3O_{7-\delta}$, 348, 405
    Cation segregation, 370
    Grain boundary dislocations, 354
    Misorientation distribution, 349
    Oxygen content and CCSL boundaries, 357
    Strain energy, 373

Heavy ion irradiation, 421
Hole concentration, 339, 361

Image simulation, 65, 144, 300, 301, 392, 457, 459, 461

Kikuchi pattern, 316

$(La_{1-x}M_x)_2CuO_4$
    Structural transformation, 42
$La_{2-x}Ba_xCuO_4$
    Structural modulation, 205
    Structural transformation, 207, 217
    Theoretical model, 226
    Twin boundary, 217

Lattice gas model, 33
Lattice parameter measurement, 342, 363
Line defects, 58
London, 10–13
    Equation, 11
    Penetration depth, 12

Meissner Effect, 9
Model for interaction potential, 31
Monte Carlo simulation
    $La_{2-x}Ba_xCuO_4$, 207
    $YBa_2Cu_3O_{7-\delta}$
        Elastic strain, 113
        Fe doping, 111
        Phase transition, 84
        Thermal quenching, 96

Neutron diffraction, 30

O-lattice, 72
Oxygen ordering, 77, 91

Planar defects
    Stacking faults, 61
Point defects, 55

RSJ model, 405

Shell model, 32
Structural modulations, 64
Superconductivity
    H_c, 7
    History, 7
    J_c, 7
    T_c, 7
    Types, 18
        Type I, 18

Index

Type II, 18

Theoretical modeling, 31
Thermal spike model, 443
Transmission electron microscopy, 27
Tweed, 65, 98, 108, 122, 133
Twin boundary, 149
    Crystallographic analysis, 179
    Displacement, 162
    DSCL treatment, 185
    Fringe Contrast analysis, 168
    Interfacial energy, 176
    $La_{2-x}Ba_xCuO_4$, 217
    Steps and dislocations, 181
    Twin tip, 190, 201
    Twinning dislocations, 181
    under electron-beam irradiation, 162
Type-II Superconductor
    Critical Current, 23
    Flux line pinning, 23

Volume defects, 62

X-ray diffraction, 30

$YBa_2Cu_3O_{7-\delta}$
    Charge distribution, 303
    Fe doping, 108
        Elastic strain, 113
        Image simulation, 144
        Lattice gas model, 108
        Reduction and reoxidation, 122
        Three-dimensional modulation, 133
        Tweedy modulation, 98
    Lattice gas model, 81
    Fe doping, 110
    Phase transition, 79
    Oxygen content and T_c, 78
    Oxygen ordering, 77
    Thermal quenching, 91
    Tweedy modulation, 98
    Twin boundary, 149
        Cation substitution, 158
        Displacement, 162
        Superconducting properties, 161
    Twin tip, 190